Civil, Structural and Constructional Engineering

Civil, Structural and Constructional Engineering

Edited by Armando Ruiz

www.clanryeinternational.com

Clanrye International,
750 Third Avenue, 9th Floor,
New York, NY 10017, USA

ISBN: 978-1-63240-623-1

Cataloging-in-Publication Data

Civil, structural and constructional engineering / edited by Armando Ruiz.
p. cm.
Includes bibliographical references and index.
ISBN 978-1-63240-623-1
1. Civil engineering. 2. Structural engineering. 3. Concrete construction. 4. Building--Technological innovations.
I. Ruiz, Armando.
TA145 .C58 2017
624--dc23

For information on all Clanrye International publications
visit our website at www.clanryeinternational.com

Printed in the United States of America.

Contents

Preface

The main aim of this book is to educate learners and enhance their research focus by presenting diverse topics covering this vast field. This is an advanced book which compiles significant studies by distinguished experts. This book addresses successive solutions to the challenges arising in the area of application, along with it; the book provides scope for future developments.

The fundamentals as well as modern approaches in the field of civil engineering have been discussed in this book. Structural and constructional engineering are important sub-fields of civil engineering. Robust development in civil engineering has been possible in the last few decades due to availability of better materials, equipment, machines and software. Civil engineering is considered to be one among the world's oldest professional engineering disciplines. Acclaimed experts of civil engineering have provided comprehensive insights in the book to assist the readers in gaining essential information. This book is a valuable compilation of topics, ranging from the basic to the most complex advancements in this field. With state-of-the-art inputs, this book targets students, researchers and professionals.

It was a great honour to edit this book, though there were challenges, as it involved a lot of communication and networking between me and the editorial team. However, the end result was this all-inclusive book covering diverse themes in the field.

Finally, it is important to acknowledge the efforts of the contributors for their excellent chapters, through which a wide variety of issues have been addressed. I would also like to thank my colleagues for their valuable feedback during the making of this book.

Editor

1

Application of genetic programming in shape optimization of concrete gravity dams by metaheuristics

Abdolhossein Baghlani[1*], Mohsen Sattari[1] and Mohammad Hadi Makiabadi[1]
*Corresponding author: Abdolhossein Baghlani, Faculty of Civil and Environmental Engineering, Shiraz University of Technology, Shiraz, Iran
E-mail: baghlani@sutech.ac.ir
Reviewing editor: Simon Smith, University of Edinburgh, UK

Abstract: A gravity dam maintains its stability against the external loads by its massive size. Hence, minimization of the weight of the dam can remarkably reduce the construction costs. In this paper, a procedure for finding optimal shape of concrete gravity dams with a computationally efficient approach is introduced. Genetic programming (GP) in conjunction with metaheuristics is used for this purpose. As a case study, shape optimization of the Bluestone dam is presented. Pseudo-dynamic analysis is carried out on a total number of 322 models in order to establish a database of the results. This database is then used to find appropriate relations based on GP for design criteria of the dam. This procedure eliminates the necessity of the time-consuming process of structural analyses in evolutionary optimization methods. The method is hybridized with three different metaheuristics, including particle swarm optimization, firefly algorithm (FA), and teaching–learning-based optimization, and a comparison is made. The results show that although all algorithms are very suitable, FA is slightly superior to other two algorithms in finding a lighter structure in less number of iterations. The proposed method reduces the weight of dam up to 14.6% with very low computational effort.

Subjects: Computer Aided Design (CAD); Structural Engineering; Water Engineering

Keywords: gravity dams; genetic programming (GP); metaheuristics; pseudo-dynamic analysis; artificial intelligence

ABOUT THE AUTHORS

Abdolhossein Baghlani is an assistant professor of civil and environmental engineering department, Shiraz University of Technology, Shiraz, Iran. His areas of research are numerical methods, optimization in civil engineering problems, and computational hydraulics and fluid dynamics.

Mohsen Sattari is an MSc graduate student of earthquake engineering from Shiraz University of Technology, Shiraz, Iran. His areas of interest are earthquake engineering, analysis of gravity dams, and optimization.

Mohammad Hadi Makiabadi is an MSc graduate student of earthquake engineering from Shiraz University of Technology and a PhD student of structural engineering in Shiraz University, Shiraz, Iran. His areas of research are structural optimization, seismic behavior of structures, and reliability.

PUBLIC INTEREST STATEMENT

Dams are among the most important hydraulic structures which are used for various purposes. Gravity dams are solid concrete structures that maintain their stability against design loads from geometric shape to mass and strength of the concrete. Hence, weight minimization of the dam can remarkably reduce the construction costs. In this paper, a procedure for finding optimal shape of concrete gravity dams with minimum weight is introduced. The procedure is computationally efficient and considerably reduces the number of structural analyses required for the design. Genetic programming (GP) along with population-based optimization approaches is used for this purpose. Optimization of the Bluestone dam is presented as a case study. By pseudo-dynamic analyses, a database is developed to find appropriate relations for design criteria of dam based on genetic programming. The developed equations are then hybridized with three different population-based optimization techniques and a comparison is made.

1. Introduction

Dams are structures built across a stream to form a reservoir. Dams are among the most important hydraulic structures used for various purposes such as water storage for domestic, industrial, and agricultural usage, flood controlling, generation of electricity, and so on. Gravity dams are solid concrete structures that maintain their stability against design loads from the geometric shape and the mass and strength of the concrete. Generally, they are constructed on a straight axis, but may be slightly curved or angled to accommodate the specific site conditions (USBR, 1976). Generally, the overall cost of dam construction is very high, and optimization methods are suitable tools for economically designing concrete gravity dams, owing to the fact that the shape of a gravity dam is so influential in its stability and that the overall cost of a gravity dam is proportional to the volume of concrete employed in dam construction. Many population-based optimization methods are recently used in engineering practice. Genetic algorithm (GA), particle swarm optimization (PSO), ant colony optimization and recently developed firefly algorithm (FA), and teaching–learning-based optimization (TLBO) are some examples. The main problem in utilizing such population-based approaches for solving real-life engineering problems is their high computational effort. First, a population of solution candidates (say 100) should be initialized and the solution is then sought iteratively (say for 100 iterations). Hence, the objective function should be evaluated several times (say 10,000). In engineering problems, the objective function is evaluated through analysis, e.g. by static or dynamic analysis of trusses, bridges, or dams which is itself time-consuming. This makes the overall procedure to be computationally expensive. A brief literature review on the application of three metaheuristics used in this study, i.e. PSO, FA, and TLBO, is presented in the next paragraph.

PSO was first presented by Kennedy and Eberhart (1995). Then, the method gained extreme popularity in solving engineering problems in many fields such as soil mechanics, optimal dam design, and structural engineering (Baghlani & Makiabadi, 2013a; Doğan & Saka, 2012; Fontana, Ndiayeb, Breyssea, Bosa, & Fernandez, 2011; Hamidian & Seyedpoor, 2010; Kang, Li, & Xu, 2012; Salajegheh & Khosravi, 2011; Seyedpoor, Salajegheh, Salajegheh, & Gholizadeh, 2011; Yagiza & Karahan, 2011). FA is a recently developed method proposed by Yang (2009). Many problems from various areas have been successfully solved using the FA and its variants (Baghlani, Makiabadi, & Rahnema, 2013; Farhoodnea, Mohamed, Shareef, & Zayandehroodi, 2014; Fister, Fister, Yang, & Brest, 2013; Gomes, 2011; Miguel & Fadel Miguel, 2012; Yang, 2013). The application of this metahurisic in optimal design of dams has not been reported yet. TLBO is a newer algorithm compared to the aforementioned ones. Rao, Savsani, and Vakharia (2011) proposed the method based on the effect of influence of a teacher on the output of learners in a class. It is reported that it outperforms some of the well-known metaheuristics regarding constrained benchmark functions, constrained mechanical design, and continuous non-linear numerical optimization problems (Rao, Savsani, & Vakharia , 2012). This relatively new optimization approach has been already employed in some engineering problems (Baghlani & Makiabadi, 2013b; Makiabadi, Baghlani, Rahnema, & Hadianfard, 2013; Niknam, Azizipanah-Abarghooee, & Rasoul Narimani, 2012; Roy, Paul, & Sultana, 2014).

Optimal design of gravity dams is a fairly complicated procedure since many factors such as fluid–structure interaction, and seismic loads should be taken into account. Furthermore, some design constraints should be satisfied to ensure safety of dams against sliding and overturning as well as fracture. Several researches can be mentioned regarding optimization of concrete gravity and arch dams. Most of these researches are focused on design of optimal shape of arch dams rather than gravity dams. Simoes and Lapa (1994) and Simoes (1995) presented optimal design of dams using quadratic programming algorithm. They investigated the optimum shape of dams under static and dynamic loadings. The objective function was the weight of dam. They reported 13% reduction in the actual weight of dam after the optimization. Ghazanfari Hashemi, Bahraninejad, and Ahmadi (2009) employed simulated annealing (SA) approach for the shape optimization of gravity dams and compared their results with the results obtained by sequential quadratic programming (SQP) approach. They reported that the weight of dam can be reduced up to 30.18% using SA and up to 30.11% using SQP. Li, Jing, and Zhou (2010) used ANSYS software for the analysis of dam models and optimized a concrete gravity dam using genetic algorithm. They could reduce the weight of the dam up to 11.9% by this approach. Salmasi (2011) employed genetic algorithm in conjunction with the spreadsheet EXCEL to find the optimal shape of gravity dams. The static analysis was performed in this study. Qi (2012) used constrained nonlinear complex optimization algorithm to optimize the shape of gravity dams. He reported 16% reduction in cross-sectional area of the dam after optimization is accomplished.

In this study, shape optimization of concrete gravity dams is carried out using a rather different approach. The proposed method is a hybridization of metaheuristics and genetic programming (GP) to reduce the computational effort in the procedure of optimization. GP is a powerful tool which has been already employed to solve various problems in science and engineering (Danandeh Mehr, Kahya, & Yerdelen, 2014; Roushangar, Mouaze, & Shiri, 2014; Silva, Santos, Matos, & Costa, 2014; Valencia, Haak, Cotillon, & Jurdak, 2014). In order to introduce the approach used in this research, shape optimization of Bluestone dam is presented as a case study. The objective function is the weight of the dam, and design constraints are necessary safety factors against sliding and overturning and satisfying behavior constraint as well. The design variables consist of four geometric variables. The open access software CADAM is used for dam models. Despite being a very effective tool for the analysis of gravity dams, the software cannot be directly employed for optimization purposes since in its present form, it cannot be linked with programming languages such as MATLAB or FORTRAN. In order to overcome this problem, GP is employed. The proposed method has also reduced the computational effort required in metaheuristics. Several models of the dam including 322 models are analyzed using CADAM in which pseudo-dynamic analysis is performed for each model. The results of the analyses are used as a database to develop appropriate formulas for design constraints by means of GP. This procedure eliminates the requirement of structural analysis in every iteration in evolutionary optimization approaches. The method is combined with three metaheuristics including PSO, FA, and TLBO, and the results are compared.

2. Problem formulation

For stability requirements, the dam must be safe against overturning and sliding. Moreover, the safe stresses in the concrete of the dam or in the foundation material shall not be exceeded. The safety factor against overturning (OSF) is defined as the ratio between the resisting moments (M_R) and overturning moments (M_O):

$$OSF = \frac{\sum M_R}{\sum M_O} \tag{1}$$

Many of the loads on the dam are horizontal or have horizontal components which are resisted by frictional or shearing forces along horizontal or nearly horizontal planes in the body of the dam, on the foundation or on horizontal or nearly horizontal seams in the foundation. A dam will fail in sliding at its base, or at any other level, if the horizontal forces causing sliding are more than the resistance available to it at that level. The resistance against sliding may be due to friction alone, or due to friction and shear strength of the joint. The stability of a dam against sliding is evaluated by comparing the minimum total available resistance to the total magnitude of the forces tending to induce sliding. Sliding resistance is a function of the cohesion inherent in the materials and at their contact and the angle of internal friction of the material at the surface of sliding. The factor of safety against sliding (SSF) can be computed from the following equation:

$$SSF = \frac{(\tan\phi . \sum F_V) + c . A_b}{\sum F_H} \tag{2}$$

in which $\sum F_V$ is the sum of all vertical forces acting on the foundation level or at the level of the assumed sliding surface, $\sum F_H$ is the sum of all the horizontal forces acting above the foundation level or at the level of the assumed sliding surface, ϕ is the internal friction angle, c is the cohesion, and A_b is the area of the dam-foundation contact surface, or of the assumed sliding surface.

In shape optimization of gravity dams, the optimal shape of the dam with minimum weight should be determined. Hence, the objective function is the weight of the dam. Moreover, some design criteria as design constraints should be satisfied. In this study, optimization of dam is performed based on USBR standard. Based on USBR, the minimum required safety factor against sliding is 1.3 and the minimum required OSF is 1.1 for seismic loads. Furthermore, for satisfying behavior constraints, the principal stresses within dam body should not exceed the allowable tensile and compressive stresses of concrete (Beser, 2005). According to these design criteria, the optimization problem can be defined as follows:

$$\text{Minimize: } W = \gamma_c V \tag{3}$$

$$\text{Subject to: SSF} \geq 1.3 \tag{4a}$$

$$\text{OSF} \geq 1.1 \tag{4b}$$

$$\sigma_{1p} \leq \sigma^{+} \tag{4c}$$

$$\left|\sigma_{2p}\right| \leq \sigma^{-} \tag{4d}$$

$$X_L \leq X \leq X_U \tag{4e}$$

in which W is the total weight of the dam, γ_c is the unit weight of concrete, V is the total volume of the dam, σ_{1p} and σ_{2p} are principal stresses, σ^{+} and σ^{-} are allowable tensile and compressive stresses, respectively, X is the vector of design variables, X_L is the lower bound of the design variables, and X_U is the upper bound. Since the value of γ_c is constant, for unit width of the dam, the cross-sectional area of the dam can be minimized instead of W.

In this study, the allowable tensile and compressive stresses of concrete are considered to be less than the tensile and compressive strength of concrete as follows:

Allowable tensile stress = $0.8\, f_t$

Allowable compressive stress = $0.8\, f'_c$

in which f_t is the tensile strength of concrete and f'_c is the compressive strength.

3. Structural analysis procedure for the case study

For analyzing gravity dams under hydrostatic and hydrodynamic pressures, gravity load, and seismic loads, various methods are available in the range of static to fully dynamic analysis. On the other hand, the structure should be analyzed many times either in the evolutionary optimization procedure or in artificial intelligence methods in order to find the optimal structure. Therefore, an effective and sufficiently accurate analysis approach should be chosen to reduce the computational effort. In this study, pseudo-dynamic analysis is selected for this purpose rather than complicated dynamic analysis. Pseudo-dynamic analysis has been found to be reliable in analyzing concrete gravity dams (Siamardi, 2007). This type of analysis is not only computationally more effective than dynamic analysis but also produces acceptable results close to those of dynamic analysis.

The pseudo-dynamic analysis is based on the simplified response spectra method. A pseudo-dynamic analysis is conceptually similar to a pseudo-static analysis except that it recognizes the dynamic amplification of the inertia forces along the height of the dam (Leclerc, Leger, & Tinawi, 2002). In this method, the dynamic magnification factor, inertial forces at the height of the dam, is considered. However, the oscillatory nature of the amplified inertia forces is not considered. That is the stress and stability analyses are performed with the inertia forces being continuously applied in the same direction. Since the pseudo-dynamic method does not recognize the oscillatory nature of earthquake loads, it is also appropriate to perform the safety evaluation in two phases: (1) the stress analysis using peak spectral acceleration values and (2) the stability analysis using sustained spectral acceleration values. It is assumed in these analyses that the dynamic amplification applies only to the horizontal rock acceleration. The period of vibration of the dam in the vertical direction is considered sufficiently small to neglect the amplification of vertical ground motions along the height of the dam (Leclerc et al., 2002).

The basic input data required to perform a pseudo-dynamic analysis, using the simplified response spectrum method proposed by Chopra (1998), are:

(1) Peak ground and spectral accelerations,

(2) Dam and foundation stiffness and damping properties,

(3) Reservoir bottom damping properties and velocity of an impulsive pressure wave in water, and

(4) Modal summation rule.

In this study, a free available code called CADAM is employed to perform pseudo-dynamic analysis of models. CADAM has been specifically developed for the analysis of gravity dams. This code was first presented by Leclerc et al. at Montreal University, Canada, in 2000 (Leclerc et al., 2002; Leclerc, Léger, & Tinawi, 2003). The main objective of the code is to analyze stability of concrete gravity dams against overturning, sliding, and breakage. It has been found that the code has great capability in analyzing gravity dams in spite of its simplicity. Because of its ability in analyzing gravity dams considering the most effective and influential forces, CADAM is one of the best codes in this area. CADAM analyzes dam models and directly reports the safety factors against sliding and overturning. It also reports the principal stresses developed within the dam body. Therefore, the implementation of the code is straightforward and no post-processing of the results is required after analysis.

The case study in the current article is devoted to optimizing the Bluestone Dam. This dam is located on the New River which is a tributary of Kanawha River 3.2 km far from Hinton city in West Virginia, USA (Edward & Stowasser, 2011). The dam was built in 1930 for the purpose of flood control and hydropower. A drainage gallery has been considered within the dam body (Enaiati, Monazzam, & GHaemian, 2011). Geometry of the dam (Ellingwood & Tekie, 2001) is shown in Figure 1 along with other specifications reported in tables. Table 1 shows material properties considered in this study for dam and foundation; in Table 2, the properties of drainage gallery are given, and Table 3 reports the parameters used in seismic analysis of the dam. Figure 2 shows one of the models developed in CADAM, schematically.

In their study on the Bluestone Dam, Tekie and Ellingwood carried out a seismic hazard study on the dam, and the result is shown in Figure 3 (Takie & Ellingwood, 2003). Based on their study, the spectral acceleration of 0.36 g is considered related to a return period of 2,475 years in which g is the gravity acceleration. Moreover, based on a hydrologic analysis, if imminent failure flood occurs, the water depth in the reservoir will be 52 m (Takie & Ellingwood, 2002). In current study, it is assumed that the dam is full (water depth equals 48.15 m in the reservoir) and all analyses are performed based on this assumption.

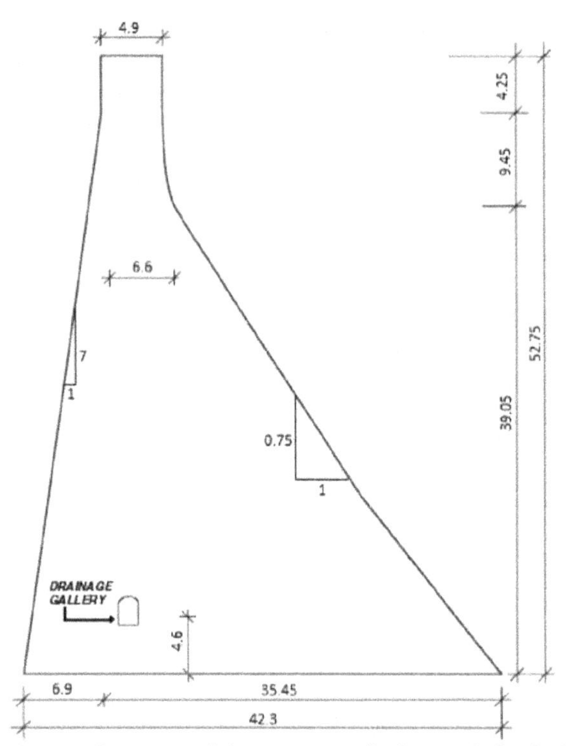

Figure 1. Geometry of Bluestone Dam (before optimization) (Ellingwood & Tekie, 2001).

Table 1. Material properties of dam and foundation (Miguel & Fadel Miguel, 2012)

Parameter	Dam	Foundation
Compressive strength of concrete (f_c')	3,4475 kp	72,397 kp
Tensile strength of concrete (f_t)	3,450 kp	6,895 kp
Cohesion (c)	998 kp	193 kp
Angle of internal friction (ϕ)	55°	46°
Density of concrete ($\rho_c = \gamma_c/g$)	2,400 kg/m³	2,643 kg/m³
Modulus of elasticity (E)	33.56×10^6 kp	27.25×10^6 kp
Poisson's ratio (υ)	0.225	0.165
Damping ratio	$\xi = 0.05$	$\eta_f = 0.1$

Table 2. Properties of drainage gallery

Guideline	Uplift reduction coefficient	Drain position of heel (m)	Drain elevation (m)
USBR	0.67	8	5

Table 3. Parameters used in pseudo-dynamic analysis of dam

Type	HPGA (g)*	VPGA (g)**	Spectral acceleration (Sa) g	Period (year)	Velocity of pressure wave (m/s)	Modal combination
Value	0.14	0.1	0.36	2,475	1,440	SRSS***

*Horizontal peak ground acceleration.

**Vertical peak ground acceleration.

***Square-root-of-the-sum-of-squares of the first mode and static correction for higher modes.

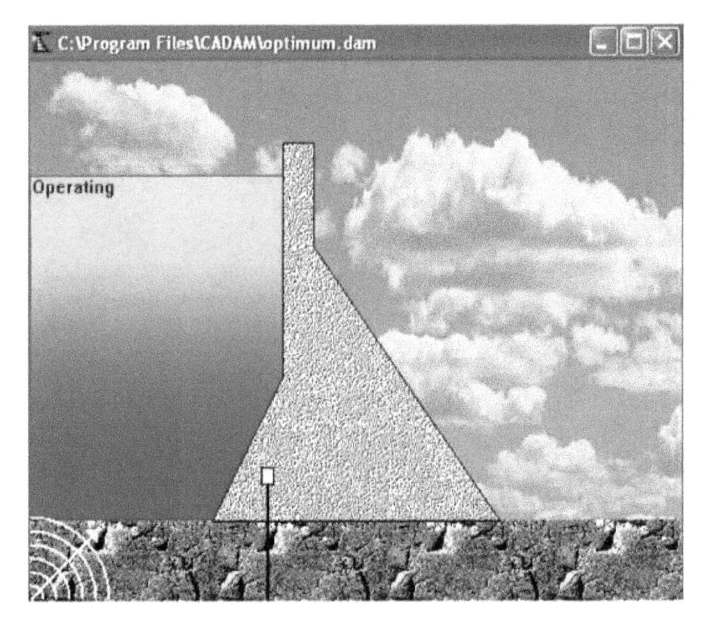

Figure 2. A schematic view of one of the dam models in CADAM.

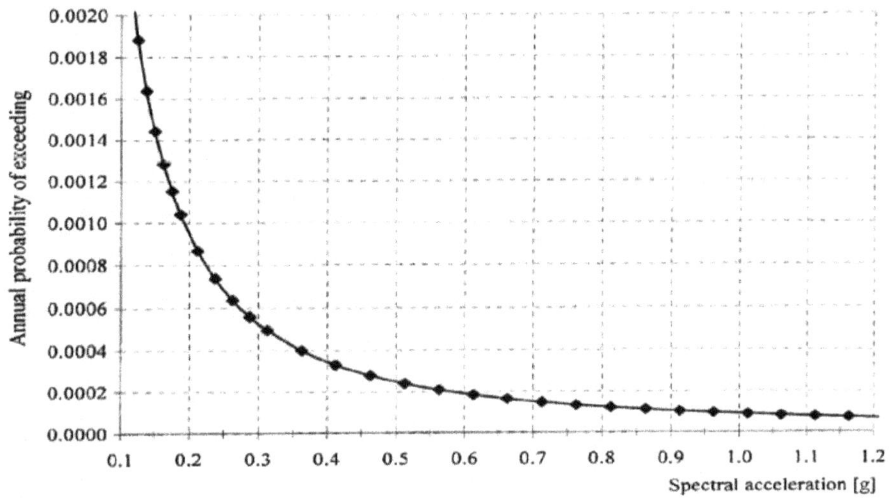

Figure 3. The results of seismic hazard study on Bluestone Dam (Tekie & Ellingwood, 2003).

4. The proposed method

The current study aims to hybridize the results found by GP with an evolutionary optimization method. For this purpose, dam models should be developed and the GP should be applied on the results in order to find appropriate relations for design constraints. Then, a suitable evolutionary algorithm can be used to determine the optimal values of design variables. The following sections illustrate this procedure in more detail.

4.1. Development of models and GP

A concrete gravity dam resists against the external loads by its own weight. For specified density of concrete, the weight of the dam is a function of its size. Hence, for other fixed material properties

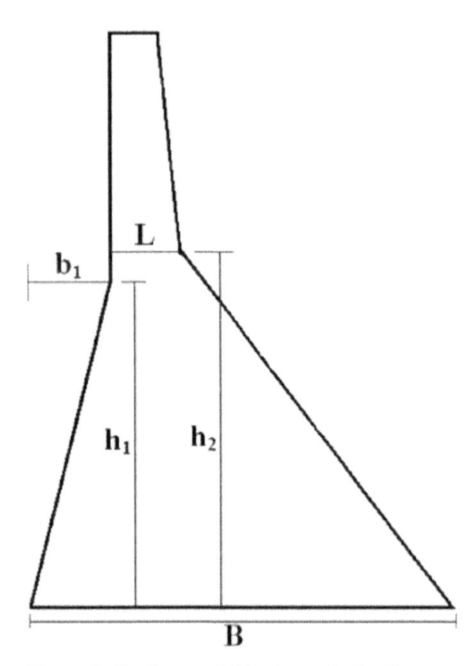

Figure 4. Design variables in optimization.

given in Table 1, the design criteria of dam including required safety factors against overturning and sliding and maximum principal stresses within the dam's body are related to the geometry of the dam.

Figure 4 shows a schematic view of Bluestone Dam with geometric design variables considered in this study. As it is clear from the figure, maximum number of geometry variables allowed in CADAM (five geometric variables) is considered and optimized. The ranges of variables are considered as follows:

$$B=[34,42],\ b_1=[0,10],\ h_1=[20,52.75],\ h_2=[30,39],\ L=[4.5,7.5] \quad (5)$$

In order to obtain a fairly complete database, 322 models of the dams with geometric values in the range of Equation 5 were developed and a complete analysis was carried out on each model by CADAM to obtain safety factors against overturning and sliding. Moreover, maximum principal stresses within dam body were obtained for all models. The results are then used to obtain appropriate formulas for these design parameters using GP. These formulas will be employed in optimization procedure in order to expedite the process of finding optimal values of design variables and to reduce computational cost of the approach. A glance on concepts of GP is presented in following paragraphs.

A branch of artificial intelligence is GP, which is an evolutionary algorithm-based methodology inspired by biological evolution to find computer programs that perform a user-defined task. In the 1960s and early 1970s, evolutionary algorithms became widely recognized as optimization methods. The first presentation of modern "tree-based" GP was introduced by Cramer (Takie & Ellingwood, 2003). This study was later greatly developed for application of GP in various complicated optimization and search problems. Recently GP has produced many novel and outstanding results in areas such as engineering, sorting, and searching, due to improvements in GP technology and the exponential growth in CPU power.

GP evolves computer programs, represented in memory as tree structures (Cramer, 1985). Trees can be easily evaluated in a recursive manner. Every tree node has an operator function and every terminal node has an operand, making mathematical expressions easy to evolve and evaluate

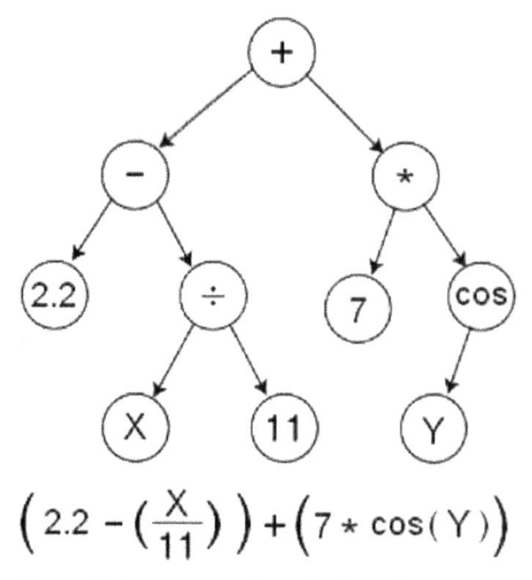

Figure 5. Representation of a function as a tree structure in G P.

(see Figure 5). Therefore, GP favors the use of programming languages such as Lisp that naturally embodies tree structures. Table 4 gives the parameter setting for GP used in this study. As it is customary, 70% of data are randomly considered for training and the rest 30% for test.

4.2. Optimization methods

The proposed method can be effectively hybridized with any evolutionary algorithm. Three different metaheuristics including, PSO, FA, and TLBO are used to find the optimal shape of Bluestone Dam and the results are compared. In the following, brief reviews on algorithms are presented.

4.2.1. PSO algorithm

For the first time, Kennedy and Eberhart (1995) proposed the standard PSO based on the swarm behavior such as fish and bird schooling in nature. This approach searches a space of an objective function by individual agents called particles. A swarm is formed by the collection of particles. The particle movement has stochastic and deterministic components. The particle is attracted toward the position of the current global best while at the same time it has a tendency to move randomly. When a particle finds a location that is better than any previously found locations, it updates it as the new current best for particle *i*. There is a current best for all *n* particles. The aim is to find the global best among all the current bests until the objective no longer improves or after a certain number of iterations are carried out. In standard PSO algorithm, the swarm is updated by the following equations:

$$V_i^{k+1} = \omega V_i^k + c_1 r_1 \left(P_i^k - X_i^k\right) + c_2 r_2 \left(P_g^k - X_i^k\right) \tag{6}$$

Table 4. Parameter setting in GP

Parameter	**Value**
Population size	600
Crossover probability	90%
Mutation probability	5%
Maximum number of generations	20,000
Maximum depth of tree after cross over	200
Maximum mutant depth	4
Selection method	Tournament selection

$$X_i^{k+1} = X_i^k + V_i^{k+1} \quad (7)$$

where X_i^k and V_i^k represent the current position and the velocity of the ith particle at time k, respectively; P_i^k is the best previous position of the ith particle (called $Pbest_i$) at time k and P_g^k is the best global position among all the particles in the swarm (called *gbest*) at time k; r_1 and r_2 are two uniform random numbers between *zero* and *1.0*; and c_1 and c_2 are two cognitive and social accelerating constants.

Shi and Eberhart (1998) introduced an inertia term ω in Equation 6 and rewrote it to the form:

$$V_i^{k+1} = \omega V_i^k + c_1 r_1 \left(P_i^k - X_i^k\right) + c_2 r_2 \left(P_g^k - X_i^k\right) \quad (8)$$

They proposed that ω be selected such that $0.4 \leq \omega \leq 0.9$. Consequently, they report improved convergence rates when ω is decreased linearly during the optimization.

In this study, c_1 and c_2 were taken to be 1.0 and the inertia term begins with a value equal to 0.9 and it is decreased linearly to a value equal to 0.4 during the iterations.

4.2.2. Firefly algorithm

One of the latest metaheuristics algorithms is FA which is a nature-inspired algorithm developed by Yang (2009), inspired by the light attenuation over the distance and fireflies' mutual attraction. FA idealizes some of the characteristics of the firefly behavior in nature. They follow three rules: (1) all the fireflies are unisex, (2) attractiveness is proportional to their flashing brightness which decreases as the distance from the other firefly increases due to the fact that the air absorbs light. Since the most attractive firefly is the brightest one, to which it convinces neighbors moving toward. In case of no brighter one, it freely moves any direction, and (3) brightness of every firefly determines its quality of solution; in most of the cases, it is proportional to the objective function.

During the loop of pair-wise comparison of light intensity, the firefly with lower light intensity will move toward the higher one. The moving distance depends on the attractiveness. The light intensity of each firefly is proportional to the quality of the solution, it is currently located at. In order to improve own solution, the firefly needs to advance towards the fireflies that have brighter light emission than is his own. Each firefly observes decreased light intensity, than the one the firefly actually emits, due to air absorption over the distance.

Attractiveness of a firefly is evaluated by the following equation (Yang, 2009):

$$\beta = \beta_0 \exp(-\gamma r) \quad (9)$$

in which β_0 is the attractiveness in distance $r = 0$ and γ is light absorption coefficient in the range $[0, \infty)$. The distance r between firefly i and j at x_i and x_j is defined as the Cartesian distance:

$$r = r_{ij} = \left\|x_i - x_j\right\| = \sqrt{\sum_{k=1}^{d} \left(x_{i,k} - x_{j,k}\right)^2} \quad (10)$$

where $x_{i,k}$ is the kth component of the spatial coordinate x_i of the ith firefly and d is the number of dimensions (Yang, 2009). Moreover, the movement of firefly i which is attracted by a more attractive or brighter firefly j is given by the following equation:

$$x_i = x_i + \beta_0 \exp(-\gamma r^2)(x_j - x_i) + \alpha(\varepsilon - 0.5) \quad (11)$$

in which α being the randomization parameter such that $\alpha \in [0,1]$, and ε is a vector of random numbers drawn from a Gaussian distribution or uniform distribution in the range [0, 1]. For most problems $\beta_0 = 1$ is considered. In the current study, the other parameters were set to $\gamma = 0.1$ and $\alpha = 0.1$.

4.2.3. TLBO algorithm

One of the latest metaheuristics is TLBO algorithm (Rao et al., 2011). Similar to most other evolutionary optimization methods, TLBO is a population-based algorithm inspired by the learning process in a classroom. The searching process consists of two phases, i.e. teacher phase and learner phase. In the teacher phase, learners first get knowledge from a teacher and then from classmates in the learner phase. In the entire population, the best solution is considered as the teacher ($X_{teacher}$). On the other hand, learners learn from the teacher in the teacher phase. In this phase, the teacher tries to enhance the results of other individuals (X_i) by increasing the mean result of the classroom (X_{mean}) towards his/her position $X_{teacher}$. In order to maintain stochastic features of the search, two randomly generated parameters r and T_F are applied in update formula for the solution X_i as:

$$X_{new} = X_i + r.(X_{teacher} - T_F . X_{mean}) \quad (12)$$

where r is a randomly selected number in the range of 0 and 1 and T_F is a teaching factor which can be either 1 or 2:

$$T_F^{\ i} = \text{round}\left[1 + \text{r and}(0,1)\{2-1\}\right] \quad (13)$$

Moreover, X_{new} and X_i are the new and existing solutions of i, (Rao et al., 2011, 2012).

In the second phase, i.e. the learner phase, the learners attempt to increase their information by interacting with others. Therefore, an individual learns new knowledge if the other individuals have more knowledge than him/her. Throughout this phase, student X_i interacts randomly with another student X_j ($i \neq j$) in order to improve his/her knowledge. In the case that X_j is better than X_i (i.e. $f(X_j) < f(X_i)$ for minimization problems), X_i is moved toward X_j. Otherwise it is moved away from X_j:

$$X_{new} = X_i + r.(X_j - X_i) \quad \text{if} \quad f(X_i) > f(X_j) \quad (14)$$

$$X_{new} = X_i + r.(X_i - X_j) \quad \text{if} \quad f(X_i) < f(X_j) \quad (15)$$

If the new solution X_{new} is better, it is accepted in the population. The algorithm will continue until the termination condition is met.

4.3. Constraint handling

The most common method for constraint handling in optimization algorithms is penalty function approach. The method has been already employed successfully to deal with constraints in other optimization problems (Ben Hadj-Alouane & Bean, 1997; Deb, 2000; Nanakorn & Meesomklin, 2001; Perez & Behdinan, 2007; Salajegheh & Khosravi, 2011). The main reason for the popularity of the method is its simplicity and its direct applicability regardless of the optimization method being used. This formulation utilizes general information of the collection of particles, such as the average of the objective function and the level of violation of each constraint in each iteration in order to define different penalties for different constraints as (Ben Hadj-Alouane & Bean, 1997):

$$f'(x) = \begin{cases} f(x) & \text{if } x \text{ is feasible} \\ f(x) + \sum_{i=1}^{m} k_i \bar{g}_i(x) & \text{otherwise} \end{cases} \quad (16)$$

in which k_i is the penalty parameter and it is calculated in each iteration by:

$$k_i = \left|\bar{f}(x)\right| \frac{\bar{g}_i(x)}{\sum_{j=1}^{m} \left[\bar{g}_j(x)\right]^2} \tag{17}$$

with $f(x)$ being the objective function and m being the number of constraints. Moreover, $g_i(x)$ is a specific constraint value so that violated constraints have values greater than zero, $\bar{f}(x)$ is the average of objective function in current particles and $\bar{g}_j(x)$ is the violation of the jth constraint averaged over the current collection of particles.

The illustration of Equation 16 is that the problem is actually solved as an unconstrained one, where in the minimization case, the objective function is designed such that non-feasible solutions are characterized by high-function values.

5. Results

The results of the implementation of aforementioned methodology including the results of GP implementation and the results of optimization are presented in following subsections.

5.1. Genetic programming

In the current study, the following operator functions and operands were used to find suitable expressions for safety factors against sliding and overturning according to the database developed by analyzing the models:

$+, -, \times, /, ^\wedge, \exp, \sin, \cos$

After evolutions in GP, the following expressions for safety factor against sliding (SSF) and OSF, and principal tensile stress were found:

$$\begin{aligned}\text{SSF} = {} & 0.118765 + 0.001535 \times b_1 \times L + 0.001053 \times B \times L + 0.000551 \\ & \times B \times h2 + 0.000329 \times b_1 \times h_1 - 1.928443e^{-6} \times B - 0.003485 \times b_1\end{aligned} \tag{18}$$

$$\begin{aligned}\text{OSF} = {} & 0.023929 \times L + 0.020932 \times B + 0.000333 \times b_1 \times hx + 0.000315 \\ & \times B \times h_2 + 0.000234 \times B \times b_1 - 0.336741 - 0.000194 \times {b_1}^2 - 0.000459 \times b_1 \times h_2\end{aligned} \tag{19}$$

$$\begin{aligned}\sigma = {} & 4021 + 646.9 \times L + 209.8 \times b_1 + 5.366 \times L \times h_1 + 1.601 \times h_1 \times h_2 - 1.147 \\ & \times {h_1}^2 - 6.851 \times b_1 \times h_1 - 32.31 \times L \times h_2\end{aligned} \tag{20}$$

Figure 6(a–c) shows the convergence history in GP for finding Equations 18, 19, and 20, respectively.

Figure 7(a–c) shows the effectiveness of Equations 18, 19, and 20 in predicting SSF, OSF and tensile stress respectively. Moreover, Table 5 reports some statistical criteria for testing goodness of the aforementioned equations in the prediction of design constraints. As Figure 7(a–c) as well as Table 5 show, the performances of Equations 18 and 19 in predicting SSF and OSF of the dam are excellent, whereas the performance of Equation 20 in predicting principal tensile stress is good. Therefore, these expressions are reliable to be used in the optimization procedure in the next step. It is worth pointing out that the analyses showed that the compressive principal stresses were found to be all well below the allowable stresses, and hence this criterion was not influential in the design of the current dam and no expressions for predicting this stress was required.

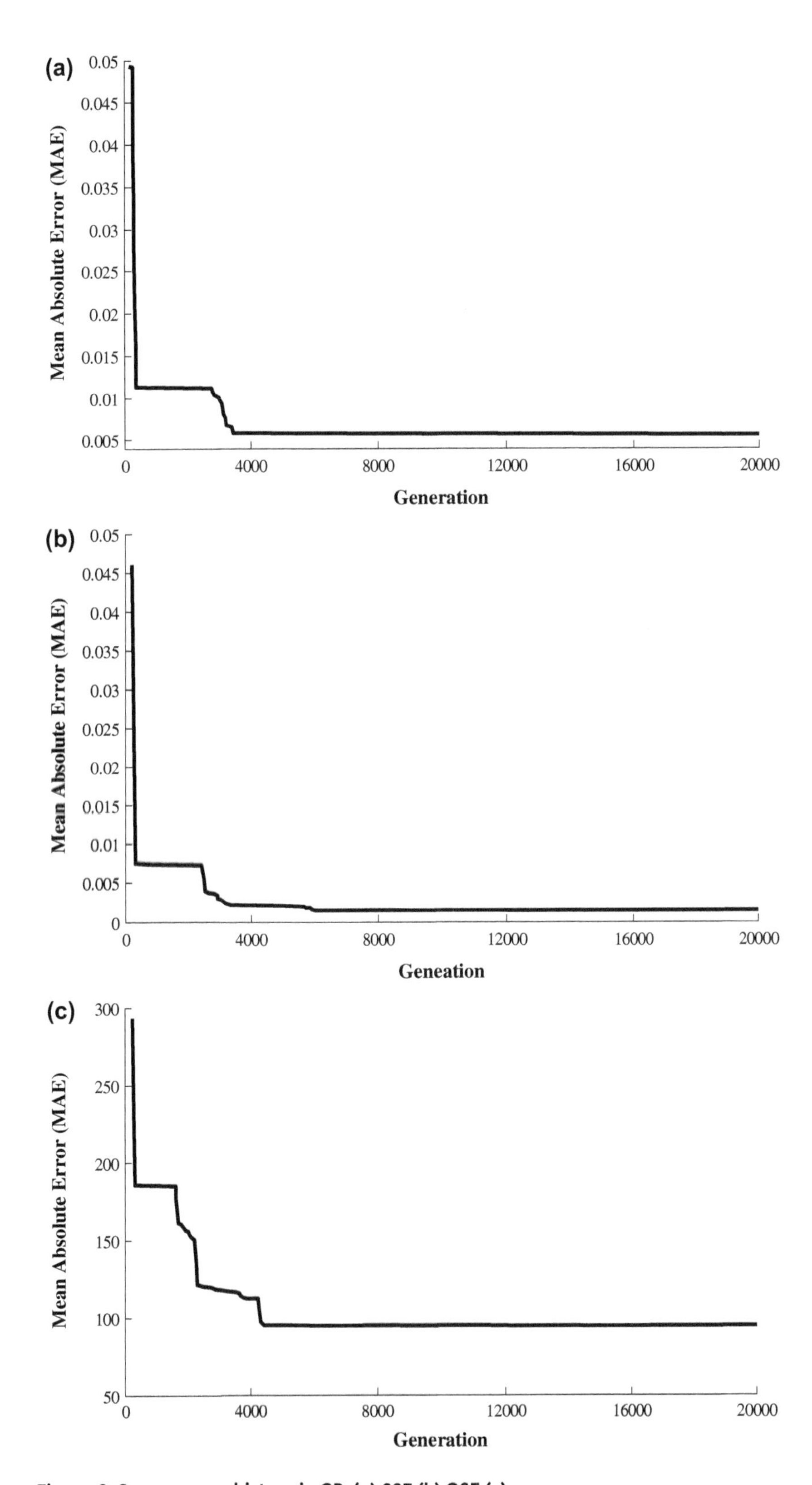

Figure 6. Convergence history in GP: (a) SSF (b) OSF (c) σ_{1p}.

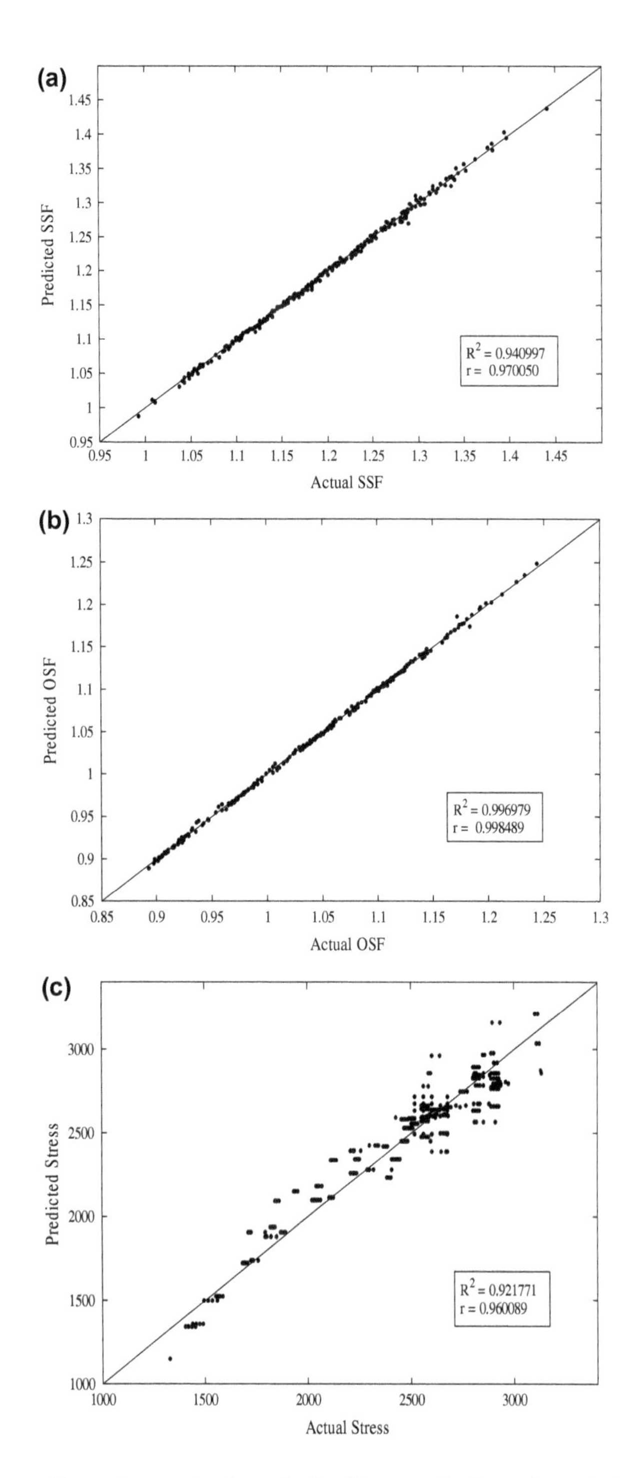

Figure 7. Actual values obtained by pseudo-dynamic analysis vs. predicted values: (a) SSF (b) OSF (c) tensile stress.

Table 5. Statistical criteria for the accuracy of obtained formulas by GP

Formula	Goodness of fit = R^2	Correlation coefficient = r	Maximum error = MA	Mean absolute error = MAE
SSF (Equation 18)	0.940997	0.970050	0.051315	0.004855
OSF (Equation 19)	0.996979	0.998489	0.054844	0.001874
σ_{1p} (Equation 20)	0.921771	0.960089	358.0287	94.76764

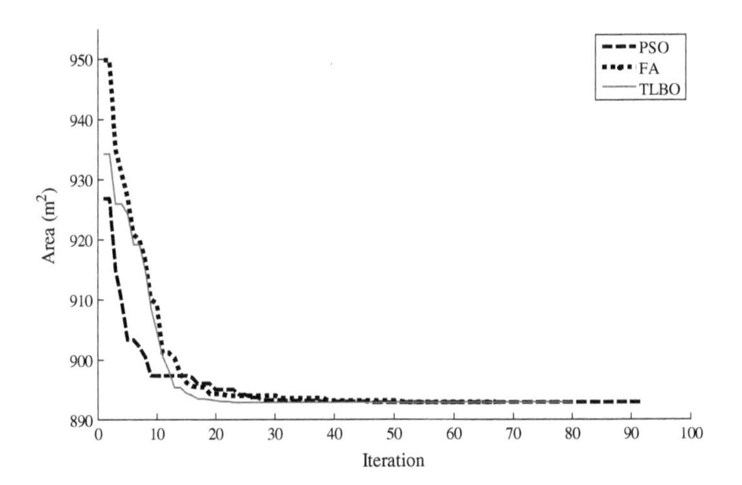

Figure 8. Comparison of convergence history of various metaheuristics.

5.2. Optimization

PSO, FA, and TLBO are used for finding the optimal values of design variables, whereas Equations 18 and 19 are used to evaluate safety factors against sliding and overturning, and Equation 20 for predicting principal tensile stress. Total numbers of 100 populations were used in each algorithm. Figure 8 compares the convergence history of minimizing cross-sectional area of dam using different metaheuristics, and Table 6 compares the results found by each algorithm. As Table 6 shows, all three methods lead to close results, indicating good performance of the algorithms. Optimal value of A = 892.8406 m^2 (W = 21,428.174 KN/m) was found by FA within 67 iterations, which shows slightly superior performance of FA compared to other two metaheuristics.

As Table 7 shows, for the design variables found by the algorithms, the values of SSF and OSF are the minimum values required for stability of the dam. After analyzing the optimal dam with CADAM, approximately the same values were found for SSF, OSF, and tensile stress, indicating accuracy of proposed formulas for these parameters. Moreover, the maximum tensile stress within the dam body was found to be σ_{1p} = 2,760 Kpa which is equal to the minimum allowable tensile stress; the maximum compressive stress is less than maximum allowable compressive stress within dam body, indicating stability of the dam against fracture as well. Compared to the actual weight of the dam

Table 6. Optimal values found for the design variables

	PSO	FA	TLBO
B	40.0078	40.0068	40.0071
b_1	10.0000	9.9997	9.9971
h_1	32.9067	32.8767	32.9069
h_2	38.4870	38.5058	38.4932
L	4.5000	4.5000	4.5000
Area (m^2)	892.7683	892.8406	892.8429
Iterations	92	67	80

Table 7. Values of the design constraints for the optimal shape of dam

SSF (Equation 18)	OSF (Equation 19)	SSF (CADAM)	OSF (CADAM)	σ_{1p}	σ_{2p}
1.3	1.1	1.308	1.104	2,760 Kpa	−5,707.519 Kpa

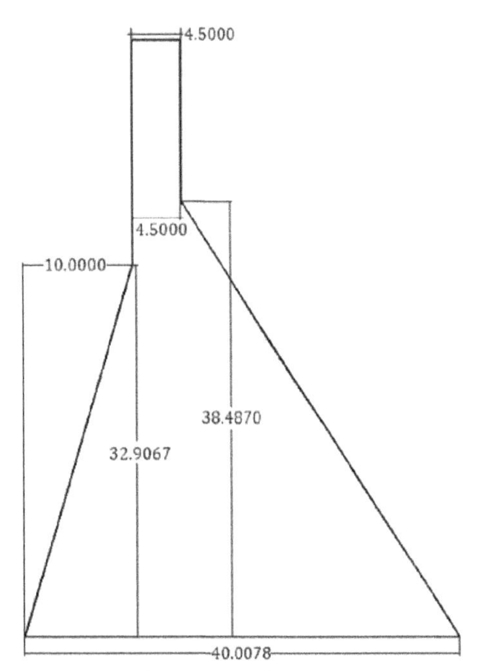

Figure 9. Optimal shape of the dam (dimensions in meter).

per unit length (W = 2,5101.44 KN/m), the optimal shape of dam shown in Figure 9 reduces the weight of dam up to a value equal to 14.6%. Figure 9 depicts the final shape of the cross section of the dam.

6. Conclusion

In this paper, the application of a powerful artificial intelligence approach, i.e. GP, in shape optimization of concrete gravity dam was introduced. A total number of 322 models of dams with regard to five design variables were developed and analyzed using pseudo-dynamic analysis. The results of the analyses were used to find appropriate formulas for design constraints. The accuracy of the formulas was verified, and they were used in an evolutionary optimization method in order to find optimal values of design variables. Three different metaheuristics, including PSO, FA, and TLBO were used for optimization and the results were compared. The procedure is computationally very effective since the necessity of time-consuming structural analysis in every evolution is eliminated. The results of applying the proposed method on a case study show that the procedure is capable of finding optimal values of design variables that satisfy all design constraints. One of the main aspects of this study is the applicability of artificial intelligence in solving real-life problems in engineering. In direct application of a structural analysis software to be used in a population-based optimization approach (say for 100 populations and 100 iterations), many analyses should be carried out to achieve the optimal design (say 10,000 analyses). Moreover, in some metaheuristics, the parameters of the algorithm should be tuned which means the overall procedure needs to be repeated several times. Hence, the population-based algorithms are very cumbersome in dealing with huge structures or in the cases in which time-consuming dynamic analysis is necessary. In hybridization of artificial intelligence techniques with population-based optimization approaches, the number of such structural analyses is remarkably reduced.

In its current form, CADAM which was used in this study as the structural analysis platform does not permit more than two slopes to be defined in the downstream face of the gravity dam. In most cases, designers prefer to use three different slopes in the geometry of the dam at downstream face. This can be mentioned as the main shortcomings of this powerful software in the design of concrete gravity dams which may be resolved in the future updated versions.

Funding
This study is a self-funded research.

Author details
Abdolhossein Baghlani[1]
E-mail: baghlani@sutech.ac.ir
Mohsen Sattari[1]
E-mail: m.sattari@sutech.ac.ir
Mohammad Hadi Makiabadi[1]
E-mail: h.makiabadi@sutech.ac.ir
[1] Faculty of Civil and Environmental Engineering, Shiraz University of Technology, Shiraz, Iran.

References

Baghlani, A., & Makiabadi, M. H. (2013a). An enhanced particle swarm optimization for design of pin connected structures. *Scientia Iranica A, 20*, 1415–1432.

Baghlani, A., & Makiabadi, M. H. (2013b). Teaching learning based optimization algorithm for shape and size optimization of truss structures with dynamic frequency constraints, IJST. *Transactions of Civil Engineering, 37*, 409–421.

Baghlani, A., Makiabadi, M. H., & Rahnema, H. (2013). A new accelerated firefly algorithm for size optimization of truss structures. *Scientia Iranica A, 20*, 1612–1625.

Ben Hadj-Alouane, A., & Bean, J. C. (1997). A genetic algorithm for the multiple-choice integer program. *Operations Research, 45*, 92–101. http://dx.doi.org/10.1287/opre.45.1.92

Beser, M. R. A. (2005). *Study on the reliability—Based safety analysis of concrete gravity dams* (Thesis). Graduate School of Natural and Applied Sciences of Middle East Technical University.

Chopra, A. K. (1998). Earthquake response analysis of concrete dams. In R. B. Jansen (Ed.), *Advanced dam engineering for design, construction, and rehabilitation* (pp. 416–465). New York, NY: Van Nostrand Reinhold.

Cramer, N. A. (1985). Representation for the adaptive generation of simple sequential programs. In *Proceedings of the 1st International Conference on Genetic Algorithms* (pp. 183–187). Pittsburgh, PA.

Danandeh Mehr, A., Kahya, E., & Yerdelen, C. (2014). Linear genetic programming application for successive-station monthly streamflow prediction. *Computers & Geosciences, 70*, 63–72.

Deb, K. (2000). An efficient constraint handling method for genetic algorithms. *Computer Methods in Applied Mechanics and Engineering, 186*, 311–338. http://dx.doi.org/10.1016/S0045-7825(99)00389-8

Doğan, D., & Saka, M. P. (2012). Optimum design of unbraced steel frames to LRFD-AISC using particle swarm optimization. *Advances in Engineering Software, 46*, 27–34.

Edward, L., & Stowasser, P. (2011). USACE modeling, mapping, & consequence center—Bluestone dam failure analysis & lessons learned. In *31st Annual USSD Conference* (pp. 1075–1088). San Diego, CA.

Ellingwood, B., & Tekie, P. (2001). Fragility analysis of concrete gravity dams. *Journal of Infrastructure Systems, 7*, 41–48. http://dx.doi.org/10.1061/(ASCE)1076-0342(2001)7:2(41)

Enalati H., Monazzam A., & GHaemian M. (2011). Performance evaluation of concrete gravity dams with dam crest displacement by non-linear analysis. In *6th International Conference on Seismology and Earthquake Engineering*. Tehran.

Farhoodnea, M., Mohamed, A., Shareef, H., & Zayandehroodi, H. (2014). Optimum placement of active power conditioners by a dynamic discrete firefly algorithm to mitigate the negative power quality effects of renewable energy-based generators. *International Journal of Electrical Power & Energy Systems, 61*, 305–317.

Fister, I., Fister Jr., Yang, X. S., & Brest, J. (2013). A comprehensive review of firefly algorithms. *Swarm and Evolutionary Computation, 13*, 34–46. http://dx.doi.org/10.1016/j.swevo.2013.06.001

Fontana, M., Ndiayeb, A., Breyssea, D., Bosa, F., & Fernandez, C. (2011, September). Soil-structure interaction: Parameters identification using particle swarm optimization. *Computers and Structures, 89*, 1602–1614.

Ghazanfari Hashemi, A., Bahraninejad, A., Ahmadi, M. (2009). Gravity dam shape optimization using simulated annealing. In *8th International Congress on Civil Engineering*. Shiraz.

Gomes, H. M. (2011). A firefly metaheuristic algorithm for structural size snd shape optimization with dynamic constraints. *Asociación Argentina de Mecánica Computacional, 76*, 2059–2074.

Hamidian, D., & Seyedpoor, S. M. (2010). Shape optimal design of arch dams using an adaptive neuro-fuzzy inference system and improved particle swarm optimization. *Applied Mathematical Modelling, 34*, 1574–1585. http://dx.doi.org/10.1016/j.apm.2009.09.001

Kang, F., Li, J. J., & Xu, Q. (2012). Damage detection based on improved particle swarm optimization using vibration data. *Applied Soft Computing, 12*, 2329–2335. http://dx.doi.org/10.1016/j.asoc.2012.03.050

Kennedy, J., & Eberhart, R. C. (1995). Particle swarm optimization. In *IEEE international conference on neural networks* (pp. 1942–1948). Piscataway, NJ.

Leclerc, M., Leger, P., & Tinawi, R. (2002). Computer aided stability analysis of gravity dams. In *4th Structural Specialty Conference of the Canadian Society for Civil Engineering*. Montreal, Canada.

Leclerc, M., Léger, P., & Tinawi, R. (2003). Computer aided stability analysis of gravity dams—CADAM. *Advances in Engineering Software, 34*, 403–420. http://dx.doi.org/10.1016/S0965-9978(03)00040-1

Li, S. H., Jing, L., & Zhou, I. J. (2010). The shape optimization of concrete gravity dam based on GA-APDL. In *International Conference on Measuring Technology and Mechatronics Automation* (pp. 982–986). China.

Makiabadi, M. H., Baghlani, A., Rahnema, H., & Hadianfard, M. A. (2013). Optimal design of truss bridges using teaching learning-based optimization algorithm. *International Journal of Optimization in Civil Engineering, 3*, 499–510.

Miguel, L. F. F., & Fadel Miguel, L. F. (2012). Shape and size optimization of truss structures considering dynamic constraints through modern metaheuristic algorithms. *Expert Systems with Applications, 39*, 9458–9467. http://dx.doi.org/10.1016/j.eswa.2012.02.113

Nanakorn, P., & Meesomklin, K. (2001). An adaptive penalty function in genetic algorithms for structural design optimization. *Computers and Structures, 79*, 2527–2539. http://dx.doi.org/10.1016/S0045-7949(01)00137-7

Niknam, T., Azizipanah-Abarghooee, R., & Rasoul Narimani, M. (2012). A new multi objective optimization approach based on TLBO for location of automatic voltage regulators in distribution systems. *Engineering Applications of Artificial Intelligence, 25*, 1577–1588. http://dx.doi.org/10.1016/j.engappai.2012.07.004

Perez, R. E., & Behdinan, K. (2007). Particle swarm approach for structural design optimization. *Computers & Structures, 85*, 1579–1588. http://dx.doi.org/10.1016/j.compstruc.2006.10.013

Qi, G. (2012). Optimized program design of gravity dam section. *International Conference on Modern Hydraulic Engineering, Procedia Engineering, 28*, 419–423.

Rao, R. V., Savsani, V. J., & Vakharia, D. P. (2011). Teaching-learning-based optimization: A novel method for constrained mechanical design optimization problems. *Computer-Aided Design, 43*, 303–315. http://dx.doi.org/10.1016/j.cad.2010.12.015

Rao, R. V., Savsani, V. J., & Vakharia, D. P. (2012). Teaching-learning-based optimization: An optimization method for continuous non-linear large scale problems. *Information Sciences, 183*, 1–15. http://dx.doi.org/10.1016/j.ins.2011.08.006

Roushangar, K., Mouaze, D., & Shiri, J. (2014, September). Evaluation of genetic programming-based models for simulating friction factor in alluvial channels. *Journal of Hydrology, 517*, 1154–1161. http://dx.doi.org/10.1016/j.jhydrol.2014.06.047

Roy, P. K., Paul, C., & Sultana, S. (2014). Oppositional teaching learning based optimization approach for combined heat and power dispatch. *International Journal of Electrical Power & Energy Systems, 57*, 392–403.

Salajegheh, J., & Khosravi, S. (2011). *Optimal shape design of gravity dams based on a hybrid meta-heruristic method* and weighted least squares support vector machine. *Iranian Journal of Optimization in Civil Engineering,* 1, 609–632.

Salmasi, F. (2011). Design of gravity dam by genetic algorithms. *International Journal of Civil and Environmental Engineering, 3*, 187–192.

Seyedpoor, S. M., Salajegheh, J., Salajegheh, E., & Gholizadeh, S. (2011). Optimal design of arch dams subjected to earthquake loading by a combination of simultaneous perturbation stochastic approximation and particle swarm algorithms. *Applied Soft Computing, 11*, 39–48. http://dx.doi.org/10.1016/j.asoc.2009.10.014

Shi, Y., & Eberhart, R. C. (1998). A modified particle swarm optimizer. In *Proceedings of IEEE International Conference on Evolutionary Computation* (pp. 303–308). Anchorage, Alaska.

Siamardi, K. (2007). Safety of concrete dams against earthquake and design criteria. In *1st National Conference of Dam and Hydrulic Structures*. Karj.

Silva, P., Santos, C. P., Matos, V., & Costa, L. (2014, October). Automatic generation of biped locomotion controllers using genetic programming. *Robotics and Autonomous Systems, 62*, 1531–1548. http://dx.doi.org/10.1016/j.robot.2014.05.008

Simoes, L. (1995). Shape optimization of dams for static and dynamic loading. In *International Course on hydroelectric power plants*. Coimbra, Portugal.

Simoes, L. M. C., & Lapa, J. A. M. (1994). Optimal shape of dams subject to earthquakes. In *The Second International Conference on Computational Structures Technology* (pp. 119–130). Athens, Greece.

Takie, P., & Ellingwood, B. (2002). *Fragility analysis of concrete gravity dams*. Atlanta, GA: US Army Corps of Engineers.

Tekie, P., & Ellingwood, B. (2003). Seismic fragility assessment of concrete gravity dams. *Earthquake Engineering and Structural Dynamics, 32*, 2221–2240. http://dx.doi.org/10.1002/(ISSN)1096-9845

USBR. (1976). *Design of gravity dams, design manual for concrete gravity dams*. Washington, DC : US Government Printing Office.

Valencia, P., Haak, A., Cotillon, A., & Jurdak, R. (2014, September 16). Genetic programming for smart phone personalisation, *Applied Soft Computing, 25*, 86–96.

Yagiz, S., & Karahan, H. (2011, April). Prediction of hard rock TBM penetration rate using particle swarm optimization. *International Journal of Rock Mechanics and Mining Sciences, 48*, 427–433. http://dx.doi.org/10.1016/j.ijrmms.2011.02.013

Yang, X. S. (2009). Firefly algorithms for multimodal optimization. In D. Hutchison, & T. Kanade (Eds.), *Stochastic algorithms: Foundations and applications, SAGA 2009, Lecture notes in computer science* (Vol. 5792 pp. 169–178). Berlin: Springer- Verlag. http://dx.doi.org/10.1007/978-3-642-04944-6

Yang, X. S. (2013). Multiobjective firefly algorithm for continuous optimization. *Engineering with Computers, 29*, 175–184. http://dx.doi.org/10.1007/s00366-012-0254-1

2

Road traffic safety perception in Jordan

Lina I. Shbeeb[1] and Wa'el H. Awad[2*]

*Corresponding author: Wa'el H. Awad, Faculty of Engineering Technology, Al Balqa' Applied University, Amman, Jordan
E-mail: whawad@awads.org

Reviewing editor: Filippo G. Pratico, University Mediterranea of Reggio Calabria, Italy

Abstract: In the past 20 years, several safety measures were taken in an attempt to reduce traffic-related fatalities. Although reduction in deaths was occasionally noticed, the sustainability of that trend has never achieved. This study explores Jordan safety profile trying to explain the traffic safety trend. It examines road traffic safety perception of both public community and road specialist. Main turning points that may have contributed in explaining prevailing traffic safety conditions have been collected. A questionnaire is administered to two distinct groups of the Jordanian society (general public and road specialist) with a total of 167 subjects. The subjects were asked to evaluate the effectiveness of a list of safety measures. Results showed that government took effective but not sustainable measures (mainly enforcement and legislative). Other ineffective measures (administrative and engineering) were taken but they were not target oriented. The main focus of interviewed subjects in order to improve traffic safety in Jordan is to establish sustainable engineering measures and improving the vehicle fleet, driver licensing, and testing procedures. As a surprising result, enforcement measures were not considered as desirable as other measures.

Subjects: Civil, Environmental and Geotechnical Engineering; Engineering Education; Transport & Vehicle Engineering; Urban Studies

Keywords: road; safety; measure; profile; effectiveness

ABOUT THE AUTHORS

Lina I. Shbeeb is an assistant professor/Faculty of Engineering/Al-Ahliyya Amman University since 2012. She has completed her PhD in traffic planning and engineering from Lund University in Sweden (2000). She served in Al Balqa' Applied University (2001–2010) and became head of civil engineering department (2006/2007). She worked in German Jordanian University as Road Safety center of excellence director (2010/2011). In 2013, she was assigned as Jordan Minister of transport for two years.

Wa'el H. Awad is an associate professor of Traffic Engineering finished his higher education in the United States (University of Colorado at Denver UCD, 1997). Worked for Colorado TransLab/UCD before returning to Jordan and worked for Al Balqa' Applied University (BAU). Awad spent five years (2008–2013) et al. Ahliyya Amman University (AAU), established a new civil engineering program and an ITS master program, and became the Dean of Engineering from 2011 to 2013.

PUBLIC INTEREST STATEMENT

Several safety measures were taken in Jordan over the last 20 years that contribute in death reduction that was not sustainable. This study explores country safety profile and attempts to explain the trend. It examines road traffic safety measures perception of both community and road specialist. Main turning points that may have contributed in explaining prevailing traffic safety trend have been collected. A questionnaire is administered to two groups (general public and road specialist). A sample of 167 subjects was interviewed.
A list of safety measures were provided and they were asked to rate their effectiveness. The study showed that government took measures (mainly enforcement and legislation) that were effective but not sustainable. Administration and engineering measures were taken but they were not target oriented. Interviewed subjects stressed on the needs for providing engineering measures and improving vehicle, driver licensing, and testing procedures. Enforcement measures were not desirable as other measures.

1. Introduction

Road traffic crashes are a global concern. It is estimated that more than 1.2 million people die worldwide as a result of road traffic crashes and some 50 million are injured per annum (Hughes, Newstead, Anund, Shu, & Falkmer, 2015). The world Bank reported that road traffic injuries ranked as the ninth leading cause of death in 2004 and expected to become the fifth leading cause of death in 2030 (Figure 1).

Road traffic injuries in Jordan caused 740 deaths in 2008 (on average two people are killed each day). Forty percent of these deaths were pedestrians, almost half of them children (less than 15-year old). The mortality rate from road traffic accidents is 12.6 deaths per 100,000 people in 2008 (Jordan Traffic Institute, 2010). A significant reduction in death rate is indicated in 2008 statistics compared to 2007, which showed a rate of 17.3 per 100,000 people (a reduction of 27%). The reduction came as a result of collective measures and initiatives that were taken at national level, which followed the occurrence of a tragic accident in the very beginning of 2008 that took a life of more than 30 persons. Private sector took some initiatives to contribute in reducing road traffic death toll. In general, the public awareness of road accidents has been raised significantly. This study document the traffic safety profile in Jordan during the past 20 years and discuss the major mile stones and turning points that had major impact on traffic safety. The study looks into the communal perception of road traffic safety.

The total length of the roadway network in Jordan increased from 6872 km (1996) to 7299 km (2013). Rural roads constitute 36.3% of the roadway network, while major roads comprise about 38%. Amman governorate, the capital of Jordan is covered by about 15% of total roadway network

Leading causes of death, 2004 and 2030 compared

TOTAL 2004

RANK	LEADING CAUSE	%
1	Ischaemic heart disease	12.2
2	Cerebrovascular disease	9.7
3	Lower respiratory infections	7.0
4	Chronic obstructive pulmonary disease	5.1
5	Diarrhoeal diseases	3.6
6	HIV/AIDS	3.5
7	Tuberculosis	2.5
8	Trachea, bronchus, lung cancers	2.3
9	Road traffic injuries	2.2
10	Prematurity and low birth weight	2.0
11	Neonatal infections and other	1.9
12	Diabetes mellitus	1.9
13	Malaria	1.7
14	Hypertensive heart disease	1.7
15	Birth asphyxia and birth trauma	1.5
16	Self-inflicted injuries	1.4
17	Stomach cancer	1.4
18	Cirrhosis of the liver	1.3
19	Nephritis and nephrosis	1.3
20	Colon and rectum cancers	1.1

TOTAL 2030

RANK	LEADING CAUSE	%
1	Ischaemic heart disease	12.2
2	Cerebrovascular disease	9.7
3	Chronic obstructive pulmonary disease	7.0
4	Lower respiratory infections	5.1
5	Road traffic injuries	3.6
6	Trachea, bronchus, lung cancers	3.5
7	Diabetes mellitus	2.5
8	Hypertensive heart disease	2.3
9	Stomach cancer	2.2
10	HIV/AIDS	2.0
11	Nephritis and nephrosis	1.9
12	Self-inflicted injuries	1.9
13	Liver cancer	1.7
14	Colon and rectum cancer	1.7
15	Oesophagus cancer	1.5
16	Violence	1.4
17	Alzheimer and other dementias	1.4
18	Cirrhosis of the liver	1.3
19	Breast cancer	1.3
20	Tuberculosis	1.1

Source: World health statistics 2008 (http://www.who.int/whosis/whostat/2008/en/index.html)

Figure 1. Expected leading causes of death worldwide between 2004 and 2030.

Source: World Health Statistics (2008). http://www.who.int/whosis/whostat/2008/en/index.html.

in the kingdom. Surprisingly, more than 75% of vehicles in Jordan are registered in Amman, the capital. Due to economic situation in the past decades, overall pavement conditions are substandard and routine maintenance is passing critical boundaries, where public criticisms occasionally pronounced.

2. Road accident situation in Jordan

The total reported number of road traffic accidents in 2010, results is 17,403 injuries and 670 fatalities (Jordan Traffic Institute, 2010). On a fatality rate scale, Jordan is ranked relatively high (Figure 2). Fatality rate as expressed per inhabitant is higher than any other selected European country, but it is lower than many reported rate of selected Middle East countries (Australia Bureau of Infrastructure, Transport and Regional Economics, 2010). Referring to WHO global reports, Jordan ranked 154 out of 185 countries (83.24%) in road fatalities per inhabitants per year, and ranked 104

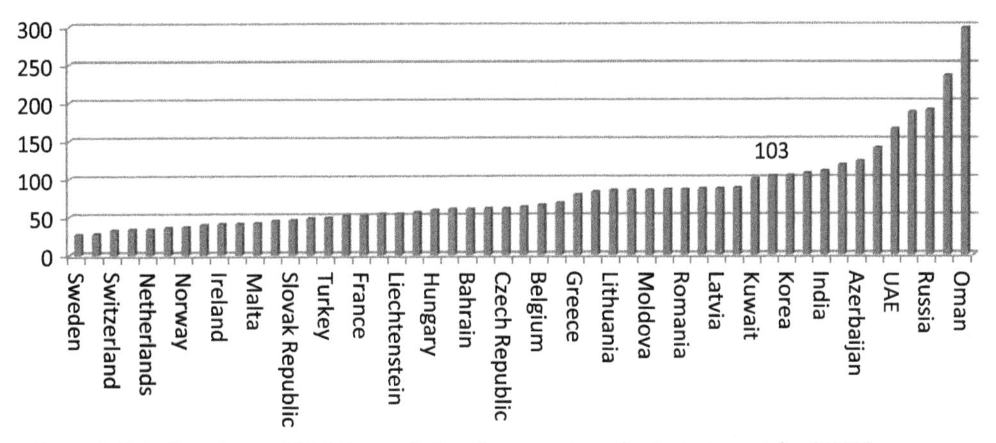

Figure 2. Fatality rate per 100,000 population for a number of selected countries (2006).

Source: International Road Safety Comparisons 2010 and National Data for Middle East countries (Australia Bureau of Infrastructure, Transport & Regional Economics, 2010).

Table 1. Road traffic accident rank among the top 10 causes of death, all ages, 2009 for selected countries around the globe

Middle East and North Africa			Africa			South America			Asia			Europe		
Country	Rank	Fatality rate	Country	Rank	Fatality rate	Country	Rank	Fatality rate	Country	Rank	Fatality rate	Country	Rank	Fatality rate
Jordan	5	16.74	Tanzania	11	6.41	Brazil	8	18.33	Malaysia	4	23.64	Greece	10	11.13
Iran	3	32.18	Kenya	10	7.71	Chile	11	13.71	Thailand	5	19.55	Czech	19	7.63
United Arab Emirates	2	24.11	South Africa	11	30.71	Mexico	9	15.96	Philippine	12	1.35	Denmark	31	4.61
Kuwait	8	17.12	Nigeria	15	3.16	Cuba	16	8.82				Estonia	18	5.82
Tunisia	5	14.50				Jamaica	11	12.90				United Kingdom	29	3.07
Syria	5	14.14										Romania	10	12.65
Libya	4	34.71										Sweden	29	4.89

Notes: World Health Organization, GLOBAL STATUS REPORT ON ROAD SAFETY TIME FOR ACTION, 2009 [1] Rank: WORLD LIFE EXPECTANCY, http://www.worldlifeexpectancy.com/ [2] Rate: WHO regional reports (Eastern Mediterranean [3], Africa [4], South America [5] the South-East Asia Region [6], and Western Pacific Region [7]).

[1] World Health Organization (2009a).

[2] World Life Expectancy, http://www.worldlifeexpectancy.com/

[3] World Health Organization (2010).

[4] World Health Organization (2009d).

[5] Organización Panamericana de la Salud (2009).

[6] World Health Organization (2009b).

[7] World Health Organization (2009c).

out of 185 countries (56.22%) in road fatalities per motor vehicles per year according to 2010 statistics (Ministry of Tourism and Antiquities/Jordan (MOTA), 2000–2009; World Health Organization, 2013).

Road accidents are one of the leading causes of death. Worldwide, it is ranked as the ninth cause of death. It is not within the top 10 causes of death in high nor in low income countries, whereas it is on ranked as the sixth cause of death in middle income countries (World Health Organization, 2009a). In Jordan, road traffic fatalities are the fifth cause of death, while it is ranked as the twenty third cause of death around the globe (Table 1). Traffic fatalities are highly ranked as main cause of death in the Middle East Countries, which is not the case in Europe except in Greece, which has high fatality rate as shown in Figure 1. Still European countries explore different means to make road traffic safer. Road deaths in South America and Africa are ranked among the top 10 causes of death but at low rank, which might be due to the presence of other epidemics and diseases that took lives of many citizens (Organizaci6n Panamericana de la Salud, 2009; World Health Organization, 2009b, 2009c, 2009d, 2010).

3. Traffic safety profile

The progression of road accident problem over the last 20 years showed a continuous fluctuation over the years. The development of vehicle ownership showed similar trend, which explains the variation in fatalities per 100,000 people over the years (Figure 3).

As elsewhere in the world as vehicle ownership increases, fatalities per 10,000 vehicles decrease. However, the trend in Jordan is not clear as in other countries in the world. On average, for every 1% increase in number of vehicles in Sweden (Eurostat, 2012), fatality rate decreases by 3% (Figure 4). The corresponding reduction in fatalities in Jordan is almost 1%.

Fatality reduction in Jordan (Figure 3) is due to the implementation of remedial measures packages which goes in parallel to political and economic circumstances that can be summarized as follows:

Jordan passed through significant economic recession during 1988–1989. The government took several measures to recover, which includes tightening the import of vehicles to save hard currency. Fatality rate (per 100,000 inhabitants) declined at the same pace as the reported declination in number of vehicles. The cabinet in 1990 took a decision mandating two-day weekend instead of one-day weekend. This decision effectively contributed to exposure reduction, which helped also in reducing the fatality rate.

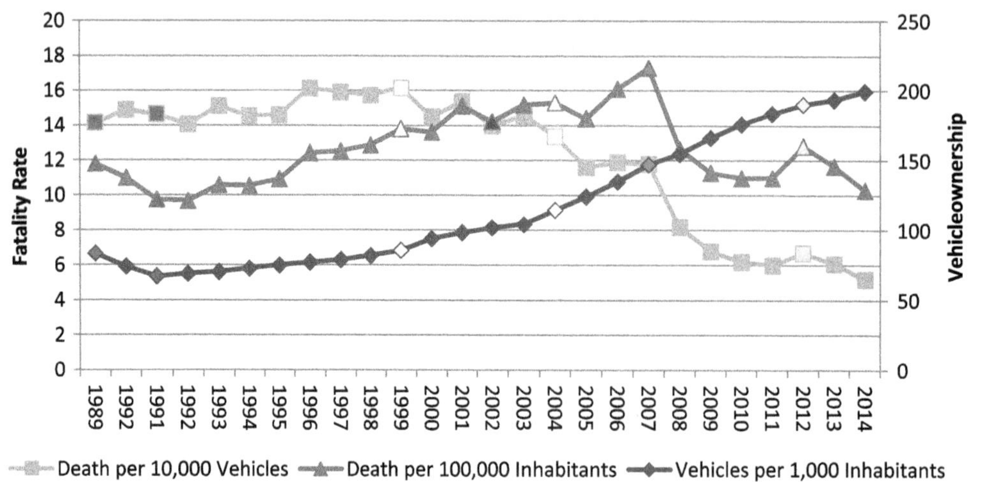

Figure 3. Jordan road traffic safety profile during 1989–2014.

Source: Jordan Traffic Institutes (2014).

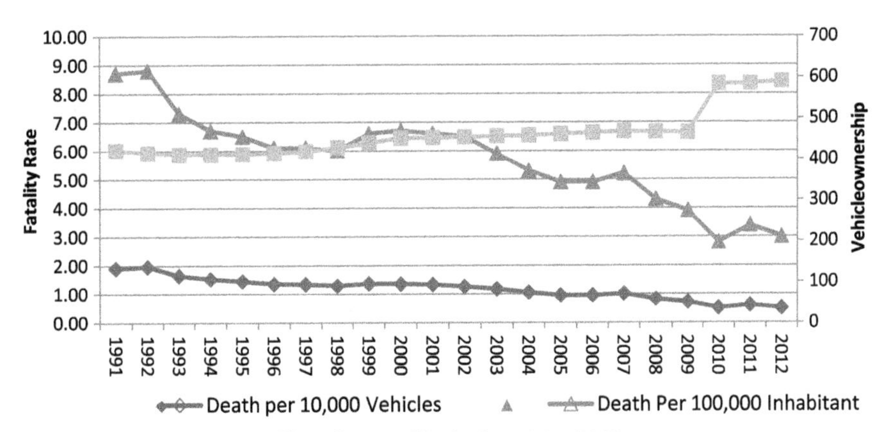

Figure 4. Sweden Road traffic safety profile during 1991–2012.

Source: Eurostat—Statistics Database (Eurostat, 2012).

In 1991, the first gulf war was launched and resulted in the reflux of many Jordanians residing in Kuwait and other Gulf countries. This helped in the recovery of the economy, the number of vehicles started to increase with time and the fatalities went up, but at a higher rate. Driving culture of the returnees was different from Jordanian driving culture, which might be the cause of the further increase in fatalities. Driving in Gulf countries is often associated with speeding that is related to the infrastructure conditions (wide and open streets) that facilitate speed and encourage speeders, which is not the case in Jordan.

Jordan signed Wadi Araba peace treaty in 1994. Jordan economy relies to some extent on foreign aids. Consequently, Jordan was granted aids and loans that improved its budget during 1995–1999 (Figure 5). Further, Jordan Parliament issued investment promotion low (low number 16/95) in 1995 to encourage private sector to invest in the country, which involved transport sector, among other sectors. Tourism industry also flourished (Figure 6), new hotels were constructed and touristic transport was an area of investment that was given incentives by Jordan Investment Cooperation (Ministry of Tourism and Antiquities, 2000–2009). Overall, number of vehicles during this period increased rapidly at 6% annual rate, leading to a 9% increase in fatality rate. The government was seriously concerned with the increase in death toll. But the emphasis was given to enforcement. The violation point system was introduced in 1997, without parallel effort in other fields, such as education or awareness campaigns. Figure 3 showed slight decrease in the progression of fatalities in the following year.

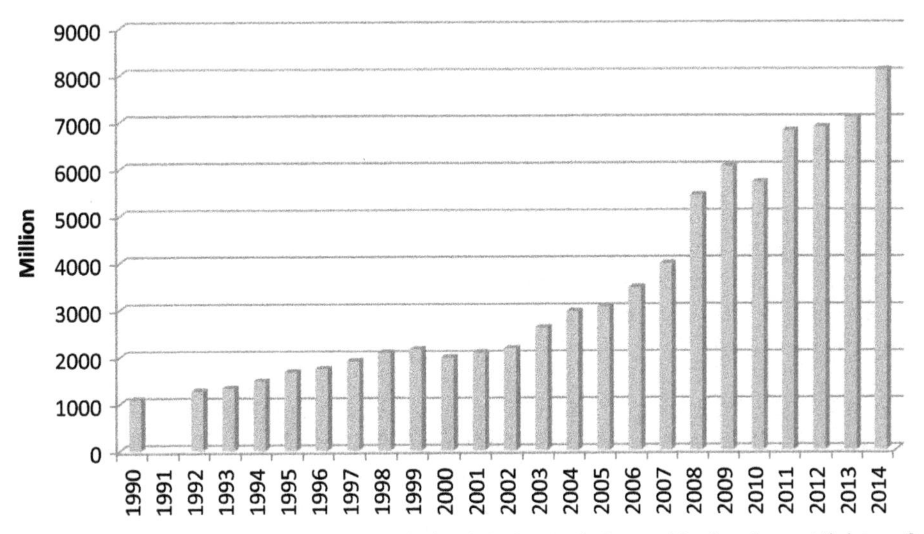

Figure 5. Budget of Jordan Government during (1990–2014). General Budget Laws, Ministry of Finance, Jordan, 1990–2014.

Safety belt law is another measure to enforce, which introduced for the first time in 1984, but not fully implemented. Following the joint cooperation between Sweden and Jordan concerning traffic safety (1998–1999), the issue of enforcing safety belt use was considered as a primary mean to reduce the fatalities. Traffic police started stringent execution of safety belt law in 1998. Jordan Traffic Institute continuously evaluated the use of safety belt showing an improvement from 23% in 1999 to 50% in 2006 (Jordan Traffic Institute, 1999, 2006).

On the legislative and administrative side, the government in 1999 reduced the customs on imported vehicles (both new and used), in addition to permitting the registration of vehicles which is more than five-year old, which has not been allowed before that date. This resulted in adding 72,000 extra vehicles on Jordan's road network in just fifteen months, without infrastructure upgrade to accommodate such an increase. Number of fatalities in 2000 increased by 13% compared to only 8% in 1999, the year that preceded the act. However, the fatality rate slightly decreased by 0.02 fatalities per 100,000 inhabitant (Jordanian Legislation System, 1988–2009).

Palestinian second uprising intifada was ignited in the fall of 2000, which negatively influenced tourism and level of investment attraction. Figure 4 showed slight budget reduction during 2000 and 2001, which affected the performance of most governmental agencies due to budgeting limitation. This implied shortage of funds to finance enforcement or engineering interventions, which may add to the severity of accident problem. The level of public frustration due to the political situation in Palestine might also have negatively contributed on driver's behavior, which became more aggressive.

In 2000 and further, Greater Amman Municipality started an implementation plan for upgrading 20 at-grade intersections to grade separated junctions in an attempt to improve traffic circulation and reduce congestion. The new setups facilitated the speed on main roads and contribute to the severity of accident problem. Pedestrian needs were not fully addressed in the new setup.

The government as a response to emergent problem issued a new traffic law in 2001 that raised the fines related to violations associated with high risk driving violations. Figure 3 showed that fatality rate in the following year (2002) decreased despite the increase in number of registered vehicles. Short-term decrease was evident that did not last and the number of fatalities went up again. A series of enforcement measures were taken such as establishing the automated surveillance program in 2003, which started in Amman, the capital. Ten cameras were installed at 10 intersections. Both speed and red light crossing were monitored. The measures were effective in reducing violation at the selected intersections but it would be illogical to assume that the implementation of the measures on such limited number of sites will have a significant reduction of fatalities at national level.

A new legislation that updated and modified existing traffic rules was issued in 2004. Again short-term effect was observed that did not last and the fatality rate increased only one year after the implementation of the new legislation. The increase in fatalities amalgamated with the significant increase in tourists, which eventually means more vehicles entering the country. Which is in a way or another is related to the increase in government expenditures. In 2004 and beyond, the increase registered vehicles was related to Gulf second war, which also brought in large numbers of Iraqis immigrants. Driving skills and culture of the Iraqis are different form Jordanian, this might add to the severity of the safety problem.

The governmental response was mainly directed to apply more strict rules and enforcement. Two traffic laws were issued in two successive years (2007 and 2008). The first law (2007) was very strict. The violation penalties were significantly increased, leading the public to complain, and as a result the law was suspended and substituted by a new law in 2008. Early 2008, a dramatic single crash occurred that took lives of many passengers. The community was shocked with the losses, the political leaders in the country felt the need to take immediate action to absorb the general public reaction to this crash. More policing and enforcement were first considered. New non-governmental agencies advocating for traffic safety were established. Their clear message was "the community can't take these loses and there is a need to take immediate actions." Since 2008, fatalities started to decrease.

4. Public perception of traffic safety measures

To examine the public perception, a questionnaire was designed and administrated. The sample consists of 167 observations. The sample includes two groups; the first group refers to general community perception hereinafter will be referred to as Public (152 subjects). The second group composes only of 15 subjects that work in road industry sector; this group will be referred to as Road Specialist.

The characteristics of the sample are given in Table 2. Female present is higher in road specialist group (47%) when compared to general public. Half of the public group hold graduated degree compared to 40% for the second group. Two-thirds of general public group do not use transit in their travel whereas two-third of road specialist commute by transit. Less than half of road specialist group own a car compared to 85% for the public group. The majority of both groups, more than 80% have private driving license.

The subjects were requested to provide information about their life style, which include income, smoking habits, availability of health insurance and the associated cost. The subjects provide also same information on their safety record, which cover accident past experience and number of committed traffic violations during the last three years.

Table 2. Sample characteristics

	General public		Road specialist	
	Frequency	Percent	Frequency	Percent
Female	14	9.2	7	47
Male	138	90.8	8	53
Less than high school	11	7.2	1	6.7
High school	22	14.5	5	36
Associate degree	36	23.7	3	20
Graduated	83	54.6	6	40
Transit user	55	36.2	10	66.7
Not transit user	97	63.8	5	33.3
Own a car	126	85.5	7	46.7
Not own a car	22	14.5	8	53.3
Hold license	145	95.4	13	86.7
Does not hold license	7	4.6	2	13.3
Private license	122	80.3	14	93.3
Smoker	81	53.3	6	40
Not smoker	69	45.4	7	46.7
No answer	2	1.3	2	13.3
Have health insurance	106	69.7	12	80
No health insurance	46	30.3	1	6.7
No Answer			2	13.3
Total	152	91%	15	9%
Average				
Average number of accidents	2.3	100%	1	58%
Average number of violations	5		2.33	

Table 2 showed that subjects in public group have higher involvement in accidents (on average, 2.3 accidents) compared to road specialist group (one accident on average). The average number of traffic violation committed of the first group (public) is twice the average reported for the second group (Road specialist).

A set of 18 remedial measures were presented to the subjects to assess their acceptance and how they perceive the effectiveness of such measures on traffic safety. The proposed measures can be grouped into five different categories of work (enforcement, engineering, education, driver/vehicle testing and licensing, administrative and rescue service). The remedial measures are briefly described as follows:

Category I—Enforcement

(1) Tightening traffic sanction: increase the presence of traffic police in the streets, whether temporally or spatially, to monitor and ticket violators.
(2) Increasing electronic surveillance in urban areas: install more electronic surveillance equipment on key locations and black spots within urbanized jurisdictions.
(3) Increasing electronic surveillance in rural areas: install more electronic surveillance equipment on key locations and black spots within rural areas.
(4) Banning smoking and mobile use: ticketing violators.

Category II—Engineering

(1) Engineering and technical measures: improve planning; design; and construction practices for highways.
(2) Improving pedestrian facilities: channelize and identify pedestrians' facilities to reduce conflict with motorized traffic.
(3) Improving public transportation means: provide competitive service with more coverage.
(4) Providing safety barriers, marking, signs: improve street furniture and traffic control devices.

Category III—Education

(1) Launching traffic safety campaigns: promote safe driving behavior in a series of scheduled and sustained program.

Category IV—Driver/vehicle testing and licensing

(1) Tightening vehicle licensing procedures: elevate the operating conditions of motor vehicle fleet.
(2) Increase driving practical training hours: improve the driving skills prior to licensing.
(3) Increase driving theoretical training hours: improve the driving knowledge of rules and regulation prior to licensing.
(4) Tightening driving test procedure: improve the competitiveness of licensed drivers.
(5) Improving the capacity of driving instructors: capacity building by certification and routine follow-up.
(6) Tightening driving license procedures: by increasing the standards and pushing up the passing thresholds.

Category V—Administrative and rescue service

(1) Establish specialized traffic courts: currently, traffic-related legal cases are referred to regular court system.
(2) Establish higher council for traffic safety: to be responsible for accomplishing the country strategy in traffic safety.
(3) Improving rescue and ambulance services: through management and coordination.

Although, the above categories are distinct theoretically, the interaction between and among each other is well investigated in the literature and beyond the scope of this paper. The infrastructure remedial measures related to geometry and pavement should be supported by traffic control remedial measures to come up with one integrated engineering perspective that can be supported by education and enforcement.

5. Data analysis

The selected analysis is aiming at testing the following hypothesis:

- General public have the same perception of traffic safety as road specialist.
- Road user involvement in traffic incidents is related to the seriousness of these events.
- People assess their experience in traffic in the same way they assess other people experience.
- The community desire to implement traffic safety measures is related to their effectiveness.

The variables included in the analysis are ordinal variables and the appropriate tests are based on cross tabulation analysis, which include both Chi Square, and Spearman correlation. Chi-Square test based on analyzing subject responses on a five-point scale whereas Spearman correlation analysis, which considers two categories which involves converting the five-scale to two categories (Frequency: Common and Less Common, Riskiness: Slight Risk and Risky). The first three-points on each scale form the first group and the remaining two-point scale forms the second group.

Road specialists believed that engineering measures are means to improve safety conditions, which is in good agreement with general publics' believe. Running traffic safety campaigns and conduct traffic education programs are ranked high in both groups priority that deemed necessary to upgrade safety condition in the country. Improving vehicle and driving licensing procedures are ranked high by road specialist group but not by the general public group. Road specialists also believe that improving rescuing procedure would reduce the severity of traffic accidents, which is to some extent in agreement with general public group view. Although both general public and road specialist stress the need of providing safe pedestrian facilities but they did not rate their effectiveness as high as providing more signs as indicated by general public or stress the need to improve

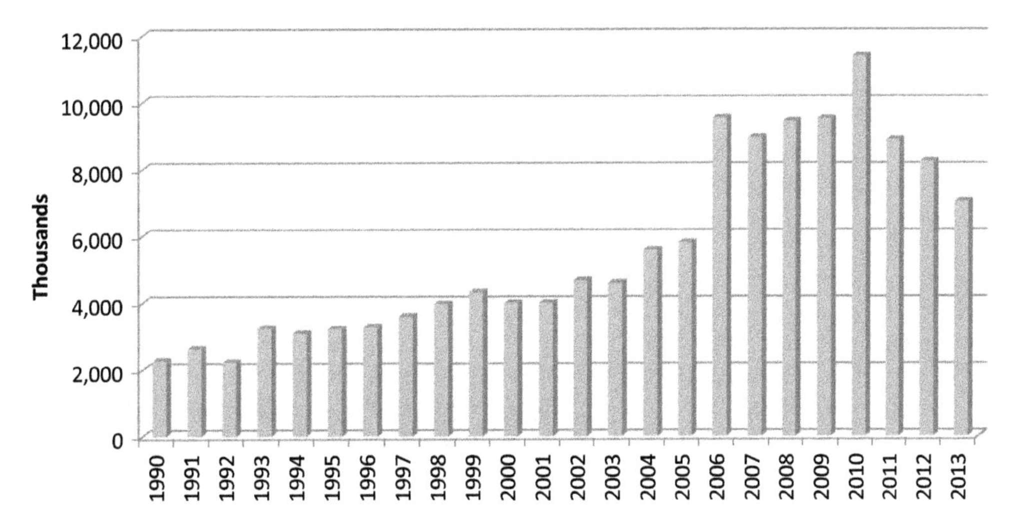

Figure 6. Number of visitors to Jordan during 1990–2013.
Source: Ministry of Tourism and Antiquities, Jordan (2000–2009).

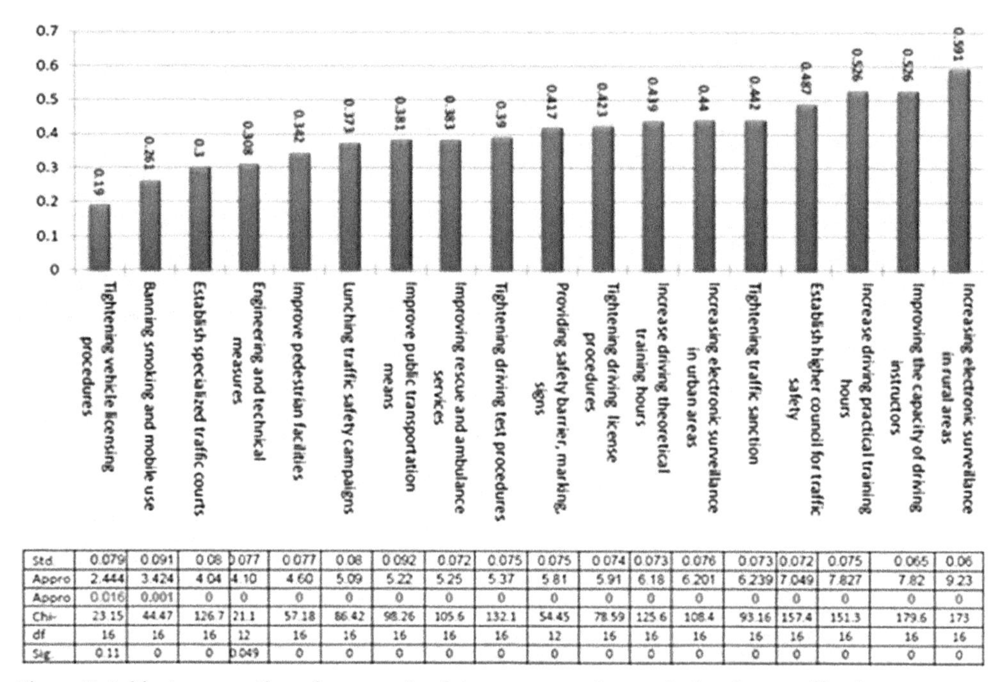

Std.	0.079	0.091	0.08	0.077	0.077	0.08	0.092	0.072	0.075	0.075	0.074	0.073	0.076	0.073	0.072	0.075	0.065	0.06
Appro	2.444	3.424	4.04	4.10	4.60	5.09	5.22	5.25	5.37	5.81	5.91	6.18	6.201	6.239	7.049	7.827	7.82	9.23
Appro	0.016	0.001	0	0	0	0	0	0	0	0	0	0	0	0	0	0	0	0
Chi-	23.15	44.47	126.7	21.1	57.18	86.42	98.26	105.6	132.1	54.45	78.59	125.6	108.4	93.16	157.4	151.3	179.6	173
df	16	16	16	12	16	16	16	16	16	12	16	16	16	16	16	16	16	16
Sig.	0.11	0	0	0.049	0	0	0	0	0	0	0	0	0	0	0	0	0	0

Figure 7. Subjects perception of proposed safety measures: the need of action vs. effectiveness.

driver instructor level as suggested by road specialists (Figure 7). Road specialists believe the improving rescue service would have the most effective measures to be taken; they think also that positive impact of up-grading of driving instructor capacity would be high on traffic safety. On the other hand, general public thinks that providing sign and marking would be the most effective means to improve road safety followed by improving the quality of pedestrian facility.

There is a significant difference in rating the needs or the desire of providing the proposed measures between both groups of subjects ($\rho = 0.661$, $p = 0.003$). However, they agree on their evaluation about the effectiveness of the presented remedial measures ($\rho = 0.313$, $p = 0.205$). There is a tremendous agreement on rating for the need of providing each measures and their effectiveness as indicated by Chi-square test (Figure 7) with a few exceptions reported for the following measures:

- The need and the effectiveness of providing more signs, marking and safety barriers
- The effectiveness of installing more cameras and other automated surveillance means

6. Discussions of results

The development of road traffic accidents in Jordan over the last 20 years showed a decrease in fatalities per 10,000 vehicles but not when considering fatalities per 100,000 people. Road accident is perceived as the main health concern for the community, which was also indicated by the results of the survey.

The measures that have been taken place over the past 20 years were always resulting in short-term effect that did not last for long or maintained continuity over the years. The measures were mostly related to enforcement and legislations aspect. Implementation of engineering measures was limited and if provided they were targeting capacity issues and localized in most cases without maintaining a systematic pattern that might support a positive behavioral pattern. Political situation in the region impose some impact on road safety, partially due to general attitude and social issues, but mainly due to financial difficulties.

The community stresses the need of considering infrastructure measures to improve safety as well as running traffic safety campaigns. Enforcement measures were not desirable as other measures. There is a significant relation between the community desires of providing safety measures and their effectiveness.

The overall international vision towards improving road safety (*Roles and responsibilities of different organisations in tackling road safety*, XXXX) requires the participation of many different organizations and sectors. No one sector working alone can effectively reduce the number of road accidents and/or deaths. This international vision is not adopted in Jordan, although many non-governmental agencies were established after the 2008 single crash to commence individual safety initiatives. No plan of coordination took place, and the unsustainable "positive" impact remained minimal.

In Jordan, no leading institution is taking responsibility to coordinate and manage the effort to establish a practical strategic plan for the future with a balanced blend of safety countermeasures. The strategic plan should incorporate room for collective efforts from all organizations (private and public) to tackle the traffic safety problem in Jordan. The international success stories in Fiji; Australia; and many other countries are leading examples (Asian Development Bank, 1997) in developing effective strategic plans to improve traffic safety.

International research (Nordfjærn, Şimşekoğlu, & Rundmo, 2014; Nordfjærn et al., 2014) is focusing on the effect of human factors and road traffic culture to improve safety, some conclusions are directing the decision-makers towards this issue, to accommodate psychological cognitions and driver behavior jointly with investments in developing the road infrastructure.

A balanced blend of safety countermeasures must be well planned for the coming 10 years to improve traffic safety in Jordan. Such plan should be based on practical strategic planning. Future research will consider the assessment of accomplished safety countermeasures in Jordan during the past 20 years to examine if Jordan is maintaining balance in measures taken to improve safety, or more emphasis were given to one measure compared to others, such as enforcement, and paying less to psychological cognitions and driver behavior.

Funding
The authors received no direct funding for this research.

Author details
Lina I. Shbeeb[1]
E-mail: l.shbeeb@ammanu.edu.jo
ORCID ID: http://orcid.org/0000-0002-1276-4011
Wa'el H. Awad[2]
E-mail: whawad@awads.org
ORCID ID: http://orcid.org/0000-0003-0639-5878
[1] Faculty of Engineering, Ahliyya Amman University, Amman, Jordan.
[2] Faculty of Engineering Technology, Al Balqa' Applied University, Amman, Jordan.

References

Asian Development Bank. (1997). *Road safety guidelines for the Asia and Pacific region*. Manila: Author. Retrieved from http://www.worldbank.org/transport/roads/saf_docs/fiji&oz.pdf

Australia Bureau of Infrastructure, Transport and Regional Economics. (2010). *International road safety comparisons*.

Eurostat. (2012). *Statistics database—Transport*. Retrieved from http://epp.eurostat.ec.europa.eu/portal/page/portal/eurostat/home/

Hughes, B. P., Newstead, S., Anund, A., Shu, C. C., & Falkmer, T. (2015). A review of models relevant to road safety. *Accident Analysis & Prevention, 74*, 250–270.

Jordan Traffic Institute. (1999/2006). *Use of safety belt* (In Arabic).

Jordan Traffic Institute. (2010). *Traffic accidents in Jordan*. Retrieved from www.jti.jo

Jordan Traffic Institute. (2014). *Traffic accidents in Jordan*. Retrieved from www.jti.jo

Jordanian Legislation System. (1988–2009). *A series of traffic, budget, transport and laws* (In Arabic). Retrieved from http://www.lob.gov.jo/ui/main.html

Ministry of Tourism and Antiquities. (2000–2009). *Tourism statistical newsletter*.

Ministry of Tourism and Antiquities/Jordan (MOTA). (2000–2009). *Tourism Statistical Newsletter*.

Nordfjærn, T., Şimşekoğlu, Ö., & Rundmo, T. (2014). Culture related to road traffic safety: A comparison of eight countries using two conceptualizations of culture. *Accident Analysis & Prevention, 62*, 319–328.

Nordfjærn, T., Şimşekoğlu, Ö., Zavareh, M. F., Hezaveh, A. M., Mamdoohi, A. R., & Rundmo, T. (2014). Road traffic culture and personality traits related to traffic safety in Turkish and Iranian samples. *Safety Science, 66*, 36–46. http://dx.doi.org/10.1016/j.ssci.2014.02.004

Organización Panamericana de la Salud[Pan American Health Organization]. (2009). *Informe sobre el Estado de la Seguridad Vial en la Región de las Américas*.

Roles and responsibilities of different organisations in tackling road safety. (XXXX). Retrieved from http://www.worldbank.org/transport/roads/saf_docs/orgs.pdf

World Health Organization. (2008). Retrieved from http://www.who.int/whosis/whostat/2008/en/

World Health Organization. (2009a). *Global Status Report on Road safety-time for action*.

World Health Organization. (2009b). *Regional Report on Status*

of road safety: The South-East Asia Region—A Call for policy direction.
World Health Organization. (2009c). *Road Safety in the Western Pacific Region—Call for action.*
World Health Organization. (2009d). *Status Report on Road safety in countries of the WHO African region.*
World Health Organization. (2010). *Eastern Mediterranean Status Report on Road safety—Call for action.*
World Health Organization (Ed.). (2013). *Global Status Report on Road Safety 2013: Supporting a decade of action* (Official Report, pp. vii, 1–8, 53ff (countries), 244–251 (Table A2), 296–303 (Table A10)). Geneva: Author. ISBN: 978 92 4 156456 4. Retrieved May 30, 2014. Tables A2 and A10, data from 2010. http://www.who.int/violence_injury_prevention/road_safety_status/2013/report/en/

Characterizing corridor-level travel time distributions based on stochastic flows and segment capacities

Hao Lei[1], Xuesong Zhou[2]*, George F. List[3] and Jeffrey Taylor[1]
*Corresponding author, Xuesong Zhou, School of Sustainable Engineering and the Built Environment, Arizona State University, Tempe, AZ 85287, USA
E-mail: xzhou74@asu.edu; xzhou99@ gmail.com
Reviewing editor: Anand J. Puppala, University of Texas at Arlington, USA

Abstract: Trip travel time reliability is an important measure of transportation system performance and a key factor affecting travelers' choices. This paper explores a method for estimating travel time distributions for corridors that contain multiple bottlenecks. A set of analytical equations are used to calculate the number of queued vehicles ahead of a probe vehicle and further capture many important factors affecting travel times: the prevailing congestion level, queue discharge rates at the bottlenecks, and flow rates associated with merges and diverges. Based on multiple random scenarios and a vector of arrival times, the lane-by-lane delay at each bottleneck along the corridor is recursively estimated to produce a route-level travel time distribution. The model incorporates stochastic variations of bottleneck capacity and demand and explains the travel time correlations between sequential links. Its data needs are the entering and exiting flow rates and a sense of the lane-by-lane distribution of traffic at each bottleneck. A detailed vehicle trajectory data-set from the Next Generation SIMulation (NGSIM) project has been used to verify that the estimated distributions are valid, and the sources of estimation error are examined.

ABOUT THE AUTHORS

Hao Lei received his PhD degree in Civil and Environmental Engineering from the University of Utah in 2013.

Xuesong Zhou received his PhD degree in Civil Engineering from the University of Maryland, College Park, in 2004. He is currently an associate professor in the School of Sustainable Engineering and the Built Environment at Arizona State University.

George F. List received his PhD degree in Civil Engineering from the University of Pennsylvania in 1984. He is currently a professor in the Department of Civil, Construction, and Environmental Engineering at North Carolina State University.

Jeffrey Taylor received his BS in Civil and Environmental Engineering from the University of Utah in 2010, and is currently a graduate student at the University of Utah. Their research interests include analytical modeling of transportation systems, and large-scale real-time traffic state estimation and prediction.

PUBLIC INTEREST STATEMENT

Travel time has long been regarded as one of the most important performance measures in transportation systems. This paper presents a theoretically sound and practically useful method for evaluating and quantifying the reliability of travel times due to the influence of random travel demand and road capacity. Freeway operating agencies and transportation planning organizations can apply and extend the model proposed in this paper to quickly estimate and improve the reliability of their systems. A number of traffic corridor management strategies can be evaluated effectively using this modeling framework to reduce congestion under recurring and non-recurring traffic conditions.

Subjects: Civil, Environmental and Geotechnical Engineering; Intelligent & Automated Transport System Technology; Transport & Vehicle Engineering

Keywords: travel time reliability; stochastic capacity; stochastic demand; queue model

1. Introduction

Travel time has long been regarded as one of the most important performance measures in transportation systems. Recently, significant attention has been devoted to evaluating and quantifying the reliability of travel times due to the influence of travel time variability on route, departure time, and mode choices. Operating agencies have also increased their efforts to monitor and improve the reliability of their systems through probe-based data collection, integrated corridor management, and advanced traveler information systems. Corridor management strategies have been designed to balance the performance of freeway and arterial networks in response to congestion and non-recurring events. Although noteworthy progress has been made in quantifying the variability in travel times, a number of challenges still need to be addressed. One is how to estimate distributions of individual vehicle travel times for both recurring and non-recurring congestion conditions, especially since both the demand and the capacity is stochastic. A framework for doing this is vital for both travelers and operating agencies (e.g. traffic management team) if they are to make informed decisions about actions that improve reliability.

Within the subject of analytical dynamic traffic network analysis, the "whole-link" model is widely adopted to describe link travel time evolution due to its simple description of traffic flow propagation through an analytical form. The link travel time function introduced by Friesz, Bernstein, Smith, Tobin, and Wie (1993) defines the travel time $\tau(t)$ on a single link at a time t as a linear function of the number of vehicles $x(t)$ on the link at time t:

$$\tau(t)=a+bx(t) \tag{1}$$

where a and b are constants in the above general linear form. A non-decreasing and continuous function is defined to calculate the cumulative number of vehicles on the link based on the time-dependent inflow and outflow rates, $u(t)$ and $v(t)$, at time t:

$$x(t)=x(0)+\int_0^t (u(s)-v(s))ds \tag{2}$$

Meanwhile, some more general non-linear travel time functions have been proposed as:

$$\tau(t)=f(x(t),u(t),v(t)) \tag{3}$$

A special case of this form, introduced by Ran, Boyce, and LeBlanc (1993), decomposes the link travel time as two different functions: g_1 accounts for flow-independent travel time and g_2 accounts for the queuing delay. A detailed mathematical representation is shown below.

$$\tau(t)=g_1[x(t),u(t)]+g_2[x(t),v(t)] \tag{4}$$

They later showed that, by assuming g_1 and g_2 are separable, i.e. $g_1=g_{1a}[x(t)]+g_{1b}[u(t)]$ and $g_2=g_{2a}[x(t)]+g_{2b}[v(t)]$, and Equation 4 can be rewritten as

$$\tau(t)=\alpha+f(u(t))+g(v(t))+h(x(t)) \tag{5}$$

where α is the free flow travel time, and $f(\cdot)$, $g(\cdot)$ and $h(\cdot)$ correspond to the functions of link inflow rate, link outflow rate, and the number of vehicles on the link, respectively.

Daganzo (1995) draws attention to problems with the general form in Equation 3, indicating that either a rapid decline in the inflows $u(t)$ or a rapid increase in outflow $v(t)$would lead to unrealistic travel time. Thus, he recommended omitting $u(t)$ and $v(t)$ from Equation 3, reducing the link travel time to a function of the number of vehicles on the link, that is, $\tau(t) = f(x(t))$. Although the link travel time function models provide some degree of simplification on travel time analysis, there is one significant drawback. Traffic congestion usually occurs at some bottleneck, and queues are produced and often grow beyond the bottleneck, which is difficult for any travel time function to capture (Zhang & Nie, 2005).

In dynamic traffic assignment and other applications, the vertical queue or point-queue model (Daganzo, 1995) was widely adopted to describe bottleneck traffic dynamics (Zhang & Nie, 2005). In a queuing-based travel time model, it is important to capture the variations of queue discharge flow rates and incoming demand to a bottleneck.

Conventionally, freeway capacity is viewed as a constant value—the maximum discharge flow rate before failure (HCM, 2000). However, the capacities vary according to different external factors in real life situations. Over the past decades, many researchers have developed a number of headway models to describe its distribution. Representatives of these models include the exponential-distribution by Cowan (1975), and normal distribution, gamma-distribution, and lognormal-distribution models by Greenberg (1966).

Incidents are one of the major contributing factors in capacity reductions, and the magnitude and duration of capacity reductions are directly related to the severity and duration of incidents (Giuliano, 1989; Kripalani & Scherer, 2007). In quantifying capacity reduction, the HCM (2000) provides guidance for estimating the remaining freeway capacity during incident conditions. Using over two years of data collected on freeways in the greater Los Angeles area, Golob, Recker, and Leonard (1987) found that accident duration fit a lognormal distribution. By extending the research of Golob et al. (1987) and Giuliano (1989) applied a lognormal distribution when analyzing incident duration for 512 incidents in Los Angeles.

It is commonly observed that travel demand fluctuates significantly within a day. During the morning and evening peak hours, surging demand may overwhelm a roadway's physical capacity and results in delays (Federal Highway Administration [FHWA], 2009). Waller and Ziliaskopoulos (2001), Chen, Skabardonis, and Varaiya (2003) and Lam, Shao, and Sumalee (2008) have used the normal distribution for modeling travel demand variation. Other researchers have modeled travel demand using the Poisson distribution (Clark & Watling, 2005; Hazelton, 2001) and the uniform distribution (Unnikrishnan, Ukkusuri, & Waller 2005).

Substantial efforts have been devoted to travel time variability estimation over the last decade. Statistical approaches (Oh & Chung, 2006; Richardson, 2003) have been widely adopted to quantify travel time variability from archived sensor data. In recent studies investigating the different sources of travel time variability, Kwon, Barkley, Hranac, Petty, and Compin (2011) proposed a quantile regression model to quantify the 95th percentile travel time based on the congestion source variables, such as incidents and weather. In their multi-state travel time reliability modeling framework, Guo, Rakha, and Park (2010) and Park, Rakha, and Guo (2011) provided connections between the travel time distributions and the uncertainty associated with the traffic states, e.g. with incidents versus without incidents.

A second approach uses numerical methods to characterize travel time distributions as a result of stochastic capacity and stochastic demand. Given a stochastic capacity probability distribution function (PDF), a Mellin transforms-based method was adopted by Lo and Tung (2003) to estimate the mean and variance of travel time distributions. Using a sensitivity analysis framework, Clark and Watling (2005) developed a computational procedure to construct a link travel time PDF under stochastic demand conditions. Ng and Waller (2010) introduced a fast Fourier transformation approach to approximate the travel time PDF from underlying stochastic capacity distributions. Although it can quantify the impacts of demand and capacity variation on the travel times, the steady-state travel time function-based approach is still unable to address the underlying time-dependent traffic dynamics.

In order to account for the inherent time-dependent traffic dynamics, some researchers (e.g. Zhou, Rouphail, & Li, 2011) have incorporated point-queue models into travel time variability estimation techniques. Using a dynamic traffic assignment simulator, Alibabai (2011) developed an algorithmic framework to investigate the properties of the path travel time function with respect to various path flow variables. While realistic simulation results require significant efforts in simulation/assignment model calibration, this approach is particularly suited for studying the effects of various uncertainty sources and assessing the benefits of traffic management strategies and traffic information systems.

The objective of this paper is to describe a model that can be used to estimate individual vehicle travel time distributions at the route level based on relatively simple information about the configuration of the corridor and its flows and capacities. Potential applications are investigated, and illustrated examples are presented. In addition, a highly detailed Next Generation SIMulation (NGSIM) vehicle trajectory data-set is used to check the validity of the travel time distributions under different traffic conditions.

The rest of the paper is organized as follows. Section 2 describes the point-queue-based travel time estimation framework with deterministic inputs. Monte Carlo simulation is then applied in Section 3 to compute the travel time distribution with stochastic inflow, outflow, and discharge rates. Highly detailed vehicle trajectory data from the NGSIM project is utilized to validate our methods in Section 4.

2. Computing route-level travel times

2.1. Problem statement

Consider a corridor with M bottlenecks, where each node in the node-link structure represents a bottleneck, and the road segments between consecutive bottlenecks are links with homogeneous capacity. Assume that node 0 is the starting point of the corridor, node m corresponds to bottleneck m, and each link between bottlenecks is denoted as link $(m-1, m)$, for $1 \leq m \leq M$. Link $(m-1, m)$ is the same as link m. Figure 1 illustrates a node-link representation for a corridor with M bottlenecks. Possible merge or diverge nodes are connected to bottleneck m and are denoted as m' or m'', respectively, so that the on-ramp before node 1 is denoted as $(1', 1)$, and the off-ramp before node 2 is denoted as $(2, 2'')$. In other words, the merge and diverge links are directly connected to the bottleneck. If there is more than one inflow or outflow between two bottlenecks, one can further decompose the link between those inflow entrances or outflow exits to several segments and merge/diverge points so as to construct the above node-link representation.

For the purposes of this analysis, the interest lies in how to estimate the travel time distribution for trips from point 0 to point m for a probe vehicle z, departing at time $t_0 = 0$. The aim is to estimate the distribution of the route-level (route) travel time, p_m^z, based on the following: (1) the number of vehicles $x_m(t_0)$ on each link m along the path at time t_0, (2) the discharge flow rate for each bottleneck c_m, and (3) the on-ramp or off-ramp flow rates f_m^{net}. The route-level travel time is defined to be the difference between the departure time at bottleneck 0 and the departure time at bottleneck m for

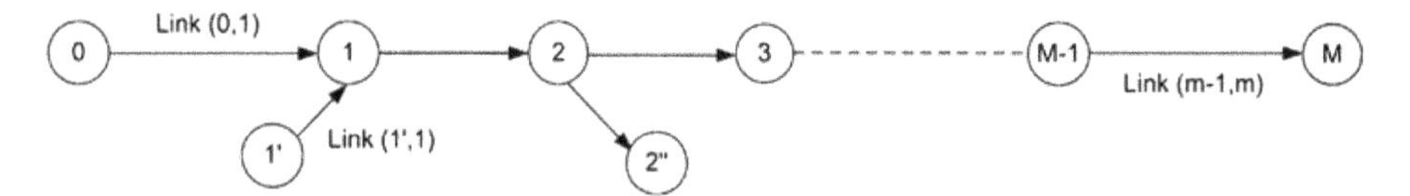

Figure 1. A node-link representation of a corridor with M bottlenecks.

probe vehicle z. Further, the departure time at bottleneck m is defined to be the time the probe vehicle leaves the queue at bottleneck m, which is the time when the number of vehicles in the queue before probe vehicle z at bottleneck m is 0.

The number of vehicles on each link is assumed to be observable from sensors, such as loop detectors, and the discharge flow rates and net flow rates on the on-ramps and off-ramps are assumed to be estimable from historical flow patterns or estimated based on prevailing traffic conditions (e.g. capacity reduction due to incidents).

The notation for the route-level travel time is described below.

Indices

z index for identifying a probe vehicle

k index for the simulation instance used in Monte Carlo simulation

m index for the bottlenecks and links along the corridor

Inputs

t_0 starting time, $t_0 = 0$

M number of bottlenecks along the corridor of interest

FFTT_m free-flow travel time over link $(m-1, m)$

c_m queue discharge rate of bottleneck m

f_m^{net} net flow rate at a merge or diverge corresponding to bottleneck m, that is, from an on-ramp to the mainline segment or from the mainline to the off-ramp

$x_m(t_0)$ number of vehicles on link $(m-1, m)$ at time t_0

$\mu_m(t)$ arrival rate of link $(m-1, m)$ at time t

$v_m(t)$ departure rate of link $(m-1, m)$ at time t

Variables to be calculated

$\tau_m^z(t)$ travel time on link m for probe vehicle z entering the link at time t

$\lambda_m(t)$ number of vehicles waiting at the bottleneck m at time t, that is, the number of queued vehicles behind bottleneck m

w_m^z waiting time in the vertical queue of bottleneck m for probe vehicle z

t_m^z arrival time for probe vehicle z at bottleneck m

p_m^z route-level path travel time from node 0 to bottleneck m

2.2. Travel time calculation

In a point-queue model, a link can be considered as two segments—the free-flow segment and the queuing segment. A vehicle travels at free-flow speed on the free-flow segment until reaching the beginning of the queuing segment, where it joins the queue waiting to be discharged. A queue is only formed if the link demand exceeds the bottleneck capacity; that is, the link arrival rate exceeds the link departure rate.

To construct a numerically tractable model for calculating route-level travel times along a corridor with multiple bottlenecks, several important assumptions are made.

(1) A point-queue model is adopted to calculate the delay on each link. On each link, a First-In, First-Out (FIFO) property is assumed to assure that any vehicles that enter the link before time t will exit the link before those entering after time t.

(2) The link traversal time is assumed to comprise a free-flow travel time and a queuing delay. The free-flow travel time is constant and flow-independent. The queuing delay is dependent on the number of vehicles in the queue when the probe vehicle arrives at the bottleneck $\lambda(t + \mathrm{FFTT})$ and the bottleneck queue discharge rate c_m. Thus, the link travel time is

$$\tau_m(t) = \text{FFTT}_m + w(t + \text{FFTT}_m) = \text{FFTT}_m + \frac{\lambda_m(t + \text{FFTT}_m)}{c_m} \tag{6}$$

where $w(t + \text{FFTT})$ is the queuing delay when vehicle z reaches the vertical queue at the bottleneck at time $t + \text{FFTT}$.

(3) The merge or diverge location is coincident with the position of the vertical queue.

(4) The bottleneck m remains congested across the estimation horizon, which extends from the current time t_0 to the arrival time of the probe vehicle z at the bottleneck m, t_m^z. The corresponding queue discharge rates c_m and net flow rates f_m^{net} in the estimation horizon are also assumed to be constant.

The first two assumptions are widely used in queuing models. The third makes it easy to incorporate the flow rate from a merge/diverge without explicitly considering the driving distance and free-flow travel time from the merge/diverge point to the bottleneck m.

Equation 6 considers the arrival time at the beginning of a link. By considering the arrival time at bottleneck m for vehicle z, t_m^z, the link traversal time can be rewritten as

$$\tau_m\left(t_m^z - \text{FFTT}_m\right) = \text{FFTT}_m + \frac{\lambda_m\left(t_m^z\right)}{c_m} \tag{7}$$

For a general queue with time-dependent arrival and departure rates, a continuous transition model can be used in Equation 6 to update the number of vehicles in the queue at any given time t.

$$\frac{d\lambda_m(t)}{dt} = \mu_m(t - \text{FFTT}_m) - v_m(t) \tag{8}$$

The number of queued vehicles $\lambda_m\left(t_m^z\right)$ at time t_m^z on bottleneck m can be derived from Equation 8, as shown in Equation 9.

$$\begin{aligned} l\lambda_m\left(t_m^z\right) &= \lambda_m(t_0) + \int_{t_0}^{t_m^z} \frac{d\lambda_m(t)}{dt} dt = \lambda_m(t_0) + \int_{t_0}^{t_m^z} \left[\mu_m(t - \text{FFTT}_m) - v_m(t)\right] dt \\ &= \lambda_m(t_0) + \int_{t_0}^{t_m^z} \mu_m(t - \text{FFTT}_m) dt - \int_{t_0}^{t_m^z} v_m(t) dt \end{aligned} \tag{9}$$

Since the fourth assumption requires the bottleneck to remain extant for the entire estimation period, the departure rate is equal to the bottleneck capacity of $\int_{t_0}^{t_m^z} v_m(t)dt = c_m \times \left(t_m^z - t_0\right)$. The remaining challenge is to estimate the unknown queue length $\lambda_m(t_0)$ at time t_0, and calculate the complex integral of $\int_{t_0}^{t_m^z} \mu_m(t - \text{FFTT}_m)dt$.

To illustrate these ideas, consider the example in Figure 1, where $m = 1$. In this case, the number of vehicles $x_1(t_0)$ and the net flow rate f_1^{net} associated with bottleneck 1 are given. For the specific starting time t_0, a probe vehicle z enters the vertical queue of bottleneck 1 at time $t_1^z = t_0 + \text{FFTT}_1$, and the number of vehicles in the queue at time t_1^z is

$$\lambda_1\left(t_1^z\right) = \lambda_1(t_0) + \int_{t0}^{t_1^z} \mu_1(t - \text{FFTT}_1)dt - c_1 \times \left(t_1^z - t_0\right) \tag{10}$$

Now consider a simpler case without merge and diverge points, i.e. $f_1^{\text{net}} = 0$. Thanks to the first-in and first-out property, we can show that $\lambda_1(t_0) + \int_{t0}^{t_1^z} \mu_1(t - \text{FFTT}_1)dt = x_1(t_0)$. The left-hand side $\lambda_1(t_0) + \int_{t0}^{t_1^z} \mu_1(t - \text{FFTT}_1)dt$ is the total number of vehicles stored in both the free-flow segment and the queuing segment before the probe vehicle z. The right-hand side is the actual number of vehicles observed on the *physical* link. One can use Figure 2 to map or "rotate" some of the vehicles from the physical link (shaded) to the vertical stack queue, and the other vehicles on the physical link (not

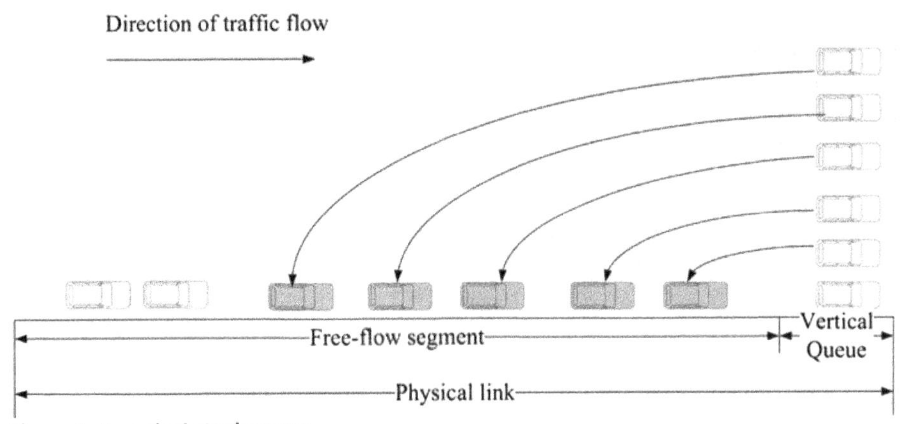

Figure 2. A vertical stack queue.

shaded) correspond to the vehicles that will arrive at the vertical queue between time t_0 and time $FFTT_1$ (that is, right before the probe vehicle). Notice that the length of the queue segment in the point-queue model is equal to zero and has unlimited storage capacity. Interested readers are referred to the paper by Hurdle and Son (2001) to examine the connection between physical queues and vertical stack queues.

The individual components of Equation 10 can be described visually using the cumulative vehicle count curves shown in Figure 3. Curve A is equivalent to the integral over the arrival rate, $A(t)=\int \mu_1(t)dt$, and the cumulative arrival curve at the vertical stack queue V is the cumulative arrival curve shifted by the free-flow travel time, $V(t)=A(t-\text{FFTT})$, and thus $V(t)=\int \mu_1(t-\text{FFTT}_1)dt$. The cumulative departure curve D is equivalent to the integral over the departure rate, $D(t)=\int v_1(t)dt=c_1\times(t_1^z-t_0)$. Substituting t with values of t_0 and t_1^z for $V(t)$, then Figure 3 shows that $V(t_1^z)-V(t_0)=\int_{t0}^{t_1^z}\mu_1(t-\text{FFTT}_1)dt$ and thus $\lambda_1(t_1^z)=\lambda_1(t_0)+\int \mu_1(t-\text{FFTT}_1)dt-c_1\times(t_1^z-t_0)$ and $x_1(t_0)=\lambda_1(t_0)+\int_{t0}^{t_1^z}\mu_1(t-\text{FFTT}_1)dt$.

By further considering the net flow rate from the merge or diverge point connected to the bottleneck, we now have

$$\lambda_1(t_1^z)=\lambda_1(t_0)+\int_{t0}^{t_1^z}[\mu_1(t-\text{FFTT}_1)-c_1]dt=x_1(t_0)+f_1^{\text{net}}\times t_1^z-c_1\times t_1^z \tag{11}$$

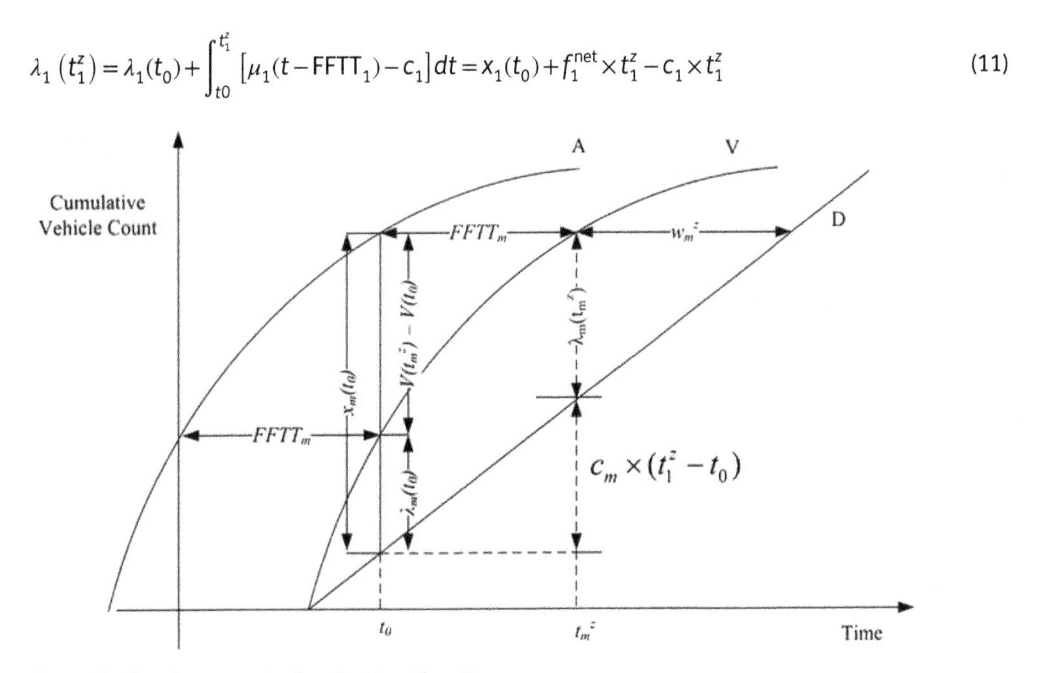

Figure 3. Visual representation for Equation 10.

Continuing to link 2 in Figure 1, the probe vehicle z will arrive at the queue of bottleneck 2 at t_2^z. Again, considering the FIFO assumption, the number of vehicles transferring from the first link to the second link before the probe vehicle z includes two terms, $\lambda_1(t_1^z)+c_1\times t_1^z$, which are the number of queued vehicles $\lambda_1(t_1^z)$ when the probe vehicle arrives at the first bottleneck at time t_1^z, and those vehicles $c_1\times t_1^z$ already entering the second link before time t_1^z. Following the derivation logic for Equation 11, the number of vehicles waiting in the queue ahead of vehicle z when it arrives at the second bottleneck at time t_2^z is

$$\lambda_2(t_2^z)=\lambda_1(t_1^z)+c_1\times t_1^z+x_2(t_0)+f_2^{net}\times t_2^z-c_2\times t_2^z \quad (12)$$

By substituting $\lambda_1(t_1^z)$ from Equations 11, 12 reduces to

$$\begin{aligned}\lambda_2(t_2^z)&=x_1(t_0)+f_1^{net}\times t_1^z-c_1\times t_1^z+c_1t_1^z+x_2(t_0)+f_2^{net}\times t_2^z-c_2\times t_2^z\\&=x_1(t_0)+f_1^{net}\times t_1^z+x_2(t_0)+f_2^{net}\times t_2^z-c_2\times t_2^z\end{aligned} \quad (13)$$

More generally, for bottleneck m:

(1) The number of vehicles waiting at the vertical queue of bottleneck m at time t_m^z can be expressed as

$$\lambda_m(t_m^z)=\sum_{i=1}^{m}x_i(t_0)+\sum_{i=1}^{m}(f_i^{net}\times t_i^z)-c_m\times t_m^z \quad (14)$$

(2) The arrival time for the probe vehicle at bottleneck m is

$$t_m^z=t_{m-1}^z+w_{m-1}^z+FFTT_m=t_{m-1}^z+\frac{\lambda_{m-1}(t_{m-1}^z)}{c_{m-1}}+FFTT_m \quad (15)$$

where $w_m^z=\frac{\lambda_m(t_m^z)}{c_m}$.

(3) Finally, the route-level travel time from bottleneck 0 to bottleneck m is

$$p_m^z=t_m^z+w_m^z=\sum_{i=1}^{m}[FFTT_i+w_i^z] \quad (16)$$

In summary, given the number of vehicles on each link, the queue discharge rate and the net flow on each bottleneck, the route-level travel time for a vehicle can be calculated by applying Equations 14–16 iteratively for links 1 through m. In each iteration, one first applies Equation 14 to obtain the number of queued vehicles at the bottleneck and then computes the queuing delay and update the route-level travel time up to the bottleneck of interest.

2.3. Illustrative example

To demonstrate how to the model can be used to calculate the route-level travel time and capture the delay propagation along a corridor, a corridor with 3 bottlenecks (Figure 4) can be used. Bottleneck 1 is on the freeway, bottleneck 2 is associated with an on-ramp, and bottleneck 3 is in conjunction with an off-ramp. The bottleneck discharge rates for those bottlenecks are 90, 90, and 60 vehicles/min, respectively. The initial number of vehicles on each link is 750, 600, and 650, respectively. The inflow rate for the on-ramp at bottleneck 2 is 20 vehicles/min (vpm), which is equivalent to 1,200 vehicles per hour. The outflow rate for the off-ramp is 18 vehicles/min (vpm). The free-flow travel time over each link is 5, 4, and 4.5 min, respectively.

For the probe vehicle in Figure 4 (starting at time 7:00 am), we now have the following calculation process for its route-level travel time.

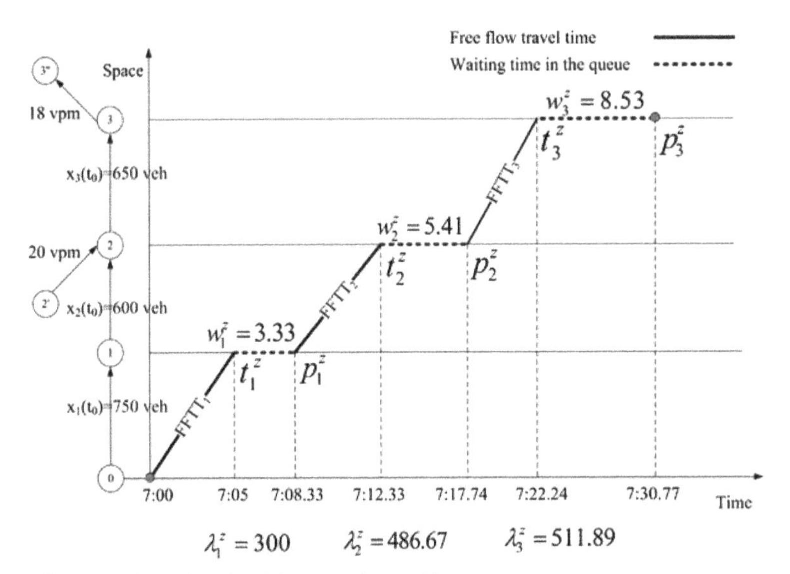

Figure 4. Three-bottleneck example corridor.

(1) Departing at 7:00, it takes 5 min (free-flow travel time) for this probe vehicle to reach the point-queue of bottleneck 1 at 7:05. At this time instance, the number of vehicles waiting in the queue is $\lambda_1\left(t_1^z\right) = 750 - (5 \text{ min} \times 90 \text{ veh/min}) = 300$ vehicles. With the discharge rate of 90 vehicles/min, this probe vehicle will spend $w_1^z = 3.33$ min waiting in the queue. Thus, the total travel time for this vehicle is 8.33 min at the end of this bottleneck.

(2) The probe vehicle enters link 2 at 7:08.33, spends 4 min traveling through the free-flow segment, and arrives at the vertical stack queue at $t_2^z = 7{:}12.33$. From 7:00 to 7:12.33, there have been 12.33 min × 20 veh/min = 246.6 vehicles entering this bottleneck from the on-ramp. The number of vehicles waiting in the queue at this time is $\lambda_2\left(t_2^z\right) = (750 + 600) + (12.33 \times 20) - (12.33 \times 90) = 486.67$. With the discharge rate of 90 vehicles/min, this vehicle leaves the queue $w_2^z = 5.41$ min later. The departure time from the second bottleneck is 7:17.74.

(3) Following the same calculation process, the number of vehicles waiting at the queue of bottleneck 3 is $\lambda_3\left(t_3^z\right) = (750 + 600 + 650) + (12.33 \times 20) + (-18 \times 22.24) - (60 \times 22.24) = 511.89$ vehicles and the waiting time in the queue is $w_3^z = 8.53$ min. This vehicle leaves bottleneck 3 at 7:30.77. The total route-level-travel time p_3^z is 30.77 min.

2.4. Travel time calculation algorithm with deterministic inputs

The algorithm for calculating the route-level path travel time for vehicle *z* entering the corridor with *M* bottlenecks at time t_0 is summarized below.

Input: The specific starting time t_0, the number of vehicles on each link $x_m(t_0)$, the net flow rate on each bottleneck f_m^{net}, and the bottleneck discharge rate c_m, at time t_0.

Route-level travel time calculation

For $m = 1$ to M

(1) Calculate the arrival time at bottleneck *m*

$$t_m^z = t_{m-1}^z + w_{m-1}^z + \text{FFTT}_m, \text{ where } t_0^z = 0, w_0^z = 0$$

(2) Use Equation 14 to calculate the number of vehicles ahead of the probe vehicle *z* in the vertical stack queue of bottleneck *m*, $\lambda_m^z\left(t_m^z\right)$, when the probe vehicle *z* reaches the beginning of the queue at time t_m^z.

(3) Use Equation 15 to calculate the delay experienced by the probe vehicle on bottleneck m, w^z_m.

(4) Use Equation 16 to update the route-level travel time over m, p^z_m

End For

Output: The route-level travel time p^z_M from bottleneck 1 to bottleneck M.

2.5. Discussions

To consider complex real-life conditions, the model must further use the following approximation methods for calculating the route-level path travel time along a corridor with multiple bottlenecks.

2.5.1. Approximating the time-dependent flow rates with average flow rates

In Equations 9 and 10, we use the maximum bottleneck discharge rates to approximate the actual discharge rates. In reality, the rates (including the queue discharge flow rates and net flow rates from and to ramps) are highly dynamic and could fluctuate significantly even in a short time interval, as shown in Figure 5. In this situation, one needs to use the average flow rate (i.e. the dashed line in Figure 5) during the interval from t_0 to the arrival time t^z_m to approximate the time-dependent volume. Although this approximation ignores traffic dynamics, in Equation 14 it still gives a reasonable estimate about the total number of vehicles leaving or entering the bottleneck before the probe vehicle. Section 4 will examine the possible impact from using this approximation method for representing travel time distributions.

2.5.2. Considering further reduced bottleneck discharge flow rate due to queue spillback

The proposed point-queue-based model captures the effects of queue spillback from a downstream bottleneck. Essentially, when a queue spillback occurs, the discharge capacity from the upstream bottleneck is then constrained by the discharge rates at the downstream bottleneck. The method detects spillback and then uses the reduced queue discharge rate to calculate the waiting time at the bottleneck with queue spillback.

As illustrated in Figure 6, the physical queue for bottleneck m spills back to bottleneck $(m - 1)$ between time t_1 and t_5 through backward waves [Interested readers are referred to the paper by Newell (1993) to learn more]. Due to the queue spillback from bottleneck m, the actual discharge rate c'_{m-1} of bottleneck $(m - 1)$ between time t_1 and t_5 is constrained by the discharge rate of bottleneck m, c_m, rather than the original discharge rate c_{m-1}. For example, at time t_2 (where $t_2 > t_1$), a probe vehicle arrives at bottleneck $(m - 1)$, if the effect of queue spillback is not taken into account, this probe vehicle in the model will leave bottleneck $(m - 1)$ at time t_3 after waiting in the queue

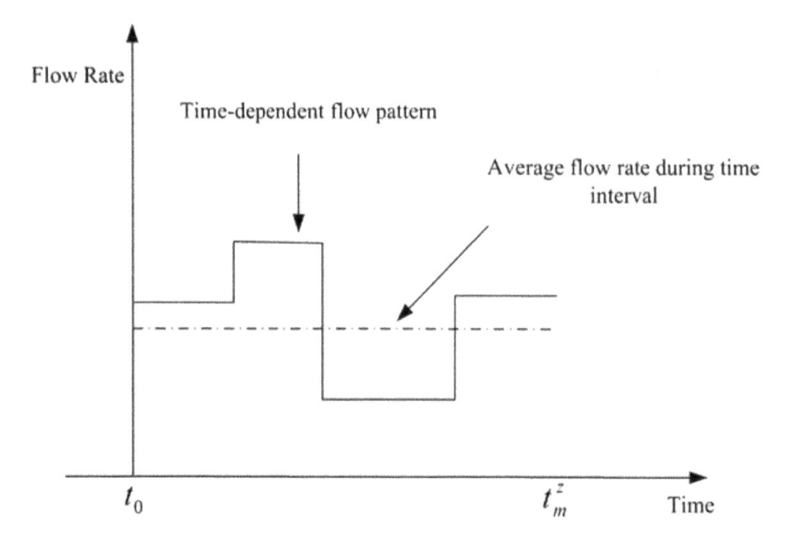

Figure 5. Time-dependent flow rate to average flow rate.

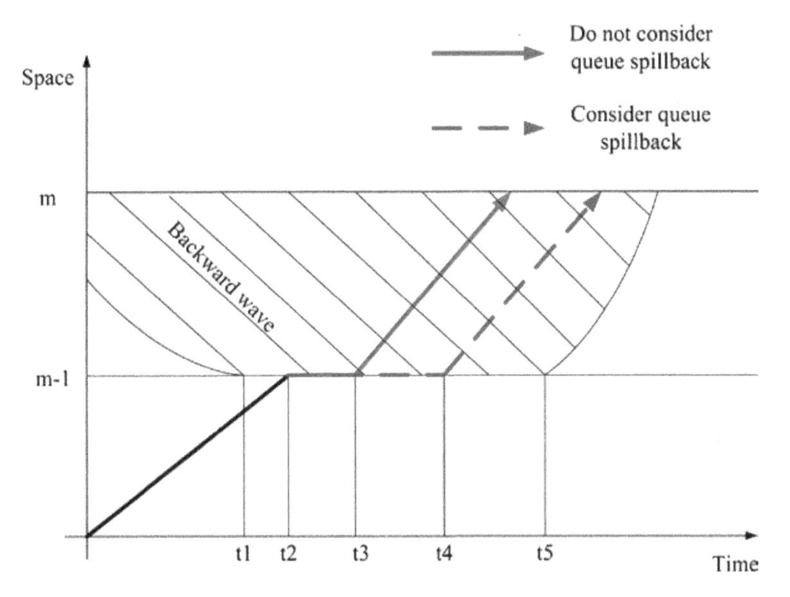

Figure 6. Queue spillback illustration in a time-space diagram.

behind bottleneck ($m-1$), using the original, unaffected queue discharge rate c_{m-1}. With the reduced discharge rate $c_m < c_{m-1}$ at bottleneck ($m-1$), the actual waiting time for the probe vehicle will be longer with a departing time of $t_4 > t_3$.

2.5.3. Calculating the net flow rate for on-ramps

When estimating the net flow at a merge or diverge location, the flow rates in previous instances are assumed to be known and time-invariant. However, special attention must be paid to conditions where the mainline and the on-ramp are both congested. In such instances, (1) the number of vehicles that can enter the bottleneck from the on-ramp and (2) the number of vehicles that can enter from the upstream segment to the bottleneck are constrained by the mainline bottleneck discharge rate. In this case, the available bottleneck discharge rate should be allocated to the upstream segment and the on-ramp proportionally, according to certain rules (Zhang & Nie, 2005). One simple rule is to split the mainline discharge rates according to the number of lanes associated with each incoming approach.

2.5.4. Considering vehicle overtaking/passing

Lastly, the FIFO property assumed on each link rules out the possibility that a vehicle can overtake and pass another vehicle. Future research will consider the impact of this condition on route-level travel time estimation using this approach.

3. Methods for calculating route-level travel time distributions

3.1. Assumptions

In the previous discussion, input parameters such as the net rates f_m^{net} at the merge and diverge points, and the bottleneck discharge rates c_m are assumed to be deterministic. In this section, we will further consider the variations or uncertainty in the input parameters, especially in the following two applications: (1) day-to-day travel time variability estimation by considering flow variations at the same time period; and (2) real-time travel time reliability estimation, where the near-future traffic flows are estimated from different sources of data with various degrees of estimation uncertainty. Emphases are placed on how to calculate the route-level travel time distribution based on the stochasticity of the random input parameters.

3.2. Important observations on path travel time

3.2.1. Simple corridor without merge and diverge

Consider a simple two-bottleneck corridor with no on-ramp and off-ramp, that is, f_1^{net} and f_2^{net} are equal to 0. According to Equations 14–16, the route-level travel time to bottleneck 1 for probe vehicle z entering link 1 at time t_0 is:

$$p_1^z = t_1^z + w_1^z = t_0 + \text{FFTT}_1 + \frac{x_1(t_0) - c_1 \times t_1^z}{c_1} = t_0 + \text{FFTT}_1 + \frac{x_1(t_0) - c_1 \times (t_0 + \text{FFTT}_1)}{c_1} \tag{17}$$

$$= t_0 + \text{FFTT}_1 + \frac{x_1(t_0)}{c_1} - (t_0 + \text{FFTT}_1) = \frac{x_1(t_0)}{c_1}$$

And the route-level travel time to bottleneck 2 is:

$$p_2^z = p_1^z + \text{FFTT}_2 + w_2^z = p_1^z + \text{FFTT}_2 + \frac{x_1(t_0) + x_2(t_0) - c_2 \times t_2^z}{c_2} \tag{18}$$

$$= p_1^z + \text{FFTT}_2 + \frac{x_1(t_0) + x_2(t_0) - c_2 \times (p_1^z + \text{FFTT}_2)}{c_2}$$

$$= p_1^z + \text{FFTT}_2 + \frac{x_1(t_0) + x_2(t_0)}{c_2} - (p_1^z + \text{FFTT}_2) = \frac{x_1(t_0) + x_2(t_0)}{c_2}$$

By comparing $p_1^z = \frac{x_1(t_0)}{c_1}$ and $p_2^z = \frac{x_1(t_0)+x_2(t_0)}{c_2}$, we can make the following important observation: the proposed formula can correctly capture the correlations between the route-level travel times p_1^z and p_2^z, as both values are dependent on the number of vehicles on link 1, $x_1(t_0)$. If $x_1(t_0)$ and $x_2(t_0)$ are assumed to be deterministic, the distributions of p_1^z and p_2^z are further dependent on the distribution of the bottleneck discharge rates, c_1 and c_2, respectively.

3.2.2. Simple corridor with merge and diverge

If we further consider situations where a merge and diverge occur at both bottlenecks, the path travel time formulas can be expressed as follows.

$$p_1^z = t_1^z + w_1^z = \frac{x_1(t_0) + f_{net}^1 \times (t_0 + \text{FFTT}_1)}{c_1} \tag{19}$$

$$p_2^z = \frac{x_1(t_0) + f_1^{net} \times \text{FFTT}_1 + x_2(t_0) + f_2^{net} \times (p_1^z + \text{FFTT}_2)}{c_2} \tag{20}$$

The above equations introduce more complex dependencies for both p_1^z and p_2^z, and no additive formula or decomposed elements can be easily constructed to simplify these equations. This observation reinforces many previous research studies which indicate that development of the route-level travel time distribution is extremely challenging.

3.3. Monte Carlo simulation

Monte Carlo simulation is widely used to simulate the behavior of various physical and mathematical systems, especially for those problems with significant uncertainty in inputs. The model presented here uses Monte Carlo simulation to investigate the route-level travel time distribution based on the proposed travel time calculation framework. In each simulation run, a realization of the random input parameters leads to a realization of the random path travel time outputs, which can be regarded as estimates of the true route-level travel time variable. A sufficient number of simulations then provide a good representation of the travel time distributions under various traffic conditions and uncertainties.

The following procedure assumes all random variables are log-normally distributed, and calculates travel time distribution through K simulation runs.

Input:

The specific starting time t_0,

The distribution of the number of vehicles on each link $x_m(t_0)$, where $x_m(t_0) \sim LN(\mu_{x_m}, \sigma^2_{x_m})$.

The distribution of the net flow rate on each bottleneck f_m^{net}, where $f_m^{\text{net}} \sim LN(\mu_{f_m^{\text{net}}}, \sigma^2_{f_m^{\text{net}}})$.

The distribution of the bottleneck discharge rate c_m on each bottleneck, at time t_0 where $c_m \sim LN(\mu_{c_m}, \sigma^2_{c_m})$.

Link free-flow travel time FFTT_m, assumed to be constant.

Number of simulations = K.

For k = 1 to K,

For m = 1 to M

1: Based on the underlying distribution parameters (μ and σ) of the individual inputs, generate a set of random samples for the following key variables: the number of vehicles on the link, the bottleneck discharge rate, and net flow rates.

2: Call the algorithm introduced in Section 2.4 to calculate the estimated route-level travel time for simulation k: $p_m^z[k]$ from this set of random samples.

End For

End For

Output: Calculate the histogram, mean, and variance for the route-level travel time from the results over K simulation runs.

3.4. Numerical experiments

3.4.1. Monte Carlo simulation

For the same example corridor in Section 2.3 (with three bottlenecks), Monte Carlo experiments were conducted to calculate the route-level travel time by assuming that the bottleneck discharge rates, inflow/outflow rates on ramps, and existing number of vehicles on the link are all lognormal variables. K = 100 simulation runs were performed with different scenarios of stochastic input parameters.

Figure 7(a–b) shows the distributions of the simulated route-level travel times p_2^z and p_3^z for probe vehicle z through bottleneck 2 and through bottleneck 3, respectively. Obviously, the mean travel time based on p_3^z is larger than that of p_2^z. In addition, a clear propagation of randomness can be observed, as p_3^z has higher variance than p_2^z. It should be noted that, by using different input distributions for flow discharge rates and the prevailing number of vehicles on the road, the resulting travel time distributions will vary. This demonstrates the advantage of the proposed model in recognizing the impact of capacity and congestion levels on travel time reliability.

3.4.2. Evaluating the improvement in reliability for traffic management strategies

In the previous section, Monte Carlo simulation was used to demonstrate the application of the proposed travel time estimation framework on route-level travel time distribution quantification. In this section, the calculation framework is further applied to evaluate the effectiveness of advanced traffic management strategies (ATMS).

To demonstrate the use of the proposed calculation framework, a 1-mile long 4-lane freeway corridor with the average bottleneck discharge rate of 2,000 vehicles per hour per lane is investigated. Before implementing ATMS, the probability of incidents on this corridor is 20% and 3 lanes are closed due to an incident if an incident occurs. After implementing ATMS (e.g. rapid incident response teams), the number of lanes closed due to incident is reduced to 2.

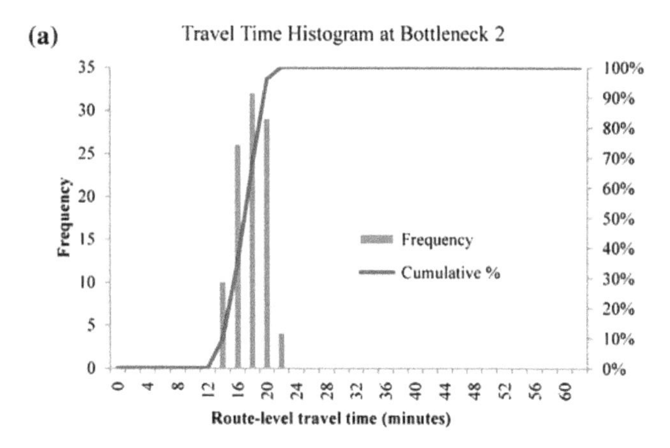

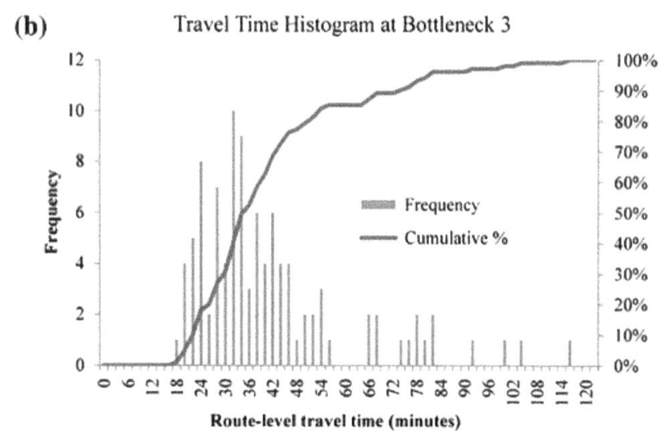

Figure 7. Route-level travel time distribution.

Figure 8(a–b) and (c–d), respectively, show the simulated distribution of the bottleneck discharge flow rate and the calculated route-level travel time, before and after the implementation of the ATMS. The first peak of Figures 8(a) and (c) represents the average capacity under incidents and the second peak represents the capacity under normal conditions. By comparing Figure 8(b) with (d), we can observe that the travel time distribution with ATMS has a shorter tail and less fluctuation. Figure 8(e) further reveals the potential benefit of ATMS in improving the travel time reliability: the 95th percentile travel time is improved from 25 min (without ATMS) to 13 min (with the implementation of ATMS), while interestingly the median travel time has not changed significantly.

4. Model validation using NGSIM data

This section uses vehicle trajectory data available from the NGSIM project (FHWA, 2006) as ground-truth data to verify the proposed methodology and examine the sources of estimation error.

4.1. Data descriptions

The NGSIM vehicle trajectory data used in this study come from the I-80 data-set, which were collected by a video camera located at Emeryville, California. This data collection point is located adjacent to I-80, as shown in Figure 9. The site was approximately 1,650 feet in length, with an on-ramp at Powell Street (indicated in Figure 9 by the circle). The freeway segment covered in the data-set includes six lanes, numbered incrementally from the left-most lane (HOV lane). Video data are available for three time intervals: 4:00–4:15 pm, 5:00–5:15 pm, and 5:15–5:30 pm, on 13 April 2005. Complete, transcribed vehicle trajectories are available with a time resolution of 0.1 s.

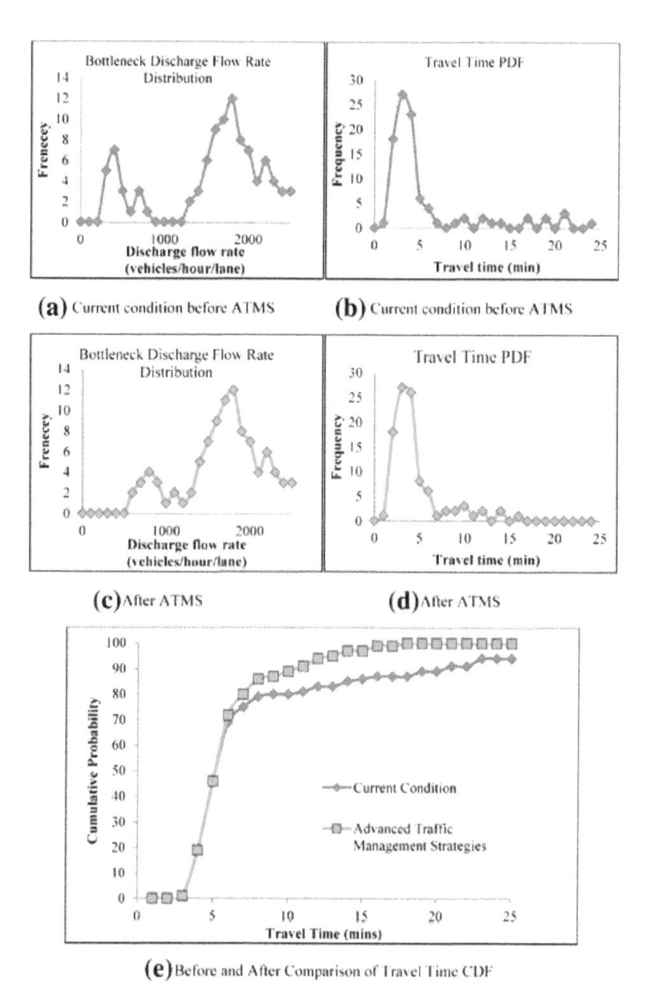

Figure 8. Before and after comparison of travel time CDF.

4.2. Data extraction from NGSIM data-set

(1) To extract vehicle flow counts data from the NGSIM data-set, we first construct a node-link structure to represent the freeway segment in Figure 9. This stretch of freeway is divided into two links, as shown in Figure 10, with the on-ramp connected with node 1.

(2) In order to obtain the flow rate at the node/bottleneck, this study introduces a set of virtual detectors at node 1 and at node 2, respectively. Meanwhile, another virtual detector is placed on the on-ramp link so that inflow vehicles from the ramp are also counted. In addition, video cameras are assumed to be installed on both links to provide link snapshots (for probe vehicle data).

(3) The vehicle trajectory data is divided into 5-min intervals for counting vehicles. An example of one 5-min span of vehicle trajectories on one lane is shown in Figure 11 to illustrate how the vehicle counts are collected. As mentioned before, two sets of virtual detectors A and B are placed at nodes 1 and 2 (shown as triangles in Figure 10), and video cameras C and D are also installed on both link 1 and link 2. Vehicles are counted along the vertical line drawn at the given time t. At time $t_0 = 0$, probe vehicle $z = 0$ enters link 1. At this time step, two vehicles are observed on link 1 and five vehicles are observed on link 2 by video cameras C and D, that is, $x_1(t_0) = 2$ and $x_2(t_0) = 5$. Similarly, probe vehicle $z = 5$ enters link 1 at time t. At this moment, $x_1(t) = 2$ and $x_2(t) = 4$.

Probe vehicle $z = 8$ is worth mentioning, which enters link 1 at time t'. However, at time t'', this vehicle changes lanes. This can be seen in Figure 11 because there is an incomplete vehicle trajectory. During this 5-min interval, 12 vehicles are counted by detector A, including two vehicles entering

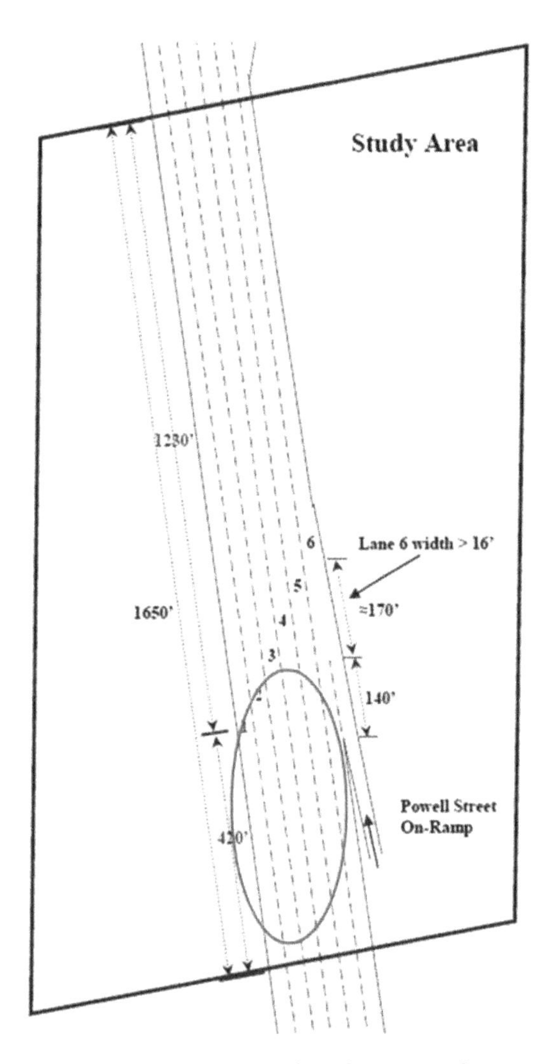

Figure 9. Schematic illustration of NGSIM study area.

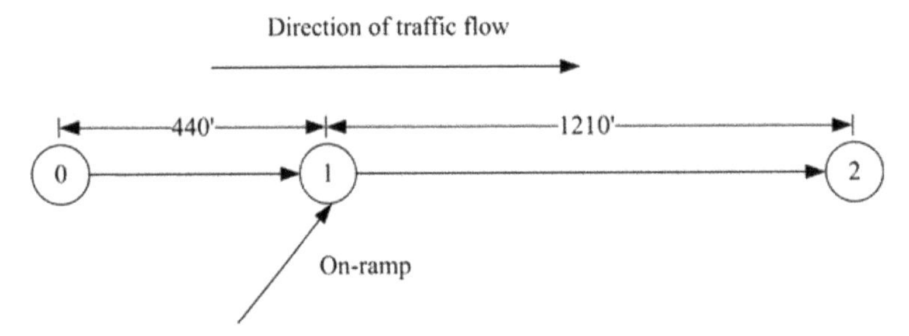

Figure 10. Node-link representation of NGSIM network.

before probe vehicle $z = 1$, but excluding probe vehicle $z = 8$. Meanwhile, 13 vehicles are counted by detector B. This count includes seven vehicles before probe vehicle $z = 1$, but excludes probe vehicles 7–11, which have not yet departed from link 2.

Figure 12 illustrates the relationship between the actual trajectories extracted from the NGSIM data-set and the number of vehicles waiting in the "modeled" vertical queue. For vehicles 1 and 2, we plot the free-flow travel times (dashed lines), the waiting times (arrowed lines) and the experienced

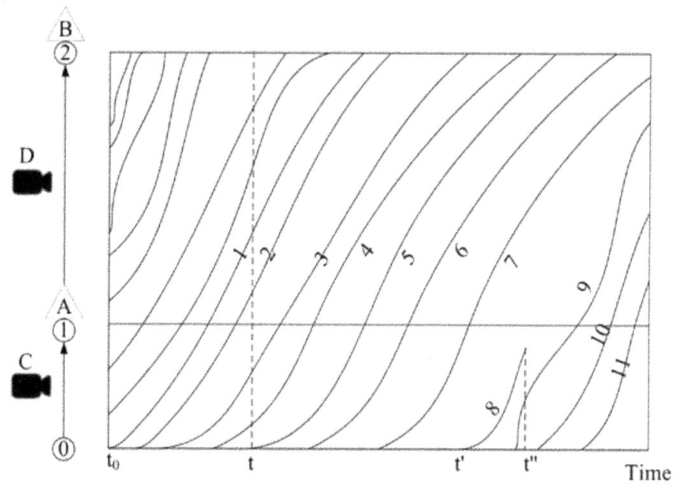

Figure 11. Vehicle trajectories on a lane.

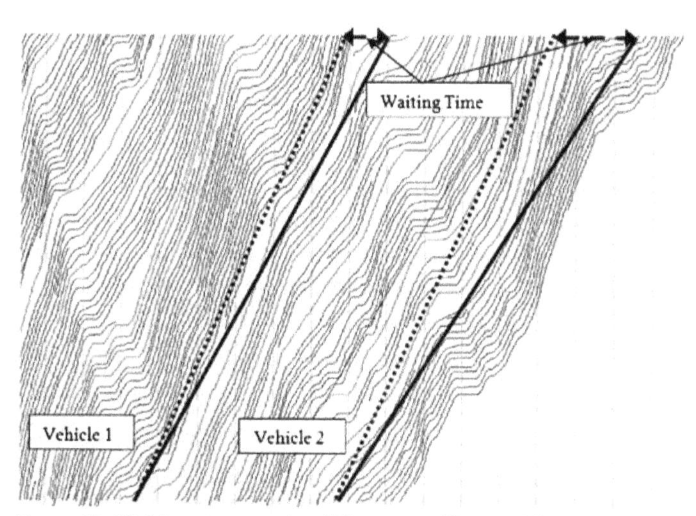

Figure 12. Waiting times under different traffic conditions.

travel times (solid lines). Essentially, the experienced travel time = free-flow travel time + waiting time at the vertical queue, where the waiting time is determined by the number of vehicles in the queue and the capacity, as stated by Equation 7.

4.3. Model validation

Two variants on the basic travel time distribution estimation approach presented here are investigated in this section. The first is lane-based; the second is link-based.

The examples require some additional notation to represent lane-specific parameters.

Z: number of probe vehicles

n: index identifying a lane

t_0^z: starting time for probe vehicle z

n^z: starting lane number for probe vehicle z

$x_m(t,n)$: number of vehicles on lane n at time t

$\lambda_m^z\left(t_m^z,n\right)$: lane n specific number of vehicles behind bottleneck m

$c_m(n)$: lane n specific discharge rate of bottleneck m

$f_m^{net}(n)$: net flow rate from or to ramps by lane n

$\theta(n)$: vehicle distribution rate from on-ramp to lane n

$w_m^z(n)$: waiting time for probe vehicle z on bottleneck m on lane n

$p_m^z(n)$: lane-based route-level travel time for probe vehicle z through lane n

The following procedure is used to calculate the lane-based travel time distribution.

For z = 1 to Z on the link

Obtain arrival time t_0^z and starting lane number n^z for each probe vehicle z.
Obtain the lane-based number of vehicles $x_m\left(t_0,n\right)$;
Obtain the lane-specific discharge rate $c_m(n)$;
Calculate net flow rate $f_m^{net}(n)$ from the on-ramp by applying $\theta(n)\times f_m^{net}$;
Calculate the number vehicles behind bottleneck m $\lambda_m^z\left(t_m^z,n\right)$;
Calculate $w_m^z(n)$ based on $c_m(n)$ and $x_m\left(t_0,n\right)$;
Update the route-level lane travel time $p_m^z(n)$.

End For

Output: Create the lane-based path travel time distribution based on $p_m^z(n)$

4.3.1. Lane-based route-level travel time estimation

The distribution of the estimated route-level travel times for each 5-min interval, calculated over the three time periods with available data (4:00–4:15 pm, 5:00–5:15 pm, and 5:15–5:30 pm) are plotted in Figure 13 with the ground truth route-level travel time obtained directly from the NGSIM data. As can be observed, the distribution of the estimated route-level travel time is very close to that of the ground truth route-level travel time. This demonstrates that our model is able to accurately estimate the route-level travel time distribution.

4.3.2. Link-based route-level travel time estimation

One common practice is to use link-based flow rates or density to estimate travel time reliability. We replace the lane-based variables in the previous approach with link-based variables $x_m(t_0)$ and c_m. That is, $x_m(t_0)$ is the existing number of vehicles on all the lanes on the link and c_m is the link discharge rate. The distribution of the estimated route-level travel time and true route-level travel time for different time intervals is shown in Figure 14.

As it can be observed, the distribution of the estimated link-based route-level travel times fails to capture the wide-spread distribution in the ground truth travel times. This is explained by the fact that link-based input variables would yield the same estimated route-level travel times for those vehicles entering the link at the same time, regardless of their driving lanes.

In order to understand the extent and sources of the lane-by-lane travel time variation, we use the time period between 4:00 pm and 4:15 pm as an example. Figure 14 shows the lane discharge rate, the existing numbers of vehicles on the lane, and the average true and estimated route-level travel times for each lane for each 5-min interval in the time period. The lane sequence is sorted by the true route-level travel time.

Several observations can be made based on Figure 15. Lane 1 (HOV lane) has the lowest existing number of vehicles on the lane and has the lowest average route-level travel time. The left-most lanes (lanes 1 and 2) usually have the highest discharge rates while lanes 3 and 4 usually have the lowest discharge rates. In most cases, lanes 3 and 4 have the highest average existing numbers of vehicles on the lane, as well as the highest average route-level travel times.

(a)

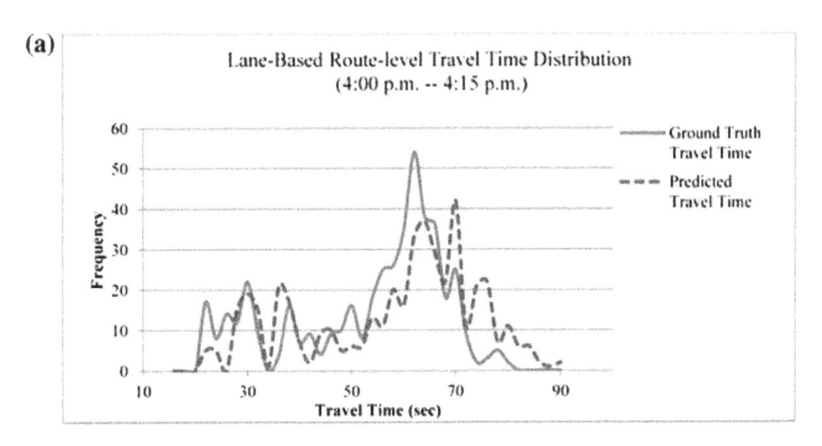

(b)

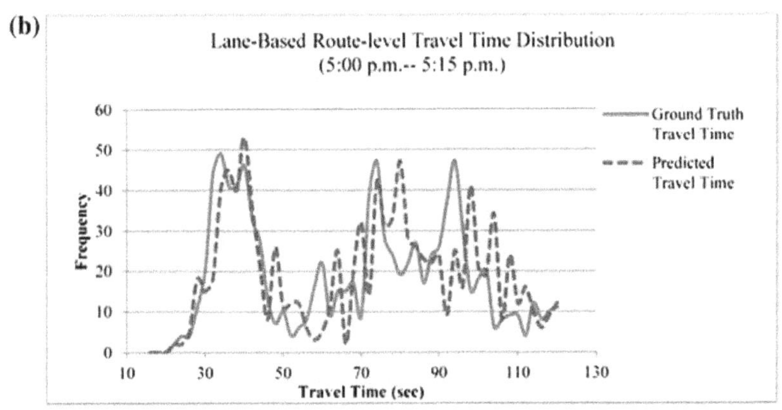

(c)

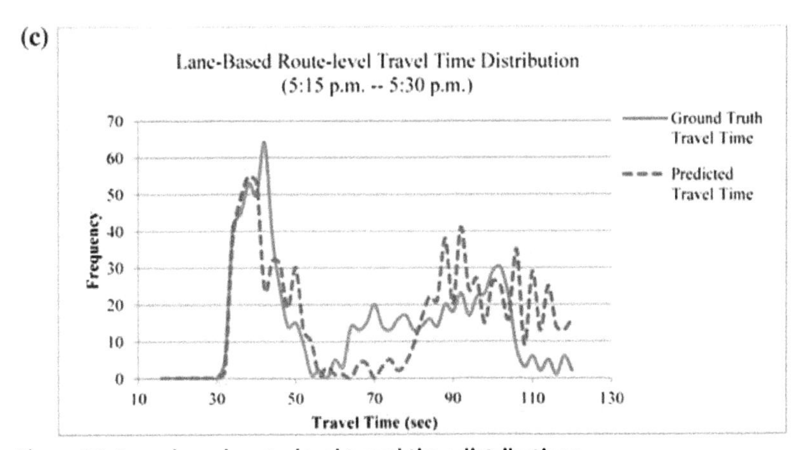

Figure 13. Lane-based route-level travel time distributions.

These observations imply that, due to the variation of the discharge rates and the number of vehicles on the lane, the route-level travel times also present strong lane-by-lane variations. As a result, we suggest using lane-based statistics to better quantify the travel time variability.

4.3.3. Estimation error sources

By comparing the estimated results with the NGSIM ground truth data, we can further uncover other possible sources of errors in the proposed travel time estimation model.

(1) Aggregation errors: The link/lane discharge rates c_m and on-ramp flow rates used in the calculations are average flow rates over a certain time interval, e.g. 5-min rates, while the existing number of vehicles on the link/lane $x_m(t_0)$ is an instantaneous value based on the entering time of a probe vehicle.

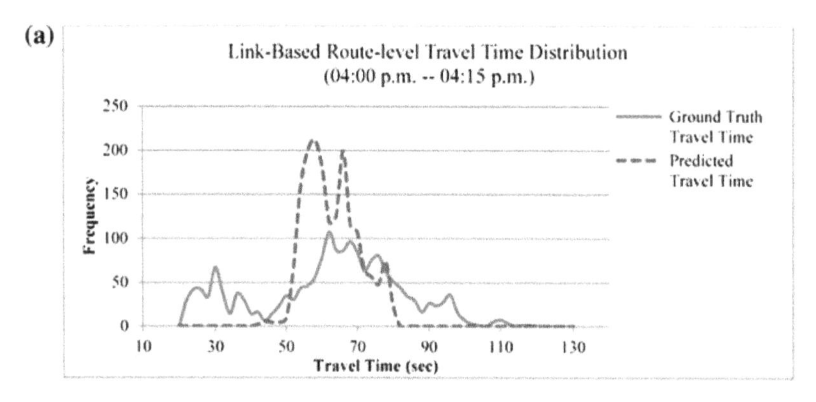

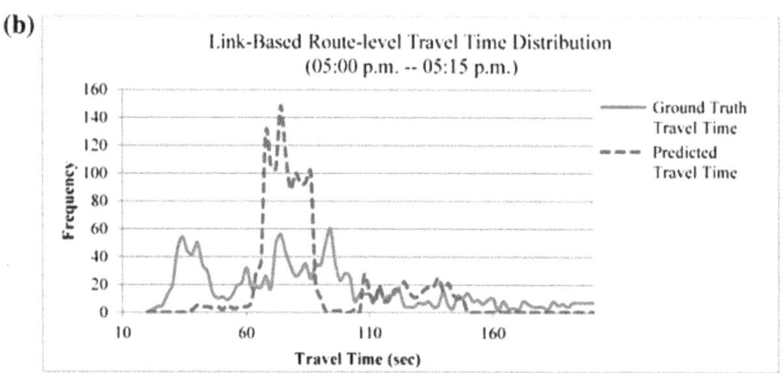

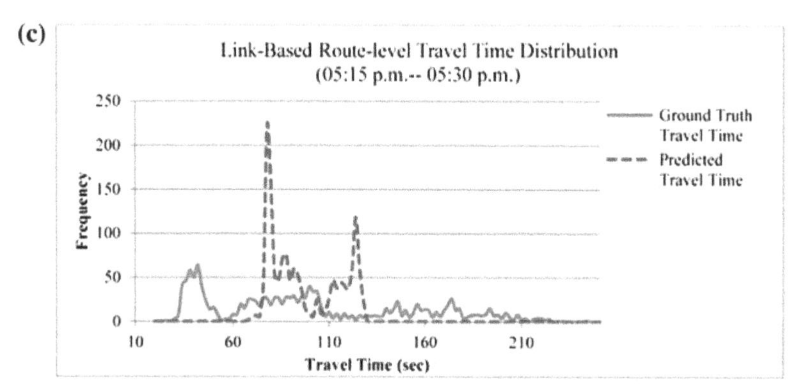

Figure 14. Link-based route-level travel time distributions.

(2) Measurement errors: The number of vehicles on the lane observed by the video camera at time t_0 is assumed to be error-free. In fact, there are always vehicle detection errors in NGSIM vehicle trajectory data associated with the underlying video recognition algorithm.

(3) Modeling errors associated with lane changing: Since the queue model incorporates the first-in-first-out principle, lane change behavior is not considered in the calculation. This will introduce two types of errors in the model:

(a) The model may underestimate or overestimate the number of vehicles behind the bottleneck, $\lambda_m\left(t_m^z\right)$. For example, some vehicles will enter the lane (from the other lanes) before a probe vehicle reaches the bottleneck, or some vehicles originally counted in $x_m(t_0)$ on the current lane will leave to one of the adjacent lanes, corresponding to a lower value of $x_m(t_0)$.

(b) When a probe vehicle changes lane from, for example, $n^z = 1$ to lane n', the discharge rate used in the calculation should be changed to the one associated with lane n'.

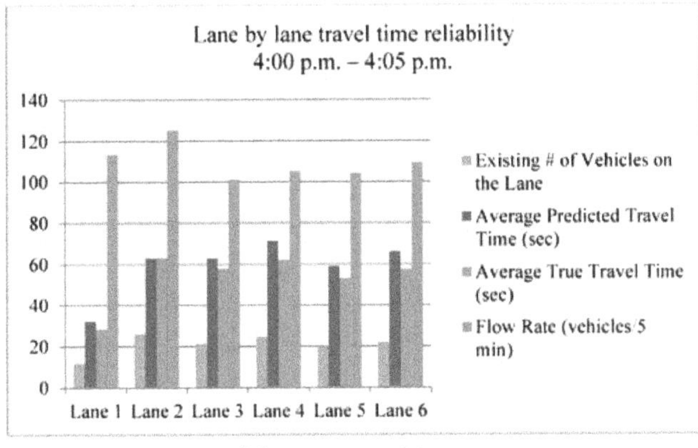

(a): 4:00 p.m. – 4:05 p.m.

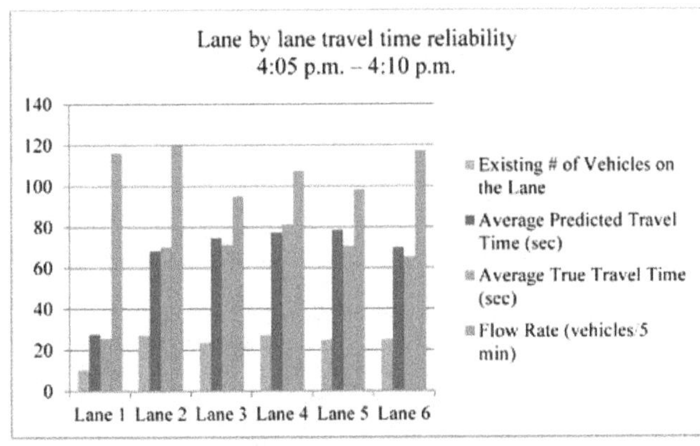

(b): 4:05 p.m. – 4:10 p.m.

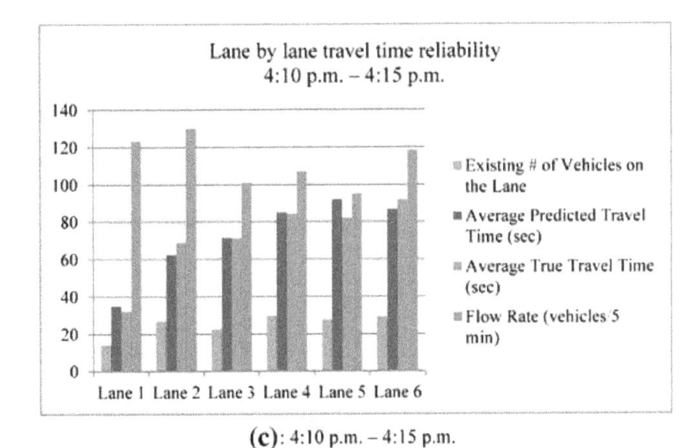

(c): 4:10 p.m. – 4:15 p.m.

Figure 15. Lane-by-lane travel time variability.

5. Conclusions

This paper has presented a model which is capable of estimating route-level travel time distributions. It does this on the basis of relatively simple information about the corridor's geometric configuration, its entering and exiting flows, its capacities, and the lane-by-lane distributions of traffic at each bottleneck. Monte Carlo simulation and mathematical approximation methods are used to calculate travel times lane-by-lane by tracking probe vehicles through the network. Ground-truth vehicle trajectory data from the NGSIM project have been used to validate the model's estimates. Several interesting observations are obtained from the research:

(1) The model offers a theoretically sound method to estimate corridor-level travel time and its distribution under different capacity and demand variations, and with possible off-ramp and off-ramp volume changes.

(2) The variation of lane-specific traffic flow parameters (such as the number of vehicles on the lanes and lane discharge rates) are significant to lane-by-lane travel time diversity.

(3) A lane-level rather than link-level representation of the system is critical in developing accurate route-level travel time distribution estimates.

Our future research will consider the impacts of downstream queue spillback on the upstream travel time. Under queue spillback, the discharge rate of the upstream bottleneck is constrained by the downstream bottleneck discharge rate, and this significantly increases dynamics and complexity in estimating the capacity for a queuing model. Moreover, it is also desirable to examine the influence of lane change frequencies on the estimated number of vehicles waiting in the queue.

Acknowledgment
This paper is based on research work partially supported through the TRB SHRP 2 L02 project titled "Establishing Monitoring Programs for Mobility and Travel Time Reliability." The work presented in this paper remains the sole responsibility of the authors.

Funding
The authors received no direct funding for this research.

Author details
Hao Lei[1]
E-mail: hlei@eng.utah.edu
Xuesong Zhou[2]
E-mail: xzhou74@asu.edu; xzhou99@gmail.com
George F. List[3]
E-mail: gflist@ncsu.edu
Jeffrey Taylor[1]
E-mail: jeff.d.taylor@utah.edu
[1] Department of Civil and Environmental Engineering, University of Utah, Salt Lake City, UT 84112, USA.
[2] School of Sustainable Engineering and the Built Environment, Arizona State University, Tempe, AZ 85287, USA.
[3] Department of Civil, Construction, and Environmental Engineering, North Carolina State University, Raleigh, NC 27695-7908, USA.

References

Alibabai, H. (2011). *Properties of simulated path travel times* (Doctoral dissertation). Northwestern University, Evanston, IL.

Chen, C., Skabardonis, A., & Varaiya, P. (2003). Travel-time reliability as a measure of service. *Transportation Research Record, 1855*, 74–79. http://dx.doi.org/10.3141/1855-09

Clark, S., & Watling, D. (2005). Modelling network travel time reliability under stochastic demand. *Transportation Research Part B: Methodological, 39*, 119–140. http://dx.doi.org/10.1016/j.trb.2003.10.006

Cowan, R. J. (1975). Useful headway models. *Transportation Research, 9*, 371–375. http://dx.doi.org/10.1016/0041-1647(75)90008-8

Daganzo, C. F. (1995). Properties of link travel time functions under dynamic loads. *Transportation Research Part B: Methodological, 29*, 95–98.

Federal Highway Administration. (2006, December). *Next generation simulation fact sheet* (Technical report, FHWA-HRT-06-135). Washington, DC: Author.

Federal Highway Administration. (2009, June). *Recurring traffic bottlenecks: A primer focus on low-cost operational improvements*. (Technical report, FHWA-HOP-09-037). Washington, DC: Author.

Friesz, T. L., Bernstein, D., Smith, T. E., Tobin, R. L., & Wie, B. W. (1993). A variational inequality formulation of the dynamic network user equilibrium problem. *Operations Research, 41*, 179–191. http://dx.doi.org/10.1287/opre.41.1.179

Giuliano, G. (1989). Incident characteristics, frequency, and duration on a high volume urban freeway. *Transportation Research Part A: General, 23*, 387–396. http://dx.doi.org/10.1016/0191-2607(89)90086-1

Golob, T. F., Recker, W. W., & Leonard, J. D. (1987). An analysis of the severity and incident duration of truck-involved freeway accidents. *Accident Analysis and Prevention, 19*, 375–395. http://dx.doi.org/10.1016/0001-4575(87)90023-6

Greenberg, I. (1966). The log-normal distribution of headways. *Australian Road Research, 2*, 14–18.

Guo, F., Rakha, H., & Park, S. (2010). Multistate model for travel time reliability. *Transportation Research Record: Journal of the Transportation Research Board, 2188*, 46–54. http://dx.doi.org/10.3141/2188-06

Hazelton, M. L. (2001). Inference for origin–destination matrices: Estimation, prediction and reconstruction. *Transportation Research Part B: Methodological, 35*, 667–676. http://dx.doi.org/10.1016/S0191-2615(00)00009-6

HCM. (2000). *Highway capacity manual*. Washington, DC: Transportation Research Board, National Research Council.

Hurdle, V. F., & Son, B. (2001). Shock wave and cumulative arrival and departure models: Partners without conflict. *Transportation Research Record, 1776*(1), 159–166. http://dx.doi.org/10.3141/1776-21

Kripalani, A., & Scherer, W. (2007, July). *Estimating incident related congestion on freeways based on incident severity* (Research Report No. UVACTS-15-0-113). Charlottesville, VA: Center for ITS Implementation Research, University of Virginia.

Kwon, J., Barkley, T., Hranac, R., Petty, K., & Compin, N. (2011). Decomposition of travel time reliability into various sources. *Transportation Research Record: Journal of the Transportation Research Board, 2229*, 28–33. http://dx.doi.org/10.3141/2229-04

Lam, W. H. K., Shao, H., & Sumalee, A. (2008). Modeling impacts of adverse weather conditions on a road network with uncertainties in demand and supply. *Transportation Research Part B: Methodological, 42*, 890–910. http://dx.doi.org/10.1016/j.trb.2008.02.004

Lo, H. K., & Tung, Y. K. (2003). Network with degradable links: Capacity analysis and design. *Transportation Research Part B: Methodological, 37*, 345–363. http://dx.doi.org/10.1016/S0191-2615(02)00017-6

Newell, G. F. (1993). A simplified theory of kinematic waves in highway traffic, part II: Queueing at freeway bottlenecks. *Transportation Research Part B: Methodological, 27*, 289–303. http://dx.doi.org/10.1016/0191-2615(93)90039-D

Ng, M. W., & Waller, S. T. (2010). A computationally efficient methodology to characterize travel time reliability using the fast Fourier transform. *Transportation Research Part B: Methodological, 44*, 1202–1219. http://dx.doi.org/10.1016/j.trb.2010.02.008

Oh, J.-S., & Chung, Y. (2006). Calculation of travel time variability from loop detector data. *Transportation Research Record, 1945*, 12–23. http://dx.doi.org/10.3141/1945-02

Ran, B., Boyce, D. E., & LeBlanc, L. J. (1993). A new class of instantaneous dynamic user-optimal traffic assignment models. *Operations Research, 41*, 192–202. http://dx.doi.org/10.1287/opre.41.1.192

Park, S., Rakha, H., & Guo, F. (2011, October). Multi-state travel time reliability model: Impact of incidents on travel time reliability. In *Intelligent Transportation Systems (ITSC), 2011 14th International IEEE Conference on* (pp. 2106–2111). Washington, DC: IEEE.

Richardson, A. J. (2003, December). *Travel time variability on an urban freeway* (TUTI Report 25-2003). Presented at the 25th Conference of Australian Institutes of Transport Research (CAITR), University of Adelaide, Adelaide, Australia.

Unnikrishnan, A., Ukkusuri, S., & Waller, S. T. (2005). Sampling methods for evaluating the traffic equilibrium problem under uncertain demand. In *Transportation Research Board 84th Annual Meeting Compendium of Papers*. Washington, DC.

Waller, S. T., & Ziliaskopoulos, A. K. (2001). Stochastic dynamic network design problem. *Transportation Research Record, 1771*, 106–113. http://dx.doi.org/10.3141/1771-14

Zhang, H. M., & Nie, X. (2005). Some consistency conditions for dynamic traffic assignment problems. *Networks and Spatial Economics, 5*, 71–87. http://dx.doi.org/10.1007/s11067-005-6662-7

Zhou, X., Rouphail, N. M., & Li, M. (2011). *Analytical models for quantifying travel time variability based on stochastic capacity and demand distributions*. Presented at Transportation Research Board 90th Annual Meeting, Washington, DC.

4

Optimization of ultra-high-performance concrete with nano- and micro-scale reinforcement

Libya Ahmed Sbia[1], Amirpasha Peyvandi[2*], Parviz Soroushian[1] and Anagi M. Balachandra[3]

*Corresponding author: Amirpasha Peyvandi, Bridge Department, HNTB Corporation, 10000 Perkins Rowe, Suite No. 640, Baton Rouge, LA 70810, USA

E-mail: Amirpasha.peyvandi@gmail.com

Reviewing editor: Ian Smith, Ecole Polytechnique Federale de Lausanne, Switzerland

Abstract: Ultra-high-performance concrete (UHPC) incorporates a relatively large volume fraction of very dense cementitious binder with microscale fibers. The dense binder in UHPC can effectively interact with nano- and microscale reinforcement, which offers the promise to overcome the brittleness of UHPC. Nanoscale reinforcement can act synergistically with microscale fibers by providing reinforcing action of a finer scale, and also by improving the bond and pullout behavior of microscale fibers. Carbon nanofiber (CNF) and polyvinyl alcohol (PVA) fiber were used as nano- and microscale reinforcement, respectively, in UHPC. An optimization experimental program was conducted in order to identify the optimum dosages of CNF and PVA fiber for realizing balanced gains in flexural strength, energy absorption capacity, ductility, impact resistance, abrasion resistance, and compressive strength of UHPC without compromising the fresh mix workability. Experimental results indicated that significant and balanced gains in the UHPC performance characteristics could be realized when a relatively low volume fraction of CNF (0.047 vol.% of concrete) is

ABOUT THE AUTHORS

Amirpasha Peyvandi is currently with HNTB Corporation as a bridge engineer. He received his PhD in civil structural engineering at Michigan State University. His research interests include application of nanotechnology in cementitious material and application of new construction materials.

Libya Ahmed Sbia received her PhD and MS in structural engineering from Michigan State University. She also received her BS and MS degrees in civil and construction management from Tripoli University, Libya. Her research interests include application of nanomaterial and structural engineering.

Parviz Soroushian is a professor of Civil and Environmental Engineering at Michigan State University. He received his MS and PhD from Cornell University. His research interests include materials science and energy-efficient construction materials and systems.

Anagi M. Balachandra is currently with Metna Co. as a senior scientist. She received her PhD in Analytical Chemistry from Michigan State University. Her research emphasizes nanocomposites and use of nanomaterials at interfaces.

PUBLIC INTEREST STATEMENT

Advances in fiber reinforced cement nanocomposites using carbon nanotube and carbon nanofiber (CNF) have allowed to enhance mechanical and durability of cementitious-based material. However, research in the area of concrete nanocomposite is rare. Ultra-high-performance which comprises with high packing density material could take advantage from introducing the nanoreinforcing system. The work reported herein; evaluate the synergistic action in mechanical properties of low-cost CNF and microscale poly vinyl alcohol (PVA) fiber in ultra-high-performance concrete. The research finally introduced the optimum values of nano/microscale reinforcement in such a concrete nanocomposite system.

used in combination with a moderate volume fraction of PVA fibers (0.37 vol.% of concrete).

Subjects: Composites; Concrete & Cement; Structural Engineering

Keywords: carbon nanofiber; ultra-high-performance concrete; polyvinyl alcohol (PVA) fiber; optimization

1. Introduction

Ultra-high-performance concrete (UHPC) materials provide compressive strengths higher than 150 MPa (22 ksi) and tensile strengths greater than 8 MPa (1.1 Ksi) (Graybeal, 2007; Habel, Viviani, Denarié, & Brühwiler, 2006; Magureanu, Sosa, Negrutiu, & Heghes, 2012; Wille, Naaman, & Parra-Montesinos, 2011).

The exceptional mechanical and durability characteristics offered by UHPC (FHWA, 2011; Van Tuan, Ye, van Breugel, & Copuroglu, 2011) are made possible by the use of high contents of cementitious binder with very low water/binder ratios (less than 0.25) (Barnett, Lataste, Parry, Millard, & Soutsos, 2010; Ding, Liu, Pacheco-Torgal, & Jalali, 2011; Peyvandi, Sbia, Soroushian, & Sobolev, 2013; Schröfl, Gruber, & Plank, 2012), dense packing (Wille, Naaman, El-Tawil, & Parra-Montesinos, 2012) particulate matter (aggregate, cement, supplementary cementitious materials, and inert powder) lowering maximum aggregate size, effective use of pozzolanic reactions to realize the binder structure and capillary pore system, and use of macro/microscale fibers (Kang & Kim, 2011; Yang, Joh, & Kim, 2011) to overcome the brittleness of UHPC.

In order to overcome the extreme brittleness of UHPC, this cement-based material usually incorporates fibers (of different types). Steel fibers are commonly used in UHPC (Kang, Lee, Park, & Kim, 2010; Wille & Loh, 2010). Slender fibers of high elastic modulus, tensile strength, and bond strength to cementitious paste improve the tensile strength and toughness of UHPC, and mitigate crack propagation under extreme loading and environmental effects. Fibers at microscale diameter at practically viable volume fractions, however, produce at relatively large fiber-to-fiber spacing (Peyvandi, Soroushian, & Jahangirnejad, 2013). Fine cracks can thus initiate and grow between microscale fibers before they reach the fiber-matrix interfaces (Foti, 2013; Konsta-Gdoutos & Metaxa, 2010; Li, Wang, & Zhao, 2005; Metaxa, Konsta-Gdoutos, & Shah, 2010; Yoo, Lee, & Yoon, 2013). Finer (nanoscale) reinforcement can be dispersed in the space between microscale fibers to effectively mitigate the inception and growth of such fine cracks.

Graphite nanomaterials, including carbon nanotube (CNT), lower cost carbon nanofiber (CNF), and graphite nanoplatelet (Musso, Tulliani, Ferro, & Tagliaferro, 2009; Nochaiya & Chaipanich, 2011; Peyvandi, Soroushian, Balachandra, & Sobolev, 2013), offer desired geometric and mechanical properties for reinforcement of cementitious materials (Gay & Sanchez, 2010; Melo, Calixto, Ladeira, & Silva, 2011; Tyson, Abu Al-Rub, & Yazdanbakhsh, 2011). Surfaces of graphite nanomaterials can also be modified for effective interaction with the cementitious matrix. In addition, nanomaterials improve the packing density of cementitious paste, tend to preserve the fresh mix workability, and provide high specific surface areas for nucleation of cement hydrates (Peyvandi, Sbia et al., 2013; Wille & Loh, 2010). Finally, the closely space nanomaterials can provide tortuous diffusion paths into UHPC, thereby improving the impermeability and durability of concrete (Kalaitzidou, Fukushima, & Drzal, 2007; Peyvandi, Soroushian, Balachandra, & Sobolev, 2013).

CNTs are not economically viable nanomaterials for use in concrete. CNFs, on the other hand, approach the desired geometric and mechanical characteristics of CNTs at a fraction of their cost. Surfaces of CNFs are also more energetic (active for convenient surface modification to introduce OH or COOH groups) than CNTs, and can thus be modified for effective interactions with cement hydrates (Galao, Zornoz, Baeza, Bernabeu, & Garcés, 2012).

When compared with steel fibers (which are commonly used in UHPC), polymer fibers offer improved stability in corrosive environments, lower diameters, and higher aspect ratios which could benefit their reinforcement efficiency in concrete. Polymer fibers, on the other hand, provide lower elastic moduli than steel, and thus offer lower reinforcement efficiency in concrete. Among the polymer fibers used in concrete (polypropylene, poly vinyl alcohol (PVA), nylon), PVA fibers offer elastic moduli (~30 GPa (4,350 Ksi)) which are about an order of magnitude higher than those of polypropylene and nylon fibers (but still below that of steel, which is 200 GPa (29,000 Ksi)). PVA fibers have a relatively simple chemical structure with a pendant hydroxyl group which benefits their interfacial interactions with cement hydrates.

The work reported herein concerned development of a new class of UHPC through complementary/synergistic use of CNF and PVA fiber. An optimization experimental program was designed and implemented in order to identify the optimum dosages of nano- and microscale reinforcement in UHPC.

2. Materials and experimental procedures

2.1. Materials

The materials used in this experimental work included: Type I Portland cement, non-densified silica fume (with ~200 nm (7.87×10^{-6} in) mean particle size, ~15 m^2/g (73,230 ft^2/Ib) specific surface area > 105% and 7-day pozzolanic activity index), superplasticizer (ADVA® Cast 575 from W.R. Grace Co., polycarboxylate-based, conforming to ASTM C494 Type F), silica sand (>99.5 wt.% SiO_2, ball milled and sieved to two particle size categories: 0.1–0.18 mm (0.004–0.007 in) and 0.18–0.5 mm (0.007–0.02 in), granite coarse aggregate (with 8 mm (0.315 in) and 3.5 mm (0.138 in) maximum and mean particle size, respectively), oxidized CNF shown in Figure 1 (with 60–150 nm ($2.36–3.94 \times 10^{-6}$ in) diameter, 40–100 μm (0.0016–0.0039 in) length, 50–60 m^2/g (2,44,100–2,92,920 ft^2/Ib) specific surface area, ~1.95 g/cm^3(121.7 Ib/ft^3) true density, and >95% purity), obtained from Applied Sciences, Inc. (brand name Pyrograf III Type PR24), and PVA fibers with 13 mm (0.52 in) length and 13 μm (0.0005 in) diameter, specific gravity of 1.3 (81.12 Ib/ft^3), and tensile strength of 1,200 MPa (174 Ksi). Surface of oxidized CNFs were treated following the procedures described later. Table 1 compares the properties of nano/microscale fibers used in this study.

2.2. Optimization experimental program

In order to find the optimum dosage of nano- and microscale reinforcement, an optimization program was designed based on response surface analysis principles, using the Central Composite Method. Thirteen different combinations of CNF and PVA fiber volume fractions were considered in this test program. The maximum PVA fiber volume fraction beyond which fresh mix workability would be compromised was identified as 0.55% by volume UHPC materials.

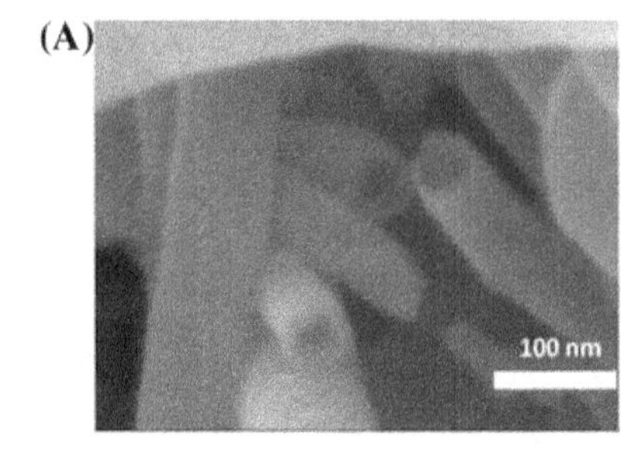

Figure 1. Scanning electron microscope image of oxidized CNFs; (A) high magnification and (B) low magnification.

Table 1. Properties of CNF and PVA fiber

Property	CNF (nanoscale reinforcement)	PVA fiber (microscale reinforcement)
Diameter nm (in)	60–150 (2.36–3.94 × 10^{-6})	13 × 10^3 (0.0005)
Length μm (in)	40–100 (0.0016–0.0039)	13 × 10^3 (0.52)
Density g/cm^3 (Ib/ft^3)	1.95 (121.7)	1.3 (81.12)
Tensile strength of MPa (Ksi)	5000 (725)	1200 (174)

Table 2. Volume percents with respect to concrete material of PVA fiber and CNF used in the optimization experimental program

Mix No.	1	2	3	4	5	6	7	8	9	10	11	12	13
Carbon nanofiber	0.04	0.04	0.04	0.00	0.08	0.04	0.08	0.04	0.04	0.00	0.00	0.10	0.04
PVA fiber	0.30	0.30	0.70	0.60	0.00	0.30	0.60	0.00	0.30	0.30	0.00	0.30	0.30

An upper limit of 0.067% by volume of UHPC materials was chosen for CNF based on preliminary studies which indicated that the CNF dispersion would be disturbed at higher volume fractions for the materials and methods used in this investigation. It should be noted that an optimization experimental program incorporates few excursions beyond these upper limits in order to test their viability. The optimization experimental program is presented in Table 2. The UHPC mix used in the optimization experimental program comprised cement: silica fume: coarse aggregate: fine aggregate: water: superplastisizer at 0.25: 0.093: 0.155: 0.38 (0.11, sand 0.1–0.18 mm (0.004–0.007 in)): 0.075: 0.018 weight ratios.

2.3. *Surface treatment of CNFs*

In order to facilitate the dispersion of CNFs in the UHPC mixing water, and also enhance their interactions with cement hydrates, hydrophilic groups were introduced on existing surface defects of nanofibers. Polyacrylic acid (PAA) was used to modify the CNF surfaces. PAA carries a high density of COOH groups, wrapping of CNFs with PAA to improve dispersion of CNFs in water, and their bonding with cement hydrates. For the purpose of modifying CNFs with PAA, CNFs were dispersed in water in the presence of PAA at PAA:CNF weight ratio of 0.1:1.0 (Peyvandi, Soroushian, Abdol, & Balachandra, 2013).

In order to disperse nanoparticle which tend to cluster due to secondary interactions over their large surface areas, a sonication technique was employed (Materazzi, Ubertini, & D'Alessandro, 2013; Raki, Beaudoin, Alizadeh, Makar, & Sato, 2010). Nanomaterials were mixed with PAA in 30% of the mixing water of UHPC, and sonicated for 30 min. The resulting dispersion was exposed to microwave radiation for 10 min at 400 W powers, and stirred for 12–15 h. The dispersion was sonicated again following the procedures described in the following section.

2.4. *Nanomaterial dispersion method*

Nanomaterials were dispersed in 30% of the UHPC mixing water using a sonicating horn following the procedure described below:

(i) The required amount of nanofiber was added to the mixing water with superplasticizer, and stirred for 12–15 h; and (ii) the mix was sonicated in a cycle comprising: (a) 10 min of sonication at 40, 40, 65, and 75% of maximum power (40 W) with 1-minute breaks in between, (b) pulse (1 min on, 30 sec off) for 10 min at 80% of maximal power, and (c) repeating the previous (pulsing) step after 2 min of rest.

2.5. *Preparation of UHPC samples*

ASTM D192 and C305 procedures were used for mixing of UHPC. This procedure involved : (i) 5 min of mixing of dry ingredients (cement, silica fume, sand, and coarse aggregate) in a Hobart Model A200F mixer at Speed 1; (ii) adding water (with dispersed nanomaterials) followed by mixing for 1 min at Speed 1, 2 min at Speed 2, and (while adding PVA fibers) 2 min at Speed 3; and (iii) casting concrete into molds per ASTM C192, and consolidating the specimens on a vibration table (FMC Syntron Power Plus) at intensity of 10. The specimens were moist-cured inside molds (ASTM C192) at room temperature for 20 h after casting, and were then demolded and subjected to 48 h of steam curing at 70 °C (158°F). The specimens were then conditioned at 50% relative humidity and ambient temperature for 7 days prior to testing.

2.6. Test procedures

Workability of fresh UHPC mixtures was evaluated using static and dynamic flow table tests (ASTM C230). Three-point flexure tests (ASTM C78) were performed on prismatic 150 × 50 × 12.5 mm (6 × 2 × 0.5 in) specimens. Impact tests (ASTM D7136) were performed on 150 mm (6 in) specimens with thickness of 12.5 mm (0.5 in). Abrasion tests (ASTM C944) were performed on cylindrical specimens with 150 mm (6 in) diameter by 12.5 mm (0.5 in) height. Compression tests (ASTM C109) were performed on 50 mm (2 in) cubic specimens. Three specimens were subjected to each test condition.

3. Experimental results and discussion

The fresh mix (static and dynamic) flow test results for UHPC mixtures with CNF and/or PVA reinforcement systems are summarized in Figures 2 and 3 versus PVA fiber and CNF concentrations, respectively. The trends depicted in Figure 2 indicate that PVA fibers have pronounced adverse effects on fresh mix workability. Fiber interactions and adsorption of water on their hydrophilic surfaces are some key factors which compromise fresh mix workability. The experimental results presented in Figure 3 indicate that fresh mix workability is not strongly influenced by the addition of CNFs (at volume fractions considered here). This can be explained by the contributions of nanofibers to the packing density of the particulate matter in the fresh cementitious matrix. Dispersed nanofibers can fill the fine space between cement and silica fume particles, thereby increasing the density of particulate matter in the binder. As a result, more water becomes available to lubricate the particles (in lieu of filling the inter particle pores). The benefits to fresh mix workability resulting from more efficient use of water towards lubricating particles compensates for any adverse effects of nanofibers on the fresh mix flow. It should be noted that the wide scatter in test results observed in Figure 3 is because the PVA fiber volume fraction (the key factor affecting flow) is not constant for different data points depicted in this figure.

Typical flexural load–deflection curves for UHPC materials with CNF and/or PVA fiber reinforcement are presented in Figure 4.

The hybrid reinforcement system comprising PVA fiber and CNF is observed to produce the least brittle material among those considered, with improved flexural strength. The instrumentation used in flexure tests suited measurement of large post-peak deformations, but not the pre-peak elastic deformations. Hence, no definitive assessments could be made of the nanomaterial effects on flexural stiffness. As far as the post-peak ductility is concerned, nanofibers alone made relatively small contributions to ductility. They were, however, effective in enhancing the ductility of PVA fiber reinforced materials. This observation points at the synergistic actions of nano- and microscale discrete reinforcement systems. This synergistic action could result from the benefits of nanofibers to the pullout behavior of PVA fibers. The particular discrete reinforcement systems considered in this investigation did not produce strain-hardening behavior. The flexural strength, maximum deflection, and energy absorption capacity as well as the impact resistance, abrasion weight loss, and compressive strength test results are presented in Figure 5. Maximum deflection corresponded to the point on flexural load–deflection curve where the flexural load-carrying capacity of the test specimen diminished. Flexural energy absorption capacity is the total area underneath the flexural load–deflection curve up to the maximum deflection. Response surface plots based on these test data, which show the trends in effects of steel fiber and CNF volume fractions on different material properties are presented in Figure 6. Each response surface plots a particular response (e.g. flexural

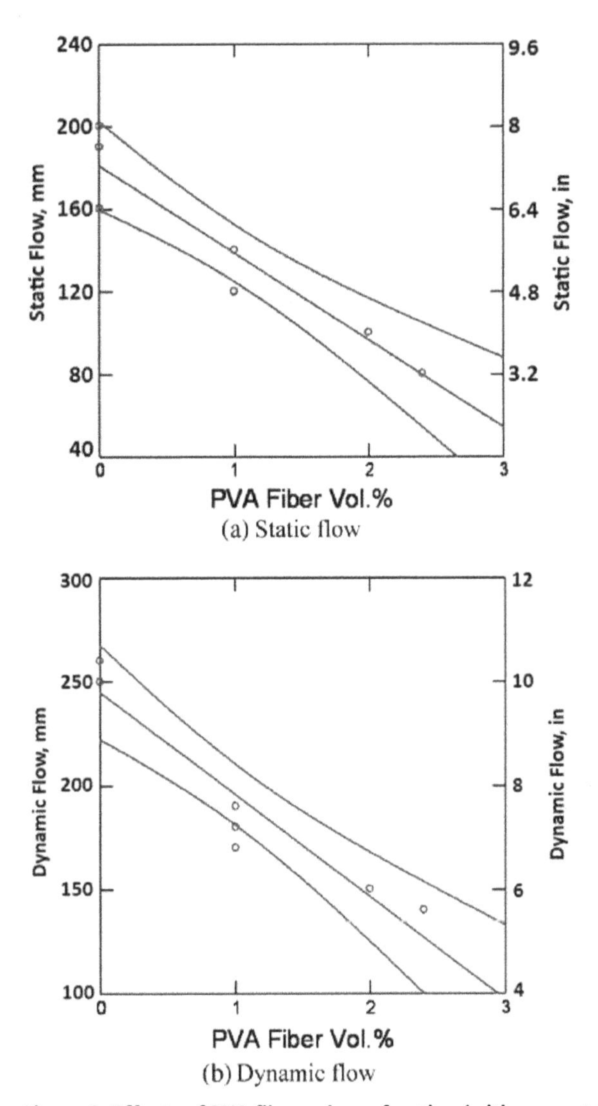

Figure 2. Effects of PVA fiber volume fraction (with respect to anhydrous cementitious materials) on fresh mix static and dynamic flow (regression lines and 80% confidence intervals are shown, CNF volume fraction varies).

strength) as a function of two variables: fiber and nanofiber volume fractions. Synergistic actions of CNF and PVA fiber towards improvement of UHPC material properties are observed in the case of impact resistance and maximum deflection test results.

Desirability (canonical) analysis of experimental results helped to determine the optimum combination of PVA fiber and CNF for achieving balanced gains in engineering properties of UHPC. The desirability function approach is used commonly for simultaneous optimization of multiple responses. In this approach, each response is first converted into an individual desirability function, and then the overall desirability function is maximized. All properties were given similar weights in the optimization process, and the objective of optimization (response surface analysis) was to identify the reinforcement condition which simultaneously maximizes. The following properties (with the targeted levels reflecting the preferred UHPC properties): flexural strength (14.3 MPa (2,070 psi) target value), maximum deflection (11.0 mm (0.43 in) target value), energy absorption capacity (1,400 N mm (398 Ib in) target value), impact resistance (4.50 mm/mm (4.5 in/in) target value), compressive strength (152 MPa (22 Ksi) target value), static flow (180 mm (7.2 in) target value) and dynamic flow (250 mm (10 in) target value), and minimize abrasion weight loss (1.12 g (0.04 oz) target value). Outcomes of this optimization process indicated that the optimum hybrid reinforcement system

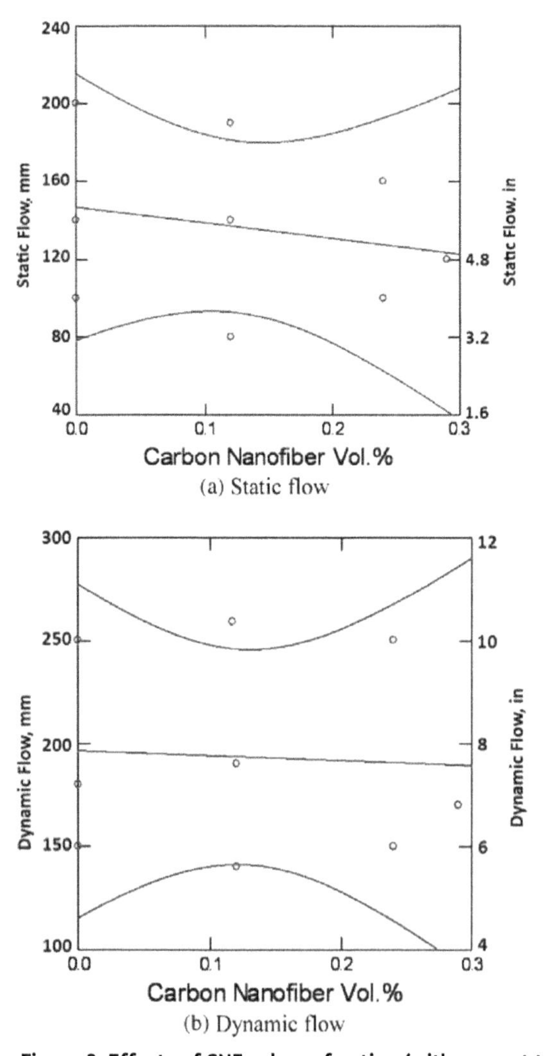

Figure 3. Effects of CNF volume fraction (with respect to anhydrous cementitious materials) on fresh mix static and dynamic flow (regression lines and 80% confidence intervals are shown, PVA fiber volume fraction varies).

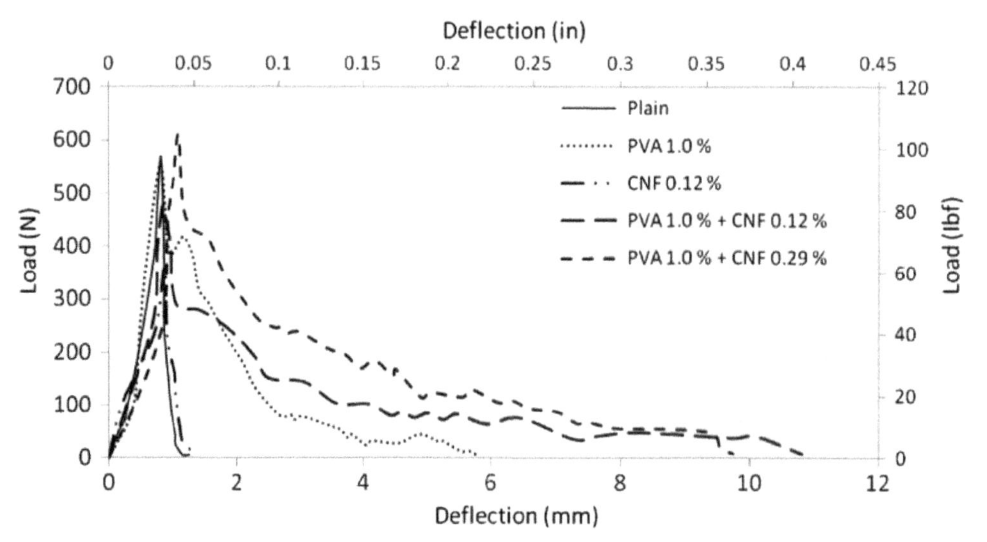

Figure 4. Typical flexural load–deflection curves of UHPC materials with CNF and/or PVA fiber reinforcement.

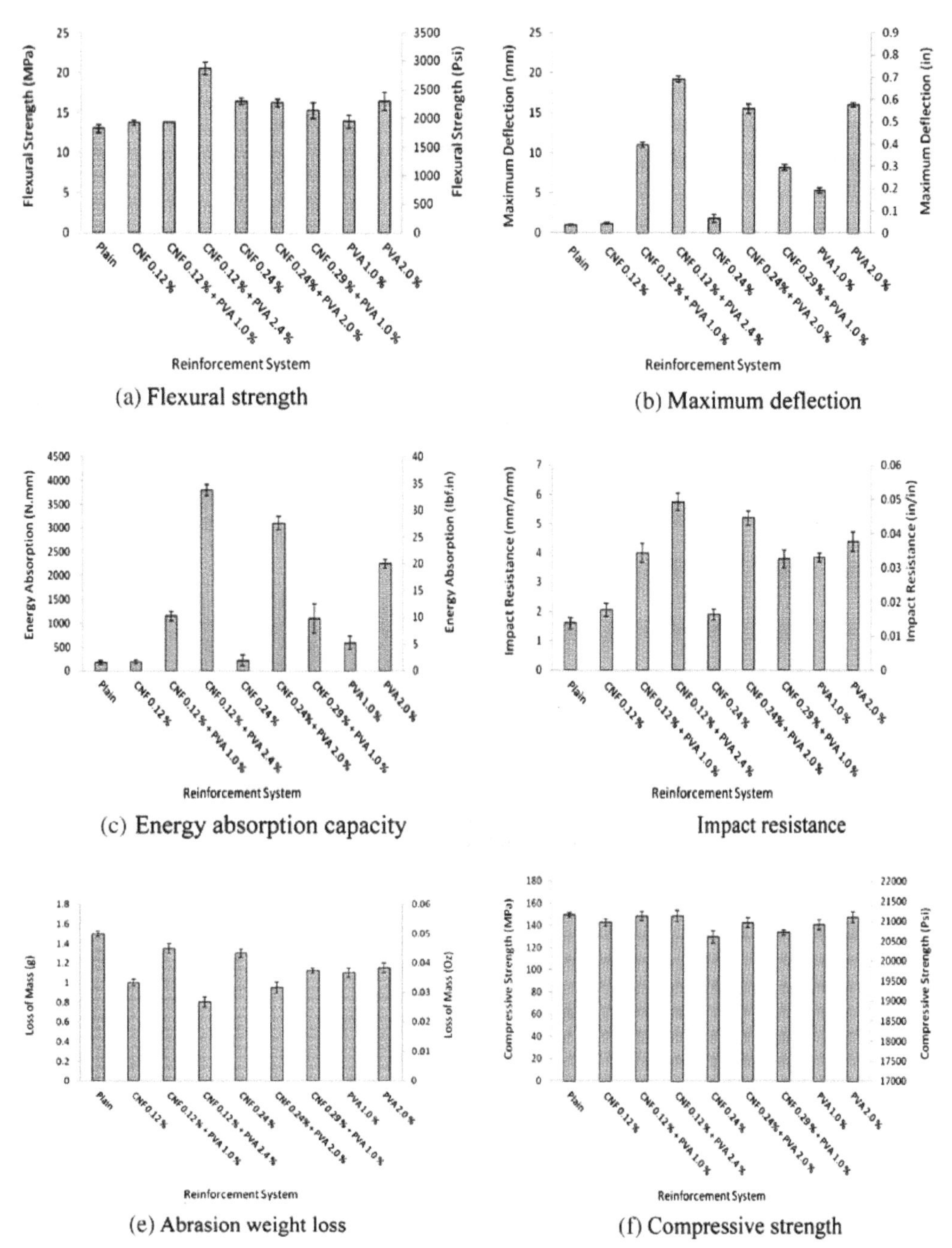

(a) Flexural strength

(b) Maximum deflection

(c) Energy absorption capacity

Impact resistance

(e) Abrasion weight loss

(f) Compressive strength

Figure 5. Experimental results on hardened UHPC material properties (means and standard errors).

comprises CNF at 0.047 vol.% of UHPC and PVA fiber at 1.20 vol.% of UHPC. Table 3 shows the wimprovements in material properties of UHPC at the predicted optimum reinforcement condition. Gains in different material properties of UHPC are also presented in Table 3. The fact that the optimized system comprises both PVA fiber and CNF points at their synergistic actions towards enhancement of the UHPC material properties. It is worth mentioning that the contribution of PVA fibers to the flexural strength of UHPC benefits from the relatively small thickness (12.5 mm, 0.5 in) of flexure test specimen compared to the fiber length (13 mm, 0.52 in). In spite of the flexibility of PVA fibers, this relatively small thickness could produce a tendency towards planar (in lieu of spatial) orientation of fibers. This geometric constraint does not alter the spatial orientation of nanofibers which have micrometer-scale lengths.

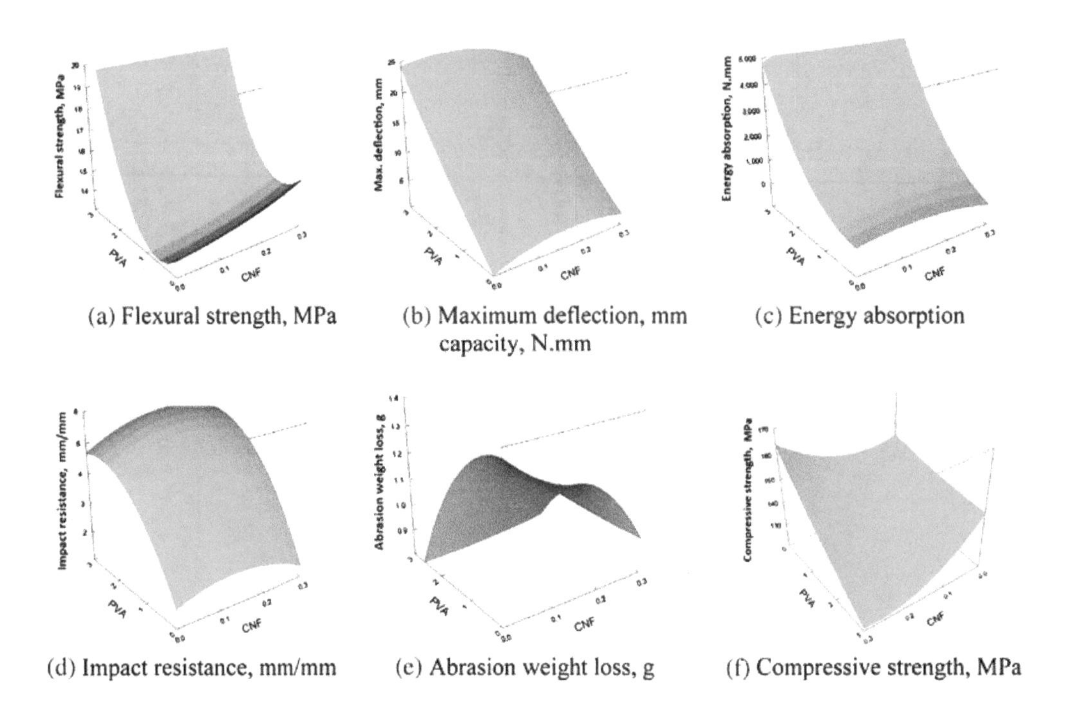

(a) Flexural strength, MPa (b) Maximum deflection, mm (c) Energy absorption capacity, N.mm

(d) Impact resistance, mm/mm (e) Abrasion weight loss, g (f) Compressive strength, MPa

Figure 6. Response surfaces developed based on the test data products for UHPC materials (PVA and CNF axes present vol.% with respect to anhydrous cementitious material).

Table 3. Effects of the optimum reinforcement condition on the material properties UHPC

	Flexural strength, MPa (psi)	Maximum deflection, mm (in)	Energy absorption capacity, N mm (Ib in)	Impact resistance, mm/mm (in/in)	Abrasion weight, loss g (oz)	Compressive strength, MPa (psi)	Static flow, mm (in)	Dynamic flow, mm (in)
	14.3 (20770)	11 (0.43)	1445 (411)	4.13 (4.13)	1.12 (0.04)	139 (20.1)	127 (5.1)	178 (7.0)
Improvement (%)	9.2	1000.0	70.0	158.2	33.9	7.5	−57.5	28.8

Effective use of nanomaterials in concrete requires thorough dispersion of nanomaterials, and their effective interfacial interactions with cement hydrates. Given the fine dimensions of nanomaterials, their dispersion and interfacial interactions in concrete benefit from increased volume fraction, fineness, and density of the binder in concrete. Therefore, nanomaterials tend to be more effective in higher performance concrete. The efforts taken in this project to ensure thorough dispersion and effective interfacial interactions of nanofibers, and also the use of UHPC have produced percent gains in material properties at relatively low nanofiber dosages which surpass those reported in the literature.

4. Summary and conclusions

A hybrid (micro/nanoscale) reinforcement system comprising CNF and PVA microfiber was optimized for balanced improvement of UHPC material properties. Oxidized CNF used in this investigation offers approach the desired geometric, mechanical, physical, and stability characteristics of CNT at about half the cost. CNF is produced at commercial scale, and is readily available for purchase. When compared with microscale (PVA) fibers, CNF offers distinct features for effective control of microcrack inception and growth. Micro- and nanoscale reinforcement offer the potential for complementary/synergistic actions in high-performance cementitious matrices because they function of different scales, and also because nanofibers can enhance the bonding and pullout behavior of microscale fibers. Surface treatment methods were developed to enhance the dispersion and interfacial interaction of CNF in cementitious matrix. An optimization experimental program was designed and implemented in order to identify the optimum dosage of PVA and modified CNF in UHPC. The material properties included fresh mix workability, flexural strength, energy absorption capacity and maximum deflection, impact and abrasion resistance, and compressive strength. Experimental results confirmed the synergistic/complementary action of nano- and microscale reinforcement in UHPC. The optimum reinforcement system, which comprised PVA fiber and CNF of 0.37 and 0.047 vol.% of UHPC improved the flexural strength, maximum deflection, energy absorption capacity, impact resistance, abrasion weight loss, and compressive strength of plain UHPC by 9.2, 1000.0, 700.0, 158.2, 33.9, and 7.5%, respectively. At the volume fractions considered here, modified CNF did not significantly alter the work abilities of fresh UHPC mixtures; PVA fiber, on the other hand, compromised the UHPC fresh mix workability. It should be noted that the discrete (nano- and microscale) reinforcement systems selected as optimum in this investigation corresponds to particular concrete materials, mix designs and production conditions, and fiber and nanofiber geometric, surface, and material characteristics. The material properties obtained in this investigation can be further improved through refinement of these parameters and identification of the corresponding optimum discrete reinforcement systems.

Acknowledgments
The authors wish to acknowledge the support of the National Science Foundation (NSF).

Funding
The research that is the subject of this paper was funded by the National Science Foundation (NSF) under grant number IIP-1142455.

Author details
Libya Ahmed Sbia[1]
E-mail: sbialiby@msu.edu
Amirpasha Peyvandi[2]
E-mail: Amirpasha.peyvandi@gmail.com
Parviz Soroushian[1]
E-mail: soroushi@egr.msu.edu
Anagi M. Balachandra[3]
E-mail: Abmetnaco@gmail.com
[1] Department of Civil and Environmental Engineering, Michigan State University, 3546 Engineering Building, E. Lansing, MI 48824-1226, USA.
[2] Bridge Department, HNTB Corporation, 10000 Perkins Rowe, Suite No. 640, Baton Rouge, LA, 70810, USA.
[3] Metna Co., 1926 Turner St., Lansing, MI 48906, USA.

References

Barnett, S. J., Lataste, J. F., Parry, T., Millard, S. G., & Soutsos, M. N. (2010). Assessment of fibre orientation in ultra high performance fibre reinforced concrete and its effect on flexural strength. *Materials and Structures, 43*, 1009–1023. http://dx.doi.org/10.1617/s11527-009-9562-3

Ding, Y., Liu, H., Pacheco-Torgal, F., & Jalali, S. (2011). Experimental investigation on the mechanical behaviour of the fiber reinforced high-performance concrete tunnel segment. *Composite Structures, 93*, 1284–1289. http://dx.doi.org/10.1016/j.compstruct.2010.10.006

FHWA. (2011). *Ultra-high performance concrete*. FHWA-HRT-11-038. McLean, VA: Research, Development, and Technology Turner-Fairbank Highway Research Center.

Foti, D. (2013). Use of recycled waste pet bottles fibers for the reinforcement of concrete. *Composite Structures, 96*, 396–404. http://dx.doi.org/10.1016/j.compstruct.2012.09.019

Galao, O., Zornoz, E., Baeza, F. J., Bernabeu, A., & Garcés, P. (2012). Effect of carbon nanofiber addition in the mechanical properties and durability of cementitious material. *Materiales de Construcción, 62*, 343–357.

Gay, C., & Sanchez, F. (2010). Performance of carbon nanofiber-cement composites with a high-range water reducer. *Transportation Research Record: Journal of the Transportation Research Board, 2142*, 109–113. http://dx.doi.org/10.3141/2142-16

Graybeal, B. A. (2007). Compressive behavior of ultra-high-performance fiber-reinforced concrete. *ACI Materials Journal, 104*, 146–152.

Habel, K., Viviani, M., Denarié, E., & Brühwiler, E. (2006). Development of the mechanical properties of an ultra-high performance fiber reinforced concrete (UHPFRC). *Cement and Concrete Research, 36*, 1362–1370. http://dx.doi.org/10.1016/j.cemconres.2006.03.009

Kalaitzidou, K., Fukushima, H., & Drzal, L. T. (2007). Multifunctional polypropylene composites produced by incorporation of exfoliated graphite nanoplatelets. *Carbon, 45*, 1446–1452. http://dx.doi.org/10.1016/j.carbon.2007.03.029

Kang, S. T., & Kim, J. K. (2011). The relation between fiber orientation and tensile behavior in an ultra high performance fiber reinforced cementitious composites (UHPFRCC). *Cement and Concrete Research, 41*, 1001–1014. http://dx.doi.org/10.1016/j.cemconres.2011.05.009

Kang, S. T., Lee, Y., Park, Y. D., & Kim, J. K. (2010). Tensile fracture properties of an ultra high performance fiber reinforced concrete (UHPFRC) with steel fiber. *Composite Structures, 92*, 61–71. http://dx.doi.org/10.1016/j.compstruct.2009.06.012

Konsta-Gdoutos, M. S., Metaxa, Z. S., & Shah, S. P. (2010). Multi-scale mechanical and fracture characteristics and early-age strain capacity of high performance carbon nanotube/cement nanocomposites. *Cement and Concrete Composites, 32*, 110–115.

Li, G. Y., Wang, P. M., & Zhao, X. (2005). Mechanical behavior and microstructure of cement composites incorporating surface-treated multi-walled carbon nanotubes. *Carbon, 43*, 1239–1245. http://dx.doi.org/10.1016/j.carbon.2004.12.017

Magureanu, C., Sosa, I., Negrutiu, C., & Heghes, B. (2012). Mechanical properties and durability of ultra-high-performance concrete. *ACI Materials Journal, 109*, 177–184.

Materazzi, A. L., Ubertini, F., & D'Alessandro, A. (2013). Carbon nanotube cement-based transducers for dynamic sensing of strain. *Cement and Concrete Composites, 37*, 2–11. http://dx.doi.org/10.1016/j.cemconcomp.2012.12.013

Melo, V. S., Calixto, J. M. F., Ladeira, L. O., & Silva, A. P. (2011). Macro- and micro-characterization of mortars produced with carbon nanotubes. *ACI Materials Journal, 108*, 327–332.

Metaxa, Z. S., Konsta-Gdoutos, M. S., & Shah, S. P. (2010). Carbon nanofiber-reinforced cement-based materials. *Transportation Research Record: Journal of the Transportation Research Board, 2142*, 114–118. http://dx.doi.org/10.3141/2142-17

Musso, S., Tulliani, J. M., Ferro, G., & Tagliaferro, A. (2009). Influence of carbon nanotubes structure on the mechanical behavior of cement composites. *Composites Science and Technology, 69*, 1985–1990. http://dx.doi.org/10.1016/j.compscitech.2009.05.002

Nochaiya, T., & Chaipanich, A. (2011). Behavior of multi-walled carbon nanotubes on the porosity and microstructure of cement-based materials. *Applied Surface Science, 257*, 1941–1945. http://dx.doi.org/10.1016/j.apsusc.2010.09.030

Peyvandi, A., Sbia, L. A., Soroushian, P., & Sobolev, K. (2013). Effect of the cementitious paste density on the performance efficiency of carbon nanofiber in concrete nanocomposite. *Construction and Building Materials, 48*, 265–269. http://dx.doi.org/10.1016/j.conbuildmat.2013.06.094

Peyvandi, A., Soroushian, P., Abdol, N., & Balachandra, A. M. (2013). Surface-modified graphite nanomaterials for improved reinforcement efficiency in cementitious paste. *Carbon, 63*, 175–186. http://dx.doi.org/10.1016/j.carbon.2013.06.069

Peyvandi, A., Soroushian, P., Balachandra, A., & Sobolev, K. (2013). Enhancement of the durability characteristics of concrete nanocomposite pipes with modified graphite nanoplatelets. *Construction and Building Materials, 47*, 111–117. http://dx.doi.org/10.1016/j.conbuildmat.2013.05.002

Peyvandi, A., Soroushian, P., & Jahangirnejad, S. (2013). Enhancement of the structural efficiency and performance of concrete pipes through fiber reinforcement. *Construction and Building Materials, 45*, 36–44. http://dx.doi.org/10.1016/j.conbuildmat.2013.03.084

Raki, L., Beaudoin, J., Alizadeh, R., Makar, J., & Sato, T. (2010). Cement and concrete nanoscience and nanotechnology. *Materials, 3*, 918–942. http://dx.doi.org/10.3390/ma3020918

Schröfl, C., Gruber, M., & Plank, J. (2012). Preferential adsorption of polycarboxylate superplasticizers on cement and silica fume in ultra-high performance concrete (UHPC). *Cement and Concrete Research, 42*, 1401–1408. http://dx.doi.org/10.1016/j.cemconres.2012.08.013

Tyson, B. T., Abu Al-Rub, R. K., & Yazdanbakhsh, A. (2011). Carbon nanotubes and carbon nanofibers for enhancing the mechanical properties of nanocomposite cementitious materials. *Journal of Materials in Civil Engineering, 23*, 1028–1035. http://dx.doi.org/10.1061/(ASCE)MT.1943-5533.0000266

Van Tuan, N., Ye, G., van Breugel, K., & Copuroglu, O. (2011). Hydration and microstructure of ultra high performance concrete incorporating rice husk ash. *Cement and Concrete Research, 41*, 1104–1111. http://dx.doi.org/10.1016/j.cemconres.2011.06.009

Wille, K., & Loh, K. J. (2010). Nanoengineering ultra-high-performance concrete with multiwalled carbon nanotubes. *Transportation Research Record: Journal of the Transportation Research Board, 2142*, 119–126. http://dx.doi.org/10.3141/2142-18

Wille, K., Naaman, A. E., El-Tawil, S., & Parra-Montesinos, G. J. (2012). Ultra-high performance concrete and fiber reinforced concrete: Achieving strength and ductility without heat curing. *Materials and Structures, 45*, 309–324. http://dx.doi.org/10.1617/s11527-011-9767-0

Wille, K., Naaman, A. E., & Parra-Montesinos, G. J. (2011). Ultra-high performance concrete with compressive strength exceeding 150 MPa (22 ksi): A simpler way. *ACI Materials Journal, 108*, 46–54.

Yang, I. H., Joh, C., & Kim, B. S. (2011). Flexural strength of large-scale ultra high performance concrete prestressed T-beams. *Canadian Journal of Civil Engineering, 38*, 1185–1195. http://dx.doi.org/10.1139/l11-078

Yoo, D. Y., Lee, J. H., & Yoon, Y. S. (2013). Effect of fiber content on mechanical and fracture properties of ultra high performance fiber reinforced cementitious composites. *Composite Structures, 106*, 742–753. http://dx.doi.org/10.1016/j.compstruct.2013.07.033

5

Performance of conceptual and black-box models in flood warning systems

Mohammad Ebrahim Banihabib[1]*

*Corresponding author: Mohammad Ebrahim Banihabib, Department of Irrigation and Drainage Engineering, University college of Aburaihan, University of Tehran, 20th km Imam Reza Road, P.O. Box 11365/4117, Pakdasht, Tehran, Iran

E-mail: banihabib@ut.ac.ir

Reviewing editor: Roberto Revelli, Politecnico Di Torino, Italy

Abstract: Flood forecasting is a core of flood forecasting and flood warning system which can be implemented by both conceptual rainfall–runoff (CRR) model and black-box rainfall–runoff (BBRR) model. Dynamic artificial neural network (DANN) as an innovative BBRR model and HEC-HMS as a traditional CRR model were used for flood forecasting. The aim of this paper is to compare the efficiency of HEC-HMS and DANN for the determination of flood warning lead-time (FWLT) in a steep urbanized watershed. A framework is proposed to compare the performance of the models based on four criteria: type and quantity of required input data by each model, flood simulation performance, FWLT and expected lead-time (ELT). Finally, the results show that FWLT and ELT were estimated longer by DANN than by HEC-HMS model. In brief, because of less required data by BBRR model and its longer ELT, future research should be focused on better verification of it.

Subjects: Civil, Environmental and Geotechnical Engineering; Water Engineering; Water Science

Keywords: flash floods; ANN; HEC-HMS; flood forecasting; flood warning system; lead-time; black-box; conceptual rainfall–runoff

ABOUT THE AUTHOR

Mohammad Ebrahim Banihabib holds a PhD of water resource engineering from Kyushu University, Japan. Currently, he is an associate professor and head of Department of Irrigation and Drainage Engineering, University of Tehran, Iran. He had studied debris flow experimentally and numerically and proposed empirical equations for its roughness and sediment transport, and a mathematical model for deposition of debris flow in floodplain. In recent decades, he has focused on three major research topics: experimental and numerical research on flash flood and debris flood control, innovative models (such as ANN and hybrid models) for flood and streamflow forecasting and strategic planning of water resources. In this paper, he focused on producing sophisticated a model for flood forecasting and flood warning. Flood forecasting and flood warning system is a non-structural measure and an effective flood control technique for mitigating flood consequences. Furthermore, it can improve the efficiency of flood management plans.

PUBLIC INTEREST STATEMENT

Population growth increases in urbanized areas and urbanization produces more floods. Also, developing urban areas in floodplains intensifies flood risk for residents. Earlier warning of floods can be used for evacuation from flood-prone areas and thus we always need appropriate models to forecast and warn floods. Frequent flash floods are reported from around the world. The extreme achievements in flood forecasting are commonly attained on large rivers. However, flash urban floods linked with dense storms in highly populated areas in urbanized watersheds need sophisticated models. The aim of this research is to compare the efficiency of HEC-HMS (a traditional model) with proposed DANN model for the determination of flood warning lead-time. This research shows that because of less required data for flood forecasting by DANN model and its longer lead-time estimation, it can be proposed for future researches.

1. Introduction

Urbanized watersheds in the mountainous areas of the world are impacted by flash floods which cause casualties and property losses and thus always need appropriate models to forecast and warn them. The utmost achievement in flood forecasting is commonly achieved for large rivers. Nevertheless, flash urban floods linked with dense storms in highly populated areas are often very tentative and are more problematical to forecast due to multifaceted dynamic phenomena tangled (Chang, Chen, Lu, Huang, & Chang, 2014). Frequent events of flash floods were reported in Southern Africa (Du Plessis, 2004), Malaysia (Alaghmand, Abdullah, Abustan, & Eslamian, 2012), the USA (Johnson, 2000), Oman (Al-Rawas, 2009), Korea (Kim & Choi, 2012), Europe (Gaume et al., 2009; Lesage & Ayral, 2007) and Iran (Golian, Saghafian, & Maknoon, 2010), which require measures to diminish their impacts (Du Plessis, 2004). In addition, because of the climate change, floods are considered one of the most significant rising natural threats of the world (Choi, 2004; Hegedus, Czigany, Balatony, & Pirkhoffer, 2013). These studies show that flash floods threat lives and properties in urbanized areas in downstream of steep watersheds; thus, appropriate models are needed for flood forecasting and flood warning system (FFFWS).

FFFWS is a non-structural measure and an efficient flood control technique for reducing flood consequences. Furthermore, flood forecasting can be used to improve the effectiveness of flood management plans (Andjelkovic, 2001; Li, Chau, Cheng, & Li, 2006; Liu & Chan, 2003). Yet, the effect of urbanization on FWLT should be deliberate to appraise their efficiency in FFWSs. FWLT is the time period between the detection of flood exceedance over specific flow threshold to the starting of flood damage. During this period, authorities should be able to apply action plan in order to reduce flood consequences. In principle, the longer the FWLT is, the higher the chance it has to decline the flood negative consequences. Studies show that urbanization will slightly raise flood-negative consequences in the next 30 years (De Roo, 1999; De Roo, Schmuck, Perdigao, & Thielen, 2003). Because of the short time for emergency action related to short watershed time-of-concentration, these negative consequences are more substantial for the highly populated areas like Tehran, Iran, with steep watersheds in the north. Therefore, existing models for flood forecasting should be examined to determine their ability in forecasting flash floods in such highly populated areas.

Flood forecasting is the central part of a FFFWS which can be implemented by a hydrological rainfall–runoff model. The hydrological rainfall–runoff model can be classified into two categories based on incorporating physical phenomenon into rainfall–runoff modelling: conceptual rainfall–runoff (CRR) model and black-box rainfall–runoff (BBRR) model. There are numerous presented FFFWSs that usually include some interacting modules which typically include at least a rainfall–runoff model. The rainfall–runoff models are mostly for real-time flood forecasting by either black-box or conceptual model in small watersheds. Black-box models do not incorporate the hydrological processes within the catchment, even in a simplified manner. However, they can be trained and verified easily in a flexible context which has made them attractive in flood forecasting (Brath, Montanari, & Toth, 2002). Additionally, conceptual models permit an explanation of spatial and time-based conservation and response laws in the watershed. Various scholars simulated floods of different watersheds via diverse rainfall–runoff forecasting models (Kafle et al., 2007; Vieux & Moreda, 2003). Numerous rainfall–runoff models were developed based on conceptual models (Feldman, 1995; Kafle et al., 2007; Vieux & Moreda, 2003). However, the necessity for a huge number of field data to calibrate and verify the models has limited their application. Clearly, there is still a strong need to provide alternate models (Chiang, Chang, & Chang, 2004). Therefore, further investigation is needed to find appropriate models for forecasting flash floods among existing black-box and conceptual models.

HEC-HMS, as a CRR model, was widely applied for forecasting flood (William & Matthew, 2010). Yet, its effectiveness in determining FWLT should be examined in steep urbanized watersheds. CRRs have been extensively applied since 1961 (Kafle et al., 2007; Sugawara, 1961; Vieux & Moreda, 2003; William & Matthew, 2010; Zhijia, Xiangguang, & Chuwang, 1998). A CRR is developed based on spatial and time-based conservation and response laws for relating rainfall and the watershed's features. In developing a CRR model, the model factors are usually watershed specific and obtained by calibrating observed floods. Conventionally, a single set of factors is expected to be related with a watershed and be appropriate to various sorts of flood occasions (Minglei, Bende, Liang, & Guangtao, 2010). Some scholars investigated run-off of a watershed by a CRR model. For instance, Vieux and Moreda (2003) simulated flash floods and debris floods of a watershed in Taiwan by Veflo™ model. Kafle et al. (2007) also investigated the flood of Bagmati Watershed in Nepal using a CRR model, HEC-HMS. HEC-HMS is developed by the US Army Corps of Engineers (William & Matthew, 2010). Their results show that there is no precise arrangement between observed and simulated low flow, but the model properly predicted the peak flow. CRR models such as HEC-HMS were considered appropriate for flood forecasting (Kafle et al., 2007). Still, the effect of urbanization on FWLT is needed to be tackled to judge flood mitigation efficiency. Therefore, HEC-HMS, as a widely used program for flood forecasting, is a typical CRR model and its efficiency can be compared with the efficiency of BBRR model in determining FWLT.

ANN can be used for forecasting floods as an innovative BBRR. However, its efficiency in determining FWLT of flash floods in steep urbanized watersheds should be examined. ANN models can capture nonlinearity of phenomenon, and thus are suitable tools for rainfall–runoff modelling (Hung, Babel, Weesakul, & Tripathi, 2009; Jeevaragagam & Simonovic, 2012; Kerh & Lee, 2006; Tawfik, 2003). Due to the ability of simulating complicated nonlinear problems, ANN is able to simulate a rainfall–runoff processes (Cameron, Kneale, & See, 2002; Chang & Chen, 2001; Chang, Liang, & Chen, 2001; Chang, Chang, & Huang, 2002; Chiang et al., 2004; Govindaraju, 2000a, 2000b; Maier & Dandy, 2000; Sivakumar, Jayawardena, & Fernando, 2002). Signal delay idea was engaged to the development of DANN in neural networking. DANN uses time delay units by recurrent connections and is computationally more strong than feed-forward artificial neural networks, and therefore recently was considered desirable for predicting (Assaad, Boné, & Cardot, 2005; Chang, Chen, & Chang, 2012; Coulibaly & Baldwin, 2005; Coulibaly & Evora, 2007; Muluye, 2011; Qian-Li, Qi-Lun, Hong, Tan-Wei, & Jiang-Wei, 2008; Serpen & Xu, 2003). Therefore, DANN is an innovative BBRR model and its efficiency can be compared with the efficiency of CRR model in determining FWLT.

HEC-HMS, as a widely used CRR, and DANN, as an innovative BBRR, were used for flood forecasting. However, their effectiveness should be evaluated by contracting their abilities in determining FWLT of the flash floods of steep urbanized watersheds. Consequently, the aim of this paper is to compare the efficiency of HEC-HMS and DANN for the determination of FWLT in north of Tehran, as a steep urbanized watershed.

2. Case study

Tajrish watershed is a steep urbanized watershed and located in the north of metropolitan city, Tehran, Iran. It was considered as the case study in this research (Figure 1). The watershed with a gross slope of 25.6% and area of 3,285 ha is a steep basin and the main flash flooder watershed in north of Tehran. Figure 2 shows the land cover of the watershed. According to Figures 1 and 2, a part of the watershed is urbanized in recent decades. The watershed is one of the major flooders of Tehran and is an appropriate case to test the performance of CRR and BBRR models in a steep urbanized watershed.

3. Calibration, verification and flood forecasting phases and required data

Characteristics of the watershed were utilized for calibration, verification and flood forecasting by the models. Table 1 illustrates data for calibration/training and verification. In addition, it shows required data for flood forecasting. HEC-HMS model needs the watershed's physical characteristics such as land cover data (as shown in Figures 1 and 2), area and slope of sub-basins, observed rainfall, and cumulative rainfall hyetograph (as shown in Figure 3) and observed flood hydrographs (as shown in Figure 6((a) and (c)). The DANN model does not require the watershed's physical characteristics such as land cover data (as shown in Figures 1 and 2), the area and slope of sub-basins.

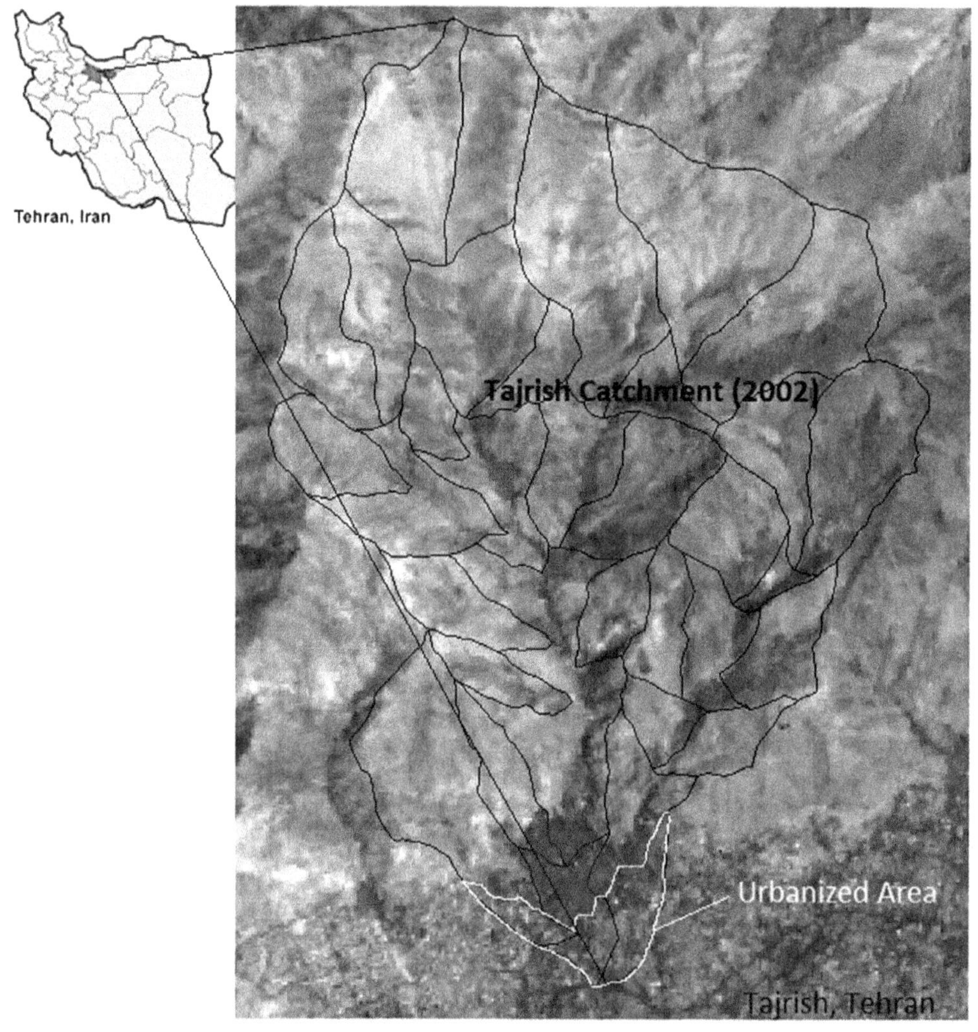

Figure 1. Watershed and its position in Iran.

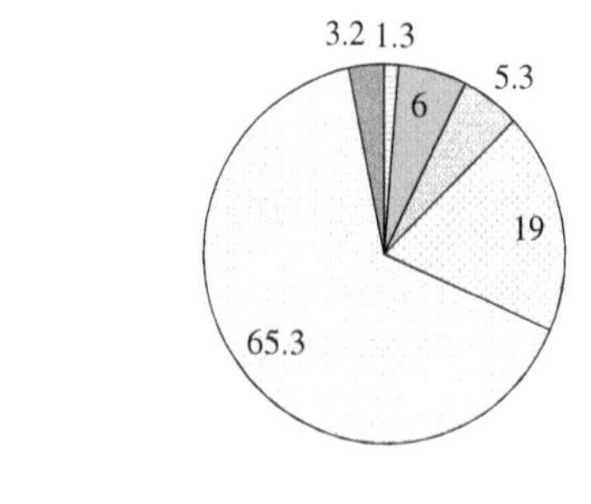

□ Farmland (%) ■ Forest/Garden (%) □ Rich (%) pasture
□ Poor pasture (%) □ Poor land (%) ■ Urbanized (%)

First, HEC-HMS model was calibrated and verified by the observed data of flow and rainfall data. Then, the simulated 10,000-year flood by HEC-HMS model was used for the training of DANN (determining neurons' weights). In addition, the simulated 25-, 50-, 100-, 200- and 1,000-year floods by HEC-HMS model were utilized for the verification of DANN. Length and width of the channels of the sub-basins and their cross sections were the physical parameters of the watershed which was used in the calibration, verification and flood forecasting by HEC-HMS model. Besides, watershed land

Figure 2. Land cover percentage in the studied watershed.

Table 1. Required data for calibration/training, verification and flood forecasting

Models	Required data for calibration/ training	Required data for verification	Required data for forecasting
HEC-HMS	Watershed's physical characteristics, observed rainfall and flood hydro-graphs on 18 March 2002	Watershed's physical characteristics, observed rainfall and flood hydro-graphs on 29 March 2002	Watershed's physical characteristics, observed rainfall
DANN	10,000-year storm temporal pattern and its simulated flood by HEC-HMS	25-, 50-, 100-, 200- and 1,000-year storm temporal pattern and its simulated flood by HEC-HMS	Observed rainfall

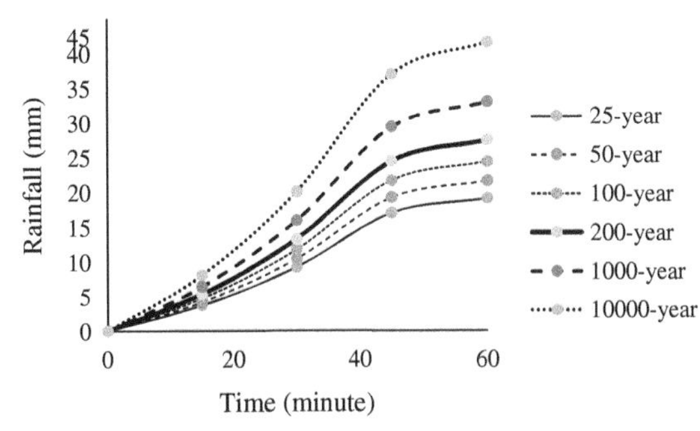

Figure 3. Cumulative rainfall hyetograph.

cover was used to approximate initial curve number (CN) of soil conservation service (SCS) as given in Figure 2. The initial CNs were calibrated using observed flood hydrographs and rainfall data of 18 March, 2002. Then, the model was validated using observed flood hydrographs of 29 March 2002. Then, 10,000-year flood was simulated by HEC-HMS using cumulative rainfall hyetograph of Niyavaran Meteorological Station as shown in Figure 3 (Banihabib, 1997). Finally, the 10,000-year flood was used for training the DANN model and the trained DANN was verified as mentioned above and was used for the assessment of FWLTs. Accordingly, using these data, the models were calibrated, verified and prepared for flood forecasting.

4. Models

4.1. Framework for comparing the models

Figure 4 shows the flowchart of this research for comparing HEC-HMS and DANN models for determining FWLT. According to the flowchart, the main stages of this research were: calibration and verification of HEC-HMS model, training and verification of DANN model, flood forecasting by the models, determination of comparing criteria, determination of FWLTs and comparing the models' performance based on criteria. Subsequently, using the proposed flowchart, the capability of HEC-HMS and DANN models was assessed by comparing their ability in determining FWLT for flash floods in a steep urbanized watershed.

HEC-HMS is used as a CRR model to determine FWLT. HEC-HMS is a new generation of models developed for rainfall-runoff simulation by the US Army Corps of Engineers (William & Matthew, 2010). First, HEC-HMS model was calibrated using the observed data as shown in Table 1. Since estimation peak of flow is very important for flood control, after determining the initial CNs of sub-basins, they were calibrated by minimizing the error of peak flow (*EPF*) using Equation (1) as follows:

$$EPF = \left| \frac{Q_{p0} - Q_{ps}}{Q_{p0}} \right| \quad (1)$$

where, Q_{p0} is observed peak flow and Q_{ps} is the simulated peak flow. Since Muskingum-Cunge has a hydraulic base and produces more accurate results, it was used for river flood routing in the

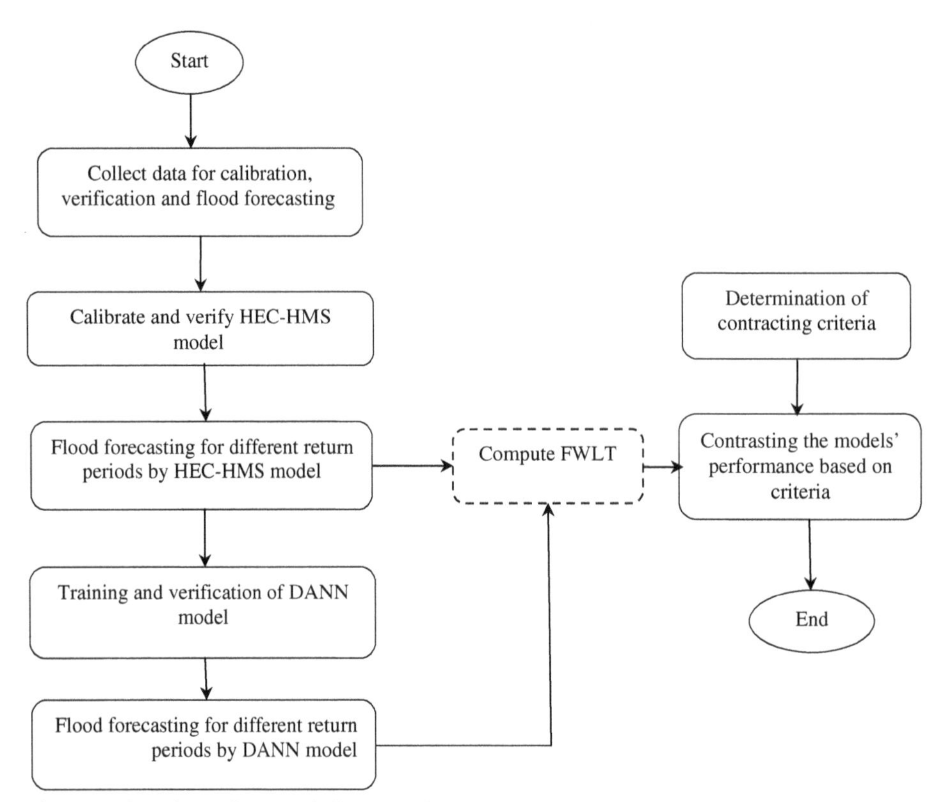

Figure 4. Flowchart of research framework.

simulation flood by the HEC-HMS model. The calibrated model is used for flood forecasting and determining FWLT as explained in Section 4.

The DANN is proposed as a BBRR in this paper which uses a recurrent mechanism to handle the memory (Figure 5). The recurrent mechanism means that the output of neurons in the output layer can also be applied as the input of the DANN. Characteristics of the structure of the proposed DANN model can be categorized into two classes: general characteristics (GCs) and tested characteristics (TCs). GCs are number of layers and activation functions which were decided based on previous researches' suggestions (Chang et al., 2014). A three-layered DANN (input, hidden and output) is flexible enough to capture any nonlinear function and was used in this paper. The activation functions of hidden and output layers of the proposed DANN were sigmoid and linear types, respectively (Chang et al., 2014). The proposed structure was completed by selecting TC of DANN structure. TC were number of delayed input, number of recurred output to input and number of neurons in the hidden layer, which were deliberated based on minimization of flood forecasting error (*FFE*) by trial-and-error process (Equation 2) (Banihabib, Arabi, & Salha, 2015):

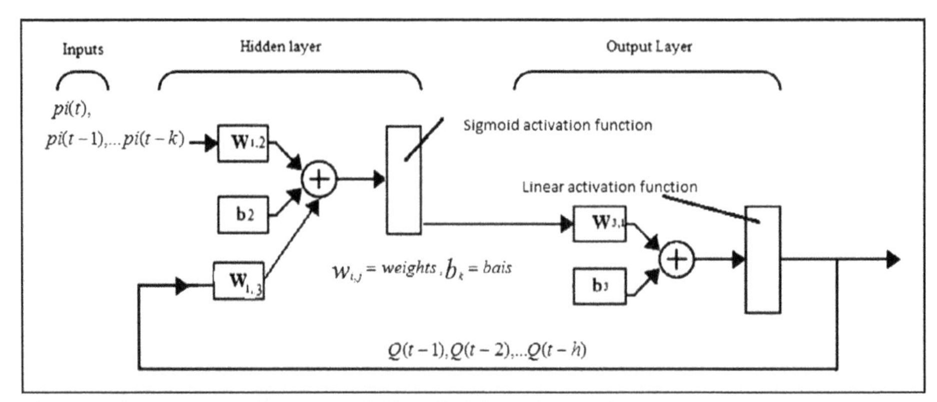

Figure 5. The structure of DANN model (Banihabib et al., 2015).

$$FFE = \text{REPE}_a \times RMSE_a \quad (2)$$

where, $RMSE_a$ and REPE_a are average of root mean square error and average of relative error of peak flow for return periods: 25-, 50-, 100-, 200-, 1,000- and 10,000-year. Number of delayed input (*K*), number of recurred output to input (*H*) and number of neurons in the hidden layer were determined by minimizing *FFE* as 37, 30 and 28, respectively (Banihabib et al., 2015). A DANN model forecasts flood using Equation (3):

$$Q(t) = F(pi(t), pi(t-1), ...pi(t-K), Q(t-1), Q(t-2), ...Q(t-H)) \quad (3)$$

where, *F* is the function of DANN, and *pi(t)* and *Q(t)* are rainfall and flow data in *t* time step, respectively. *K* and *H* are number of delayed input and number of recurred output to input which are TCs of DANN.

4.2. Determination of FWLT

Next stage of the framework was to estimate the FWLT and expected lead-times (ELTs) of flood warning. FWFFS is able to reduce flood damages and protect lives if appropriately designed and performed. The main propose of FWFFS is to announce advanced warning of a flood. Accordingly, action plan may be applied to evacuate in risk areas (Pingel, Jones, & Ford, 2005). Specifically, the duration between the first forecastable or observed rainfall and the time at which the flood flow exceeds the threshold for a flood risk to life or properties at a critical location is the maximum possible warning time (Dotson & Peters, 1990). If a warning is known, the remaining time until the exceeded threshold is the forecast lead-time. Initially, time is required for system operators to gather, evaluate and forecast based on the available data. This data collection and evaluation time are considered as forecasted recognition time (T_F). The forecast lead-time (*FWLT*) is the difference of time between risk recognition (T_F) and the time of flow that exceeds the threshold limit (T_E).

FWLT estimated for different return periods i.e. 25, 50, 100, 200, 1,000, 10,000 years. As noted, trained DANN and calibrated HEC-HMS models were applied as part of the process. First, according to Iranian regulation for rivers, the minimum discharge for the starting flood damage which is equal to peak flow of 25-year flood (Q_{25}) was defined as the threshold flow for warning (TFW) (Standard and Technical Criteria Office, 1997). Second, to determine the rainfall for flood forecasting, a time step (*dt*) was selected. Classically, this value was nominated based on the time increment of the DANN and HEC-HMS models (in this case, it equals to one minute). In the next step, T_F was set (duration of rainfall) equal to *dt* and the rainfall that occurs during that *dt* was applied for the flood forecast. Once the forecast was complete, the forecasted peak flow was compared to TFW. Then, if the forecasted flow was less TFW, the precipitation during T_F was not enough to be risk for downstream of the watershed. Consequently, T_F was increased by adding another *dt*. In the next step, the flood was forecasted again. This illustrated process was repeated until the peak flow was equal to or greater than the TFW. If the flow threshold exceeded, a warning should be issued and FWLT was determined. An expected lead-time (ELT) which uses (1/*T*) as weight of each FWLT was used for contracting the performance of the models in flood warnings.

5. Criteria

The performance of the models can be examined based on four criteria: type and quantity of the required input data by each model, flood simulation performance, time-length of FWLT and ELT. These criteria were used for contracting the models' capabilities in determining FWLT and ELT as follow:

Type and quantity of required data for HEC-HMS and DANN models in calibration/training, verification and forecasting phases are easy-accessed data and less number is preferred for both phases. Therefore, availability and numbers of data are the indices used for assessment of the required data in calibration/training and forecasting phases. Flood simulation performance can be examined by error of peak flow (EPF) and root mean square error (RMSE). The former index is the indicator for the performance of model in the simulation of the peak flow of a flood. The recent index can be used to assess the ability of the simulated hydrograph in following the observed or target hydrograph. In addition, flood simulation performance can be tested by comparing simulated flood hydrographs with observed ones. Therefore, application of these indices and the comparison can denote the performance of the model in forecasting flood hydrographs. FWLT and ELT are used for assessing the performance of models used in FFFWS. These indices show the lead-time for warning in each return period and in average using HEC-HMS and DANN models.

6. Results and discussion

6.1. Assessment of the models based on required input data

The review of the required data for calibration/training verification of and flood forecasting by HEC-HMS model and DANN model shows that DANN model requires more flood hydrographs data for verification than HEC-HMS model. As DANN model used five flood hydrographs of Figure 6((d)–(h)), HEC-HMS model just used one flood hydrograph of Figure 6(c). However, in the forecasting phase, DANN needs less data than HEC-HMS model (Table 1). As described in part 3 of this paper, HEC-HMS model requires the watershed's physical characteristics such as land use data (as shown in Figures 1 and 2), area and slope of sub-basins, where DANN model does not requires those data. Whereas HEC-HMS model needs one observed hydrograph for calibration (as shown in Figure 6(a)), our tests show that DANN model has to use extreme flood, 10,000-year flood (as shown in Figure 6(b)), for training to be able to simulate other smaller floods. Moreover, the DANN model needs more simulated flood hydrographs (five flood hydrographs) for verification than HEC-HMS model. The reason for that is DANN model doesn't incorporate physical process of rainfall-runoff and should be verified for a range of floods to confirm its ability. On the other hand, HEC-HMS model needs watershed's physical characteristics for calibration, verification and flood forecasting phases, while DANN model does require only rainfall data for flood forecasting (Table 1). Consequently, required input data for the preparation of DANN model is more than HEC-HMS model, but its application for forecasting is easier than HEC-HMS.

6.2. Calibration/training and verification

Comparing trained hydrographs of DANN model and calibrated ones of HEC-HMS model shows that the trained hydrographs followed target hydrographs better than following observed ones by calibrated ones. Figure 6((a) and (b)) shows the comparison of the calibrated and trained flood hydrographs by observed and 10,000-year simulated floods, respectively. These figures illustrate that the DANN model training performed better than HEC-HMS calibration. Consequently, the results of the comparison of trained hydrographs by DANN model and calibrated ones by HEC-HMS with target and observed hydrographs demonstrate the better performance DANN in training than HEC-HMS in calibration phases.

Comparison of simulated flood hydrographs of the models with observed and target flood hydrographs denotes that HEC-HMS model was verified better than DANN model. Figure 6(c) shows the verification of HEC-HMS model with observed flood hydrographs. Figure 6((d)–(h)) denotes the verification of DANN model by target flood hydrographs (25-, 50-, 100-, 200-, 1,000-year floods which were simulated by HEC-HMS). Comparing Figure 6(c) with Figure 6((d)–(h)) illustrates that simulated flood hydrographs by HEC-HMS follow observed ones better than DANN model does in its verification. Comparing the result of DANN model's verifications (Figure 6(d)–(h)) reveals that the verification becomes better as the return period of target hydrograph increases. Hence, the result of DANN model's verification by target flood hydrographs demonstrates that DANN model follows the target flood hydrograph which was in the same order to the trained one. However, generally, HEC-HMS model had better performance than DANN model in the verification phase.

Flood simulation performance of the models can be assessed by comparing the EPF and RMSE of the models. These indices show that the models perform conversely in training/calibration and verification phases. Table 2 shows these indices of the models for calibration and verification phases. Both indices are better for DANN model than HEC-HMS model in the calibration phase, whereas they

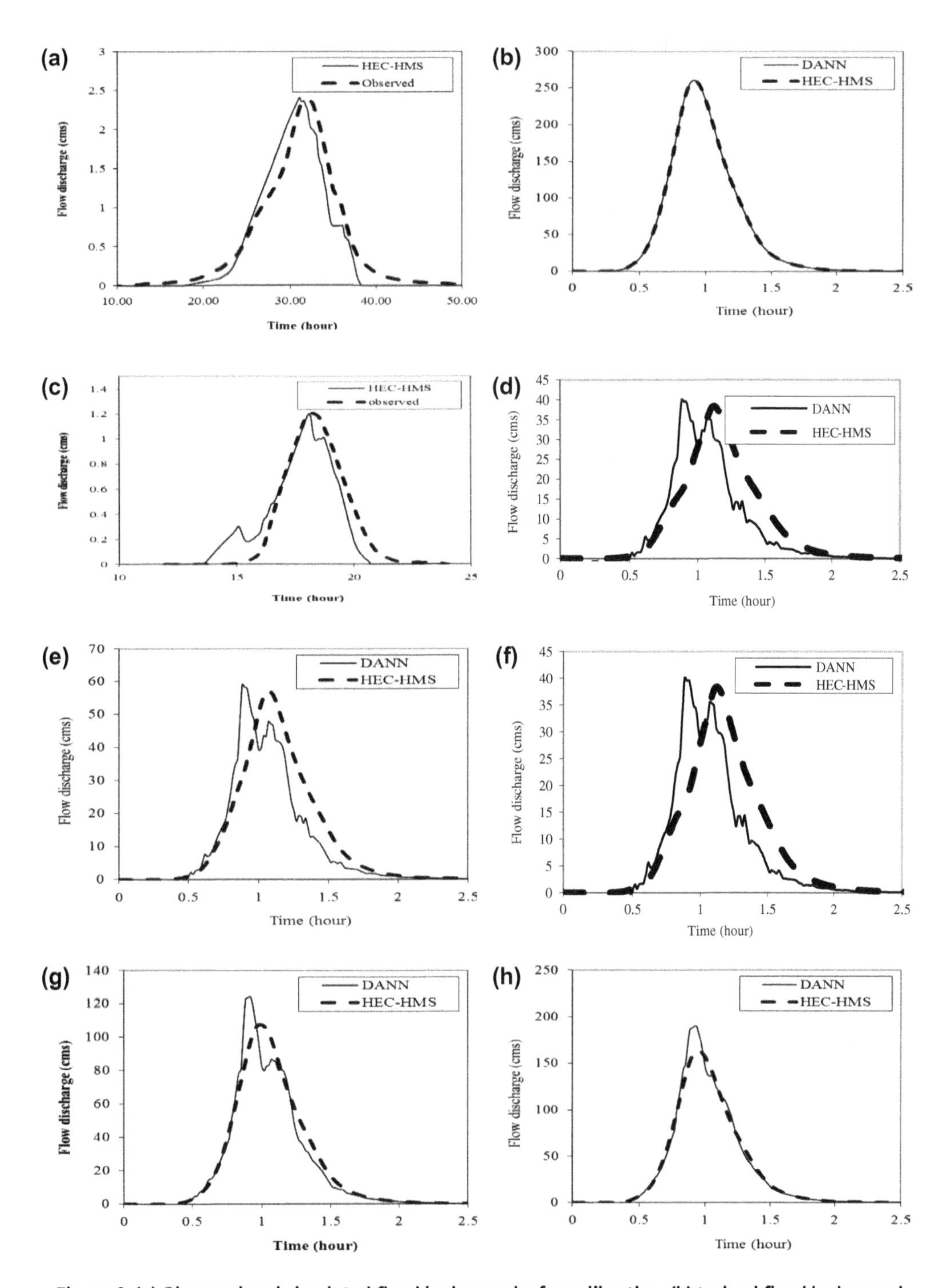

Figure 6. (a) Observed and simulated flood hydrographs for calibration, (b) trained flood hydrograph (DANN) and simulated 10000-year flood, (c) observed and simulated flood hydrographs for verification, ((d)–(h)) 25-, 50-, 100-, 200-, 1000-year-simulated flood hydrograph (DANN) and verification data.

Table 2. Error indices of the models for calibration/training and verification

Models	Calibration/training		Verification	
	RMSE	REPE	RMSE/RMSEa	EPF/EPFa
HEC-HMS	0.185768	0.03196	0.125088323	0.032006633
DANN	0.0048	0.00000000238	4.810	0.096

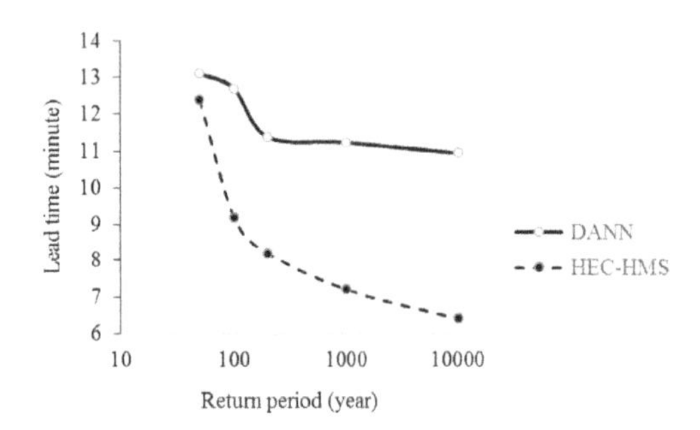

Figure 7. FWLTs of the models.

are better for HEC-HMS model than DANN model in the verification phase. The reason for that is the incorporation of watershed's physical characteristics in the verification phase by HEC-HMS model. Consequently, both indices show that DANN model performed better in the training phase than HEC-HMS model and in contrast, HEC-HMS model acted better than DANN model in the verification phase.

6.3. Assessment of the flood warning performance of the models

The results of the models show that FWLT and ELT were estimated longer by DANN than HEC-HMS model. Figure 7 indicates comparing FWLT of various return periods using DANN and HEC-HMS models. This figure shows that the FWLTs of DANN model were longer than HEC-HMS. However, the difference between FWLTs of DANN and HEC-HMS models declines by decreasing the return period. Considering relatively better verification of DANN model for higher return periods, this difference can be considered fairly precise. ELT for DANN and HEC-HMS models was 12.7 and 10.8 min, respectively. Therefore, assessment of the flood warning performance of the models based on FWLT and ELT indicated longer estimation of these indices by DANN model than HEC-HMS model.

7. Conclusion

Flood forecasting is the core of a flood warning system which can be applied by either a CRR or BBRR model. HEC-HMS, a CRR model, is used widely for forecasting flood. Yet, its effectiveness in determining FWLT was required to be examined in steep urbanized watersheds. DANN can be used for forecasting floods as an innovative BBRR. However, its efficiency for determining FWLT of flash floods in steep urbanized watersheds was essential to be examined. Thus, the aim of this paper was to compare the efficiency of HEC-HMS and DANN for the determination of FWLT in the north of Tehran, as a steep urbanized watershed. The performance of the models can be inspected based on these criteria: type and quantity of required input data by each model, flood simulation performance and duration of FWLT and ELT, and the following major conclusions are derived:

- The review of the required data for calibration/training, verification of and flood forecasting by HEC-HMS model and DANN model shows that DANN model requires more data for training and verification than HEC-HMS model. Conversely, in the forecasting phase, DANN desires less data than HEC-HMS model.
- The trained hydrographs of DANN model followed target hydrographs better than following observed ones by calibrated ones of HEC-HMS model.
- Comparison of simulated flood hydrographs of DANN and HEC-HMS models with observed and target flood hydrographs denotes that HEC-HMS model has better verification than DANN model. Furthermore, flood simulation performance of the models can be assessed by comparing the model based on EPF and RMSE indices which show that the models performed conversely in training/calibration and verification phases. DANN model acted better in the calibration phase, whereas HEC-HMS model performed better in the verification phase.
- FWLT and ELT were estimated longer by DANN than HEC-HMS model.

In conclusion, comparing CRR model, HEC-HMS, and BBRR model, DANN, shows that because of less required data for flood forecasting by BBRR model and its longer FWLT and ELT estimation, future research should be focused on better verification of this kind of model for flood forecasting.

Acknowledgements
The authors appreciate Elmira Nazar for editing the English writing of this paper and Azar Arabi for helping in research.

Funding
The author received no direct funding for this research.

Author details
Mohammad Ebrahim Banihabib[1]
E-mail: banihabib@ut.ac.ir
[1] Department of Irrigation and Drainage Engineering, University College of Aburaihan, University of Tehran, 20th km Imam Reza Road, P.O. Box 11365/4117, Pakdasht, Tehran, Iran.

References

Alaghmand, S., Abdullah, R., Abustan, I., & Eslamian, S. (2012). Comparison between capabilities of HEC-RAS and MIKE11 hydraulic models in river flood risk modelling (a case study of Sungai Kayu Ara River basin, Malaysia). *International Journal of Hydrology Science and Technology, 2*, 270–291. http://dx.doi.org/10.1504/IJHST.2012.049187

Al-Rawas, G. A. A. (2009). *Flash flood modelling in Oman wadis* (Doctoral dissertation). Calgary: University of Calgary.

Andjelkovic, I. (2001). *Guidelines on non-structural measures in urban flood management* (Technical documents in hydrology). Paris: UNESCO.

Assaad, M., Boné, R., & Cardot, H. (2005). Study of the behavior of a new boosting algorithm for recurrent neural networks. In W. Duch, J. Kacprzyk, E. Oja, & S. Zadrozny (Eds.), *Artificial neural networks: Formal models and their applications–ICANN 2005* (pp. 169–174). Heidelberg: Springer. Berlin.

Banihabib, M. E. (1997). *Guideline for determination of design flood return period in rivers works*. Tehran: Standard and Technical Criteria Office, Iran Water Resources Management Company.

Banihabib, M. E., Arabi, A., & Salha, A. A. (2015). A dynamic artificial neural network for assessment of land-use change impact on warning lead-time of flood. *International Journal of Hydrology Science and Technology, 5*, 163–178. http://dx.doi.org/10.1504/IJHST.2015.070093

Brath, A., Montanari, A., & Toth, E. (2002). Neural networks and non-parametric methods for improving real-time flood forecasting through conceptual hydrological models. *Hydrology and Earth System Sciences, 6*, 627–639. http://dx.doi.org/10.5194/hess-6-627-2002

Cameron, D., Kneale, P., & See, L. (2002). An evaluation of a traditional and a neural net modelling approach to flood forecasting for an upland catchment. *Hydrological Processes, 16*, 1033–1046. http://dx.doi.org/10.1002/(ISSN)1099-1085

Chang, F., Chang, L. C., & Huang, H. L. (2002). Real-time recurrent learning neural network for stream-flow forecasting. *Hydrological Processes, 16*, 2577–2588. http://dx.doi.org/10.1002/(ISSN)1099-1085

Chang, F.-J., & Chen, Y.-C. (2001). A counterpropagation fuzzy-neural network modeling approach to real time streamflow prediction. *Journal of Hydrology, 245*, 153–164. http://dx.doi.org/10.1016/S0022-1694(01)00350-X

Chang, F.-J., Chen, P.-A., Lu, Y.-R., Huang, E., & Chang, K.-Y. (2014). Real-time multi-step-ahead water level forecasting by recurrent neural networks for urban flood control. *Journal of Hydrology, 517*, 836–846. http://dx.doi.org/10.1016/j.jhydrol.2014.06.013

Chang, F.-J., Liang, J.-M., & Chen, Y.-C. (2001). Flood forecasting using radial basis function neural networks. *IEEE Transactions on Systems, Man and Cybernetics, Part C (Applications and Reviews), 31*, 530–535. http://dx.doi.org/10.1109/5326.983936

Chang, L.-C., Chen, P.-A., & Chang, F.-J. (2012). Reinforced two-step-ahead weight adjustment technique for online training of recurrent neural networks. *IEEE Transactions on Neural Networks and Learning Systems, 23*, 1269–1278. http://dx.doi.org/10.1109/TNNLS.2012.2200695

Chiang, Y.-M., Chang, L.-C., & Chang, F.-J. (2004). Comparison of static-feedforward and dynamic-feedback neural networks for rainfall–runoff modeling. *Journal of Hydrology, 290*, 297–311. http://dx.doi.org/10.1016/j.jhydrol.2003.12.033

Choi, W. (2004). Climate change, urbanisation and hydrological impacts. *International Journal of Global Environmental Issues, 4*, 267–286. http://dx.doi.org/10.1504/IJGENVI.2004.006054

Coulibaly, P., & Baldwin, C. K. (2005). Nonstationary hydrological time series forecasting using nonlinear dynamic methods. *Journal of Hydrology, 307*, 164–174. http://dx.doi.org/10.1016/j.jhydrol.2004.10.008

Coulibaly, P., & Evora, N. (2007). Comparison of neural network methods for infilling missing daily weather records. *Journal of Hydrology, 341*, 27–41. http://dx.doi.org/10.1016/j.jhydrol.2007.04.020

De Roo, A. (1999). LISFLOOD: A rainfall-runoff model for large river basins to assess the influence of land use changes on flood risk. *Ribamod: River basin modelling, management and flood mitigation. Concerted action, European Commission, EUR, 18287*, 349–357.

De Roo, A., Schmuck, G., Perdigao, V., & Thielen, J. (2003). The influence of historic land use changes and future planned land use scenarios on floods in the Oder catchment. *Physics and Chemistry of the Earth, Parts A/B/C, 28*, 1291–1300. http://dx.doi.org/10.1016/j.pce.2003.09.005

Dotson, H. W., & Peters, J. C. (1990). *Hydrologic aspects of flood warning-preparedness programs* (DTIC Document).

Du Plessis, L. (2004). A review of effective flood forecasting, warning and response system for application in South Africa. *Water Sa, 28*, 129–138.

Feldman, A. D. (1995). *HEC-1 flood hydrograph package. Computer models of watershed hydrology, 119*, 150.

Gaume, E., Bain, V., Bernardara, P., Newinger, O., Barbuc, M., Bateman, A., ... Daliakopoulos, I. (2009). A compilation of data on European flash floods. *Journal of Hydrology, 367*, 70–78. http://dx.doi.org/10.1016/j.jhydrol.2008.12.028

Golian, S., Saghafian, B., & Maknoon, R. (2010). Derivation of probabilistic thresholds of spatially distributed rainfall for flood forecasting. *Water Resources Management, 24*, 3547–3559. http://dx.doi.org/10.1007/s11269-010-9619-7

Govindaraju, R. S. (2000a). Artificial neural networks in hydrology. I: Preliminary concepts. *Journal of Hydrologic Engineering, 5*, 115–123.

Govindaraju, R. S. (2000b). Artificial neural networks in hydrology. II: Hydrologic applications. *Journal of Hydrologic Engineering, 5*, 124–137.

Hegedus, P., Czigany, S., Balatony, L., & Pirkhoffer, E. (2013). Sensitivity of the hec-hms runoff model for near-surface soil moisture contents on the example of a rapid-response catchment in SW Hungary. *Riscuri si Catastrofe, 12*, 125–136.

Hung, N. Q., Babel, M. S., Weesakul, S., & Tripathi, N. (2009). An artificial neural network model for rainfall forecasting in Bangkok, Thailand. *Hydrology and Earth System Sciences, 13*, 1413–1425. http://dx.doi.org/10.5194/hess-13-1413-2009

Jeevaragagam, P., & Simonovic, S. P. (2012). Neural network approach to output updating for the physically-based model of the Upper Thames River watershed. *International Journal of Hydrology Science and Technology, 2*, 306–324. http://dx.doi.org/10.1504/IJHST.2012.049188

Johnson, L. E. (2000). Assessment of flash flood warning procedures. *Journal of Geophysical Research, 105*, 2299–2313. http://dx.doi.org/10.1029/1999JD900125

Kafle, T., Hazarika, M., Karki, S., Shrestha, R., Sharma, S., & Samarakoon, L. (2007). *Basin scale rainfall-runoff modelling for flood forecasts.* Paper presented at the Proceedings of the 5th Annual Mekong Flood Forum, Ho Chi Minh City.

Kerh, T., & Lee, C. (2006). Neural networks forecasting of flood discharge at an unmeasured station using river upstream information. *Advances in Engineering Software, 37*, 533–543. http://dx.doi.org/10.1016/j.advengsoft.2005.11.002

Kim, E. S., & Choi, H. I. (2012). Estimation of the relative severity of floods in small ungauged catchments for preliminary observations on flash flood preparedness: A case study in Korea. *International Journal of Environmental Research and Public Health, 9*, 1507–1522. http://dx.doi.org/10.3390/ijerph9041507

Lesage, S., & Ayral, P.-A. (2007). Using GIS for emergency management: A case study during the 2002 and 2003 flooding in south-east France. *International Journal of Emergency Management, 4*, 682–703. http://dx.doi.org/10.1504/IJEM.2007.015738

Li, X.-Y., Chau, K., Cheng, C.-T., & Li, Y. (2006). A web-based flood forecasting system for Shuangpai region. *Advances in Engineering Software, 37*, 146–158. http://dx.doi.org/10.1016/j.advengsoft.2005.05.006

Liu, P.-S., & Chan, N. W. (2003). The Malaysian flood hazard management program. *International Journal of Emergency Management, 1*, 205–214. http://dx.doi.org/10.1504/IJEM.2003.003303

Maier, H. R., & Dandy, G. C. (2000). Neural networks for the prediction and forecasting of water resources variables: A review of modelling issues and applications. *Environmental modelling & software, 15*, 101–124.

Minglei, R., Bende, W., Liang, Q., & Guangtao, F. (2010). Classified real-time flood forecasting by coupling fuzzy clustering and neural network. *International Journal of Sediment Research, 25*, 134–148.

Muluye, G. Y. (2011). Improving long-range hydrological forecasts with extended Kalman filters. *Hydrological Sciences Journal, 56*, 1118–1128. http://dx.doi.org/10.1080/02626667.2011.608068

Pingel, N., Jones, C., & Ford, D. (2005). Estimating forecast lead time. *Natural Hazards Review, 6*, 60–66. http://dx.doi.org/10.1061/(ASCE)1527-6988(2005)6:2(60)

Qian-Li, M., Qi-Lun, Z., Hong, P., Tan-Wei, Z., & Jiang-Wei, Q. (2008). Multi-step-prediction of chaotic time series based on co-evolutionary recurrent neural network. *Chinese Physics B, 17*, 536–542. http://dx.doi.org/10.1088/1674-1056/17/2/031

Serpen, G., & Xu, Y. (2003). Simultaneous recurrent neural network trained with non-recurrent backpropagation algorithm for static optimisation. *Neural Computing & Applications, 12*(1), 1–9.

Sivakumar, B., Jayawardena, A., & Fernando, T. (2002). River flow forecasting: Use of phase-space reconstruction and artificial neural networks approaches. *Journal of Hydrology, 265*, 225–245. http://dx.doi.org/10.1016/S0022-1694(02)00112-9

Sugawara, M. (1961). On the analysis of runoff structure about several Japanese rivers. *Japanese Journal of Geophysics, 2*(4).

Tawfik, M. (2003). Linearity versus non-linearity in forecasting Nile River flows. *Advances in Engineering Software, 34*, 515–524. http://dx.doi.org/10.1016/S0965-9978(03)00039-5

Vieux, B. E., & Moreda, F. G. (2003). Ordered physics-based parameter adjustment of a distributed model. *Calibration of Watershed Models*, 267–281. http://dx.doi.org/10.1029/WS006

William, A., & Matthew, J. (2010). *Hydrologic modeling system HEC-HMS user's manual.* Washington, DC: US Army corps of engineers Hydrologic Engineering Center (HEC). 20314–21000.

Zhijia, L., Xiangguang, K., & Chuwang, Z. (1998). Improving Xin′ anjiang Model. *Hydrology, 1998*.

6

Average opportunity-based accessibility of public transit systems to grocery stores in small urban areas

Nimish Dharmadhikari[1]* and EunSu Lee[2]
*Corresponding author: Nimish Dharmadhikari, Indian Nations Council of Governments, Tulsa, OK 74103, USA
E-mail: nimish.dharmadhikari@ndsu. edu
Reviewing editor: Filippo G. Pratico, University Mediterranea of Reggio Calabria, Italy

Abstract: This research studies the accessibility of grocery stores to university students using the public transportation system, drawing from a case study of Fargo, North Dakota. Taking into consideration the combined travel time components of walking, riding, and waiting, this study measures two types of accessibilities: accessibility to reach a particular place and accessibility to reach the bus stop to ride the public transit system. These two accessibilities are interdependent and cannot perform without each other. A new method to calculate the average accessibility measure for the transit routes is proposed. A step-wise case study analysis indicates that one route provides accessibility to a grocery store in eight minutes. This also suggests that the North Dakota State University area has moderate accessibility to grocery stores.

Subjects: Cities & Infrastructure; GIS, Remote Sensing & Cartography; Transport Geography; Transportation Engineering

Keywords: public transit; accessibility; bus; geographic information systems

1. Introduction

The US Department of Agriculture defines a food desert as "a low-income census tract where a substantial number or share of residents has low access to a supermarket or large grocery store" (USDA, 2009). University students are a low-income group. It is important from the food security point of view for university students to have access to healthy food options. A grocery store (supermarket) is a place

ABOUT THE AUTHORS

Nimish Dharmadhikari is a Transportation Modeling Coordinator at INCOG, Metropolitan Planning Organization (MPO) for Tulsa region. He is currently pursuing his PhD in Transportation and Logistics from North Dakota State University (NDSU), Fargo, USA. He has completed his MS in Industrial Engineering and Management from NDSU. His research interests are GIS, Simulation, Transportation Modeling, Forecasting, and Optimization.

EunSu Lee is an associate research fellow at the Upper Great Plains Transportation Institute at North Dakota State University (NDSU). He received his PhD in Transportation & Logistics and an MS in Industrial Engineering and Management from NDSU, an MBA in Production Management from Hanyang University in South Korea, and a BSE in Computer Science from Kwandong University in South Korea. He holds GISP, CPIM, and CSCP.

PUBLIC INTEREST STATEMENT

Providing accessibility to local businesses is a critical service of public transportation systems. In this study, we used geographic information systems to measure indices that indicate the level of college students' access to grocery stores in Fargo, North Dakota. All existing public transportation routes and grocery stores in Fargo are investigated with regard to boarding time, transfer time, walking time, and travel time to calculate total traveling costs. We found that all grocery stores are accessible, but some major and specialized stores are too far away for a reasonable round trip. We found that, given the cold weather during the winter, a five-minute walking distance to a bus shelter is appropriate for designing bus routes. Public transit agencies and city planners can use this study and its approach for planning public transportation services.

where they can buy a range of healthy food products. This research studies the accessibility of public transit to grocery stores using the public bus system. A case study of the grocery stores in an area of the city of Fargo, North Dakota, is performed for the development of an average accessibility measure for the transit route.

Accessible is defined as "(of a place) Able to be reached or entered" (U. Oxford, 2012). It is from this word that accessibility is derived, which means the ability to reach a particular place or area (Litman, 2015). This definition is used in two different regards while studying transit system accessibility: (1) Accessibility to reach a particular place, such as a healthcare facility, work place, or grocery store using public transit, and (2) Accessibility to reach the public transit stop to ride the system. It is important to find the accessibility of bus stops based on the population it is serving, to make transit route decisions in urban settings. These two accessibilities are interdependent and cannot perform without each other. If a person wants to start a journey with public transport, he or she decides the place and then selects a point to board the bus to reach his or her destination. Thus, it is important to study both accessibility by measuring walking distance and opportunity to ride the public transit. Litman (2015) in his updated study provides several other meanings of accessibility with respect to different purposes such as engineering, urban economics, pedestrian planning, and social planning.

Important factors affecting accessibility are: mobility, transport options, land use, and affordability (Litman, 2015). Affordability of transit depends on the ability of a person to pay the cost for a particular trip. This personal ability is collectively expressed as the average household income. Public transit system development is focused on providing equal opportunities to all economic groups. Thus, the system should be accessible to all economic groups. Kwan (1998) studied three types of accessibility measures: (1) gravity-type, (2) cumulative-opportunity, and (3) space–time measures. They compare space–time measures with other two measures and also develop a computational algorithm using geographic information systems (GIS). Miller (1999) studied individual accessibility and connected three measures: space–time constraint-oriented approach, attraction measures, and land-use approach. This study provided a measure that reflects locations, distances and travel velocities by urban transportation network. Different classification criteria are presented in Geurs and van Wee (2004). Their criteria are used for accessibility of all modes of transportation. This research focuses on the accessibility of public transit system, thus concentrating on the literature and measures used for public transit system.

This research addresses the issue of the accessibility of the public transit modes from a residential area, combining it with the routing decisions based on the average household income. This accessibility is determined with the help of GIS. The data from the census and economic survey are joined with the spatial data to make routing decisions. A case study of the transit system in the city of Fargo, North Dakota, is performed to learn about accessibility and routing decisions based on average household income. This study focuses on the accessibility of grocery stores to a particular low-income group (university students) by the use of walking and Metro Area Transit Bus (MATBUS) travel.

This study proposes a method to calculate the average accessibility measure for the transit routes. Section 2 describes related efforts in earlier literature. Section 3 provides the methodology and development of the average accessibility measure. Section 4 provides the case study analysis. Section 5 presents results and discusses them in detail. Conclusion and future research are explained in Section 6.

2. Literature review

The literature regarding accessibility to transit system can be divided into three parts: (1) Need for an accessibility study, (2) methods to determine the accessibility of various places using the public transit system, and (3) methods to determine the accessibility to the public transit system.

2.1. Need for an accessibility study

Connecting people to their places of work using public transit has long been a topic of conversation in the transit industry. A vast study of the metropolitan area transit systems in the United States and jobs in those areas was performed by Tomer, Kneebone, Puentes, and Berube (2011). They analyzed jobs in metropolitan regions and connectivity of those jobs with the help of public transit systems. They provided findings about the residents living in areas with access to public transit. The large metropolitan areas of New York, Los Angeles, Houston, and Washington, DC provide frequent service (every 10 min) through a variety of route combinations. About 30% of jobs can be reached with the public transit within 90 min, which was their threshold of study.

Tomer (2011) stated that even though zero-vehicle households' areas are well served by public transit systems, they cannot provide ample connectivity with job opportunities. About 90% of 7.5 million zero-vehicle households from the metropolitan United States live in areas served by some kind of public transit system, but only 40% of them have job access within 90 min via transit. Accessibility is also important to reach different places such as healthcare facilities, grocery stores, and educational institutions. Also, it is important to be able to access a public transit system that is within walking distance from a household.

Curl, Nelson, and Anable (2011) review the literature on accessibility with the perspective of planning practitioners in the United Kingdom. They classify accessibility measures into five categories: infrastructure-based, cumulative, gravity-based, utility-based, and activity-based measures. They conclude that though accessibility is given importance in transportation planning, there is a greater need for work based on comparing objective measures and people's perspective of accessibility.

2.2. Determination of accessibility of various places using public transit system

Different studies have been performed to determine the accessibility of various locations from the public transit system. An important issue of access to healthy food is addressed by Burns and Inglis (2007). They used the GIS accessibility program to evaluate access to healthy and unhealthy foods in the Melbourne metropolitan area of Australia. They found that the areas with low-income households had closer access to fast-food locations, which they considered unhealthy options, while higher income households had better access to grocery stores to buy healthy foods. They state that only 50% of the population dependent on the bus had access to grocery stores within 8–10 min of travel time. They used an extension called accessibility analyst developed by the Centre of International Agricultural Tropica for ArcView GIS 3.2.

In their research regarding cancer patients in the UK, Jones et al. (2008) found an important relationship between accessibility to the transit system and cancer treatment urgency. This study highlights the importance of accessibility in the field of healthcare. Information from the patient register was combined with road network and transit information, and GIS was used to calculate accessibility to surgeries (Lovett, Haynes, Sunnenberg, & Gale, 2002). They found that around 13% of the population could not reach hospitals for surgery using the bus service. They suggested using the patient register with GIS as a planning tool for further research.

Shannon et al. (2006) presented results of an online survey about commuting patterns and transit decisions by the university population in Australia. They found that 30% of students and staff could switch to public transport if the barriers are reduced, thereby providing evidence of potential ridership. They only used online survey data and conducted statistical analysis to acquire the results.

2.3. Determination of accessibility to the public transit systems

The previous section talks about the accessibility to different destinations using public transit system, while this section reviews the literature about the accessibility to public transit system. These studies talk about how people can reach public transit system in order to use it. Salon (2009) used survey data in her study about commuting in the city of New York. She used discrete choice econometrics to estimate a model of the choices of car ownership and commute mode while also modeling the related choice of residential location. A multinomial logistic regression model is used to relate choice of residential neighborhood, car ownership status, and commute transport mode. She proposed if the relative travel time is less for transit than cars, auto ownership and car commuting can be reduced.

An application to perform measurement of public transit supply and needs in the socially disadvantaged area of Melbourne, Australia, was proposed by Currie (2010). He recognized gaps between social needs and transport service. Thus, different policy changes are suggested for transit planning. A study of GIS-based accessibility in Auckland, New Zealand, was performed by Mavoa, Witten, McCreanor, and O'Sullivan (2012). They created a multimodal transit network and combined it with the transit service frequency to calculate actual accessibility. They put forward the importance of transit frequency use in future studies and stated that frequency affects transit usage decisions.

A new method of using GIS to produce accessibility maps is provided by Langford, Fry, and Higgs (2011). They avoided using the criticized method of using buffers and demonstrated a modified two-step floating catchment method as an accessibility measure. A different approach to solve the transit route problem was used by Curtin and Biba (2011). They used a mathematical model to maximize the service value of a route, rather than minimizing cost, called Arc-Node Service Maximization. They performed a case study on the street network in Richardson, TX, and found that their method increases the service of the transit system. Zhao, Chow, Li, Ubaka, and Gan (2003) estimated transit walk accessibility using population, employment data, transit routes, and street configuration. They found that the network ratio method with actual walking distance shows better results than the simple buffer method because of natural barriers in the study region.

Biba, Curtin, and Manca (2010) proposed a new method to determine the population with walking access to transit. They offered a parcel-network method, which used the spatial and aspatial attributes of parcels and the network distances from parcels to bus stop locations. This method also avoids the criticized and unrealistic buffer method. They compared their method with the buffer analysis and network-ratio methods in the case study of transit in the metropolitan area of Dallas, TX.

The current literature highlights the need for accessibility measures for healthy food options. Though there are various efforts for defining accessibility measures based on different perspectives, there is a need for developing a measure based on the actual need of the people and their walking perspective. Thus, we try to develop a modified average accessibility measure for a transit system based on walking distance to the bus stop and income group.

3. Methodology

3.1. Data sources

This study uses two groups of data sets: census and transportation networks. Population and household income data are used for census data. Population shapefiles provided by the U.S. Census Bureau for the year 2010 are used for census blocks (U.S. Census Bureau, 2011, https://www.census.gov/geo/maps-data/data/tiger-line.html). A census block is the smallest geographical unit for which data are presented by U.S. Census Bureau. Household income data of the year 1999 are available from a database provided by the University of Wisconsin at Milwaukee (Employment and Training Institute, 2008, https://www4.uwm.edu/eti/PurchasingPower/ETIshapefiles.htm).

Cass County provides a GIS shapefile of roads and other geographic boundaries from the county's website (Cass County Geographic Information Systems, 2012, https://www.casscountynd.gov/county/depts/GIS/download/Pages/shapefiles.aspx). MATBUS, the transit provider in the Fargo area, operates routes and bus stop locations. As destinations for grocery shopping in the study, eight grocery stores were identified and digitized. The stores are three Hornbacher's (North, South, and 13th Ave.), two Sun Marts (North and South), and two Wal-Marts (13th Ave. and 55th Ave.), and a Cash Wise. Some other grocery stores were excluded in Moorhead, MN, the neighboring city in Minnesota.

3.2. Accessibility measures

The components of riding public transit for grocery shopping are composed of walking time from home of origin *o* to the closest bus stop *i* en route (W_{oi}) and riding time from a bus stop *i* to a grocery store of destination *j* (R_{ij}). The riding time can be decomposed into three components: riding time from a starting bus stop i to a transit station k (R_{ik}), transfer time at the transit station for transferring from one bus to another bus (Q_k), and traveling from the transit station k to the destination bus stop j (R_{kj}). W_{jd} presents the walking time from the bus stop j to the grocery store d as destination. This time is represented as $\text{Travel}_{\text{outbound}}$, the traveling time toward destination j in Equation 1:

$$\text{Travel}_{\text{outbound}} = W_{oi} + R_{ij} + W_{jd} = W_{oi} + (R_{ik} + Q_k + R_{kj}) + W_{jd} \tag{1}$$

Therefore, as it is possible to reach the destination with only one trip without transfer, the component $(R_{ik} + Q_k + R_{kj})$ can be zero in that situation. Waiting time at the transfer point will be included as transfer time (Q_k), while the first waiting time can be added to the walking time W_{oi}. However, the return journey, which is an inbound trip from a grocery store to home, is not always the same as the outbound trip to the grocery store due to route strategies, one-way roads, and other barriers. Thus, the total travel time will be represented as a summation of time toward destination and time from destination to reach origin i. This time is given as $\text{Travel}_{\text{total}}$ in Equation 2:

$$\text{Travel}_{\text{total}} = \text{Travel}_{\text{outbound}} + \text{Travel}_{\text{inbound}} \tag{2}$$

The students at North Dakota State University (NDSU) and other higher education institutions in the Fargo–Moorhead area ride the buses fare-free through a U-Pass program. The travel time and convenience likely have significant impacts on the use of transit for grocery shopping. In comparison to the public transit traveling time ($\text{Travel}_{\text{total}}$), the driving time can be estimated as walking time from home to the parking location (W_{op}) and traveling from the parking location to the grocery store (R_{pd}). Driving is faster than riding transit in this region and is also convenient for carrying groceries. However, the cost of driving is approximately $0.771 per mile for composite average of small sedan, medium sedan, and large sedan considering fuel, maintenance, tires, insurance, license, depreciation, and financing (AAA Association Communication, 2012, https://exchange.aaa.com/wp-content/uploads/2012/04/Your-Driving-Costs-20122.pdf). Thus, this study assumes the students are sensitive to cost, and students plan to ride public transit despite the cost of time and convenience, so this study focuses on walkability and opportunity-based accessibility.

The accessibility measure referred to as 'average opportunity accessibility' is derived from the opportunity measure proposed by Wachs and Kumagai (1973). Opportunity-based measures compute the number of opportunities that someone could have within a certain distance or during some period. A larger number indicates a greater level of accessibility. This model could be written as Equation 3:

$$A_i = \sum_j D_{j(t \le t_0)} \tag{3}$$

where A_i is the accessibility of origin i, D_j is the number of opportunities at destination j. Destination j is within a threshold travel time t_0 from origin i.

This study modifies the accessibility to incorporate the public transit route available and grocery stores served by the transit route. For example, some census block might have one or more service routes with at least one grocery store through one route, while some might not have any public transit service available. Even if some census blocks belong to one or more public transit routes, the routes might not have any grocery store along them. Equation 4 addresses such issues. The public transit routes (BR_m), which are located within the threshold walking distance from each census block as origin i to the closest bus stops o (t_0), are found. The number of grocery stores (G) through the routes ($BR_{m'}$) is summed. Then, the total number of grocery stores is divided by the population in the census block (POP_i). In other words, the accessibility for origin i (a census block) equals the sum of grocery store on a route that can be accessed in less than n minutes, divided by the population for origin i:

$$A_i = \frac{\sum_{m(w_{oi} < t_0)} G_{d(d \in BR_{m'})}}{POP_i} \tag{4}$$

Equation 4 gives the value of average opportunity-based accessibility. It is given by dividing the total number of grocery stores through one bus route with total population in the census blocks through that route within walking distance.

This study connects the average accessibility measure, which is a technical measure, with the population perspective in terms of paying for service and use of the public transit system. This is an important addition as it will be useful for transportation planners.

4. Analysis

The analysis is performed with the help of GIS map development and statistical study. This is presented in three steps. The first step is analysis of the average household income and bus routes in Fargo. The second step is analysis of the total population in census blocks and transit routes 13A and 15. The third step is performing network analysis for generating service areas. A statistical analysis to find accessibility is also performed in this step.

Step 1: Analysis of the average household income and bus routes

The average annual household income is calculated by adding the total annual incomes of households in a given area and dividing it with the total number of households in that particular census area. These data are joined with the shapefile of the census area for visualization. The data for average annual household income is presented in Figure 1. This figure also includes the road network of Fargo and selected bus routes. The GIS map is used to analyze the bus route distribution based on the average annual household income in census block groups. According to U.S. Census Bureau, a census block group is the unit clustering census blocks and contains around 600 to 3,000 people. This map shows the average annual household income is the lowest around the bus routes 13A and 15.

Route 13A travels around North Dakota State University. The majority of the population around this route is made up of students, which explains the smaller average annual household income. This route connects the university to the Fargo downtown area and the Ground Transportation Center (GTC), which is a hub for public transit. Route 15 connects the GTC with the new business developments like West Acres Mall and Wal-Mart. Large numbers of university students frequently use these routes for travel. Thus, this study is focused on the accessibility analysis near routes 13A and 15.

Step 2: Analysis of the total population in census block

The second step analyzes the total population in the census blocks and important MATBUS routes. The main bus stop serving the NDSU campus is located at the south side of the Memorial Union building, where there is a large shelter that most students use when waiting for the bus.

The area grocery stores are located and digitized using their latitude and longitude. The MATBUS routes connecting NDSU and the grocery stores are identified. Figure 2 shows routes 13A, 11, 14, 16, 23, 25, and 15 and the total population in census blocks. The student population use MATBUS to travel to different grocery stores.

Step 3: Network analysis

The analyses in Steps 1 and 2 found the seven important routes to be 11, 13A, 14, 15, 16, 23, and 25. There is a moderate or high population in the census blocks around NDSU. This study assumed that a majority of this population is the student population. The average annual household income of this population is also smaller. Thus, it is assumed that this student population needs transit access to travel to different places.

The walking accessibility of this population to reach a transit stop is studied in this step. The Network analyst is the default functionality in the ESRI ArcGIS software. This functionality has different methods

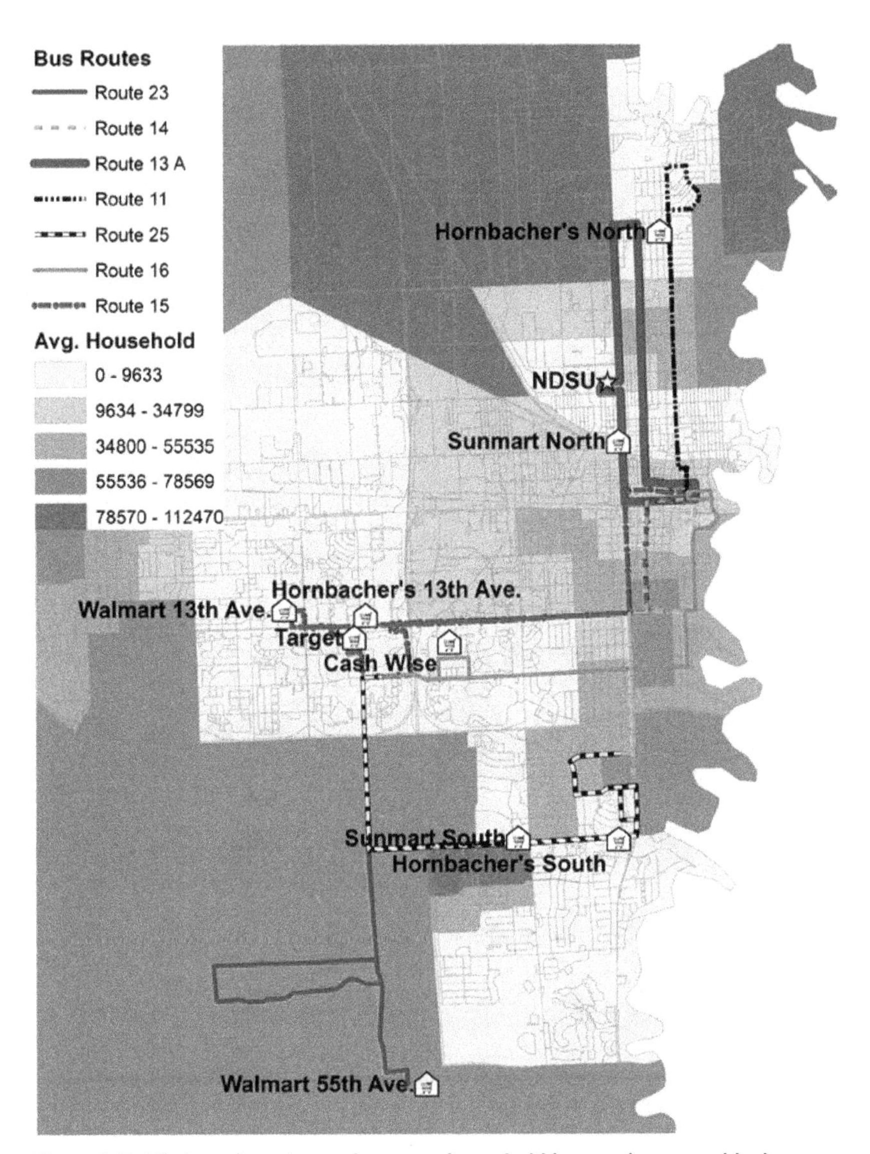

Figure 1. Public transit routes and average household income in census block groups.

such as closest facility analysis, service area, and vehicle routing problem. This study uses service area and closest facility methods. The service area method is used to study the accessibility based on walking distance, thereby presenting WT in the total travel time. In the city of Fargo, MATBUS allows flagged stops at all intersections on their routes, so they are assumed as the intermediary destinations. The service area method uses the actual route information for drawing the service area from the specific facility. The service area is the polygon around the facility, which is accessible from within that polygon.

Gates, Noyce, Bill, and Van Ee (2006) stated that the four feet per second (feet/s) walking speed can be used near the college campus. It is assumed for this study that the pedestrian walking speed is 4 feet/s. Two different times are assumed as the different willingness to walk to the transit stop. Five minutes and slightly more than 10 min are considered as two walking times for this case study. This gives two different distances to use in the GIS analysis. If the walking time is 5 min, 1,200 feet is the walking distance (5 min × 60 × 4 feet/s = 1,200 feet), and the walking distance 2,500 feet (10.5 min × 60 × 4 feet/s ≈ 2,500 feet) is for slightly more than 10 min.

Distances of 1,200 feet and 2,500 feet are used to plot the service area. For the service area, maximum distances traveled by the available path are 1,200 feet and 2,500 feet. The road network in the

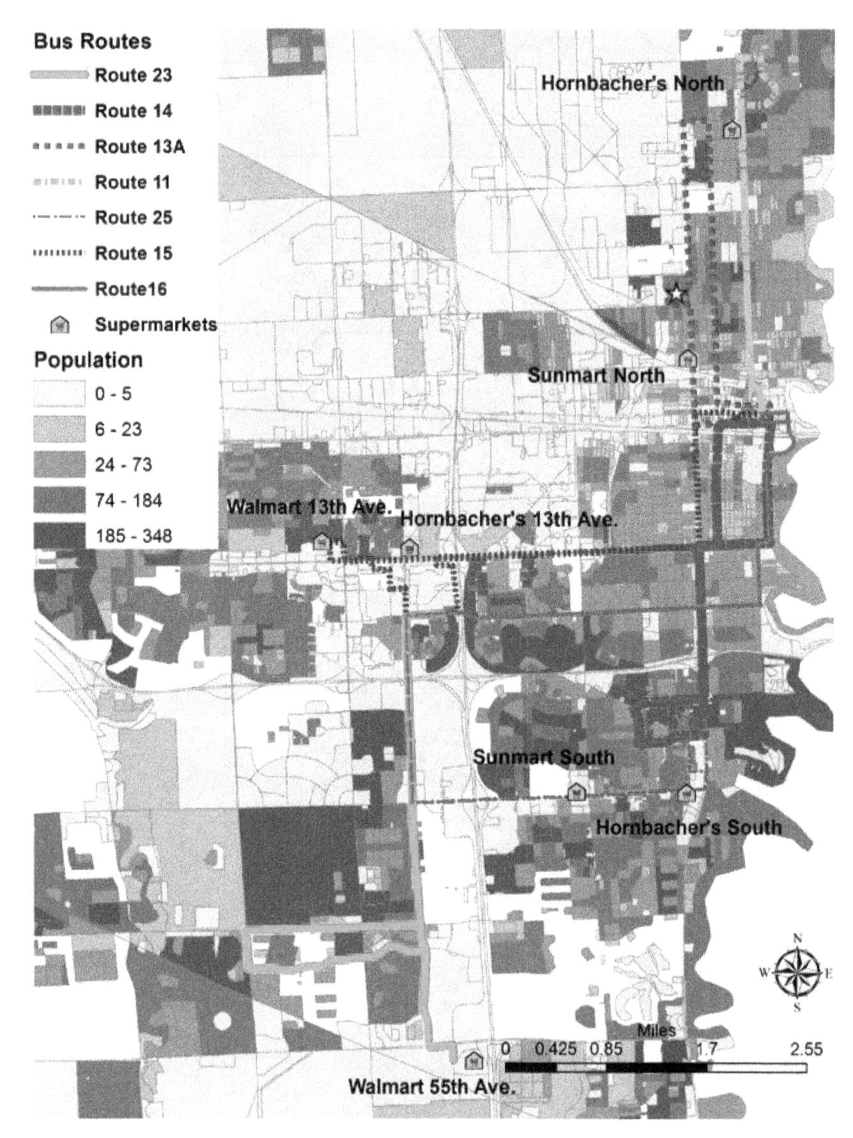

Figure 2. MAT bus routes with population.

city of Fargo is assumed to have sidewalks for the pedestrians to reach the transit stops. In reality, some streets may not have sidewalks, reducing pedestrian accessibility, but for the area being studied, most of the streets have sidewalks. Instead of using straight-line distances, a service area method based on actual route by avoiding natural and man-made barriers is proposed to be more suitable for the accessibility study. The service area method uses the sidewalk network as the route network. Thus, it has fewer obstacles and the actual distance and time traveled can be calculated.

5. Results and discussion

5.1. Results

It is found that there are four groups of routes, which connect NDSU to the following area grocery stores: Hornbacher's, Sun Mart, Wal-mart, and Cash Wise. Service areas around these routes are formed using walking distances of 1,200 and 2,500 feet. For example, Figure 3 shows the service areas around routes 15 and 13A in (a) and (b), respectively. A major portion of Route 15 is along 13th Avenue South for both directions (Figure 3(a)). The route provides service between the GTC downtown and the West Acres Shopping Center on the west side of the route.

Route 13A starts from the GTC and provides direct service to the NDSU campus and Sun Mart North (Figure 3(b)). Route 13A, when using a 1,200-feet walking distance, serves 14,577 people through 94 possible stop locations. Each stop might serve 155 people. When using the 2,500-feet walking distance, Route 13A serves 21,915 people, and each stop of the route might serve 233 people (Table 1).

For the walking distance of 1,200 feet, the accessibility of the census blocks along the routes for different service areas are calculated with Equation 4 developed in the methodology. For example, Route 13A with service area for 1,200-feet walking distance serves a total population of 14,577. The average accessibility can be calculated using GIS analysis. For example, one census block near NDSU through route 13A has a population of 95 and it reaches 1 grocery store. Thus, using Equation 4 this block has accessibility of 0.0105 (1/95). A summation of all census blocks average accessibilities for all census blocks along Route 13A equals 0.129, which is higher than that along Route 11 but lower than those along Routes 14 and 15. Route 14 provides the highest accessibility among the routes; however, this route does not have grocery stores along the route using a combination of two or more routes.

For the walking distance of 2,500 feet, the average accessibility of the census blocks along Route 13A is the highest (0.226) since both Sun Mart North and Hornbacher's North are located in the service area.

This accessibility is visualized with the help of GIS maps in Figure 4: (a) for the 1,200-feet walking distance and (b) for the 2,500-feet walking distance. It is difficult to connect Hornbacher's South and Sun Mart South with NDSU by transit, showing low accessibility for both of 1,200-feet and 2,500-feet walking distances. Students have to take three different routes to reach these two grocery stores through 13A–15–23 routes. Figure 4 shows the transferring at GTC from Route 13A to 15 and at West Acres Shopping Center from Route 15 to 23.

Wal-Mart 13th Ave. plays an important role in grocery shopping for the lower income level in the city. The store is accessible with routes 13A and 15 for NDSU students by transferring from route 13A to 15 at the GTC. Route 13A is highly accessible to students to the stores, but the route must be linked to Route 15 to reach Wal-Mart 13th Ave. and Target (Figure 4(a)).

The newest grocery store in the Fargo area is Wal-Mart 55th Ave. at the south edge of the city. The probability that NDSU students will ride a transit route, which is far from the university campus, or travel for grocery shopping by walking long distances is very low. Though Wal-Mart 55th Ave. is accessible by connecting routes of 13A–15–23 from the university, the probability of students going there for shopping is very low.

Routes 13A and 11 connect NDSU to Sun Mart and Hornbacher's North.Sun Mart North is the closest grocery store to the students, taking 12 min without stopover at the transit center (Table 2). The routing decisions are made with the help of bus speed and distance in miles for the routes. The ArcGIS® Network Analyst tool was used to find routes from NDSU to grocery stores with the shortest travel times. The average effective bus speed in Fargo is assumed as 15 miles/h. This is assumed with the standard delays for the stoppage for passengers. Total travel time and distance traveled is presented in Table 2.

The closest grocery store is Sun Mart North. It takes five to ten minutes walking and approximately three minutes by bus without any transfer. The farthest store is Wal-Mart 55th Ave., which requires five to ten minutes walking plus 10-min transfer time, and 38-min riding in bus; thereby it takes 52.78 min in total.

Routes 13A and 11 run parallel from downtown to the north of the city. Hornbacher's North is located near the edge of route 13A (Figure 5). If only the route 13A is used to reach Hornbarcher's North, it takes five minutes more on foot in a 2,500-feet walking distance (see Figure 5(b)). From the analysis, Route 13A

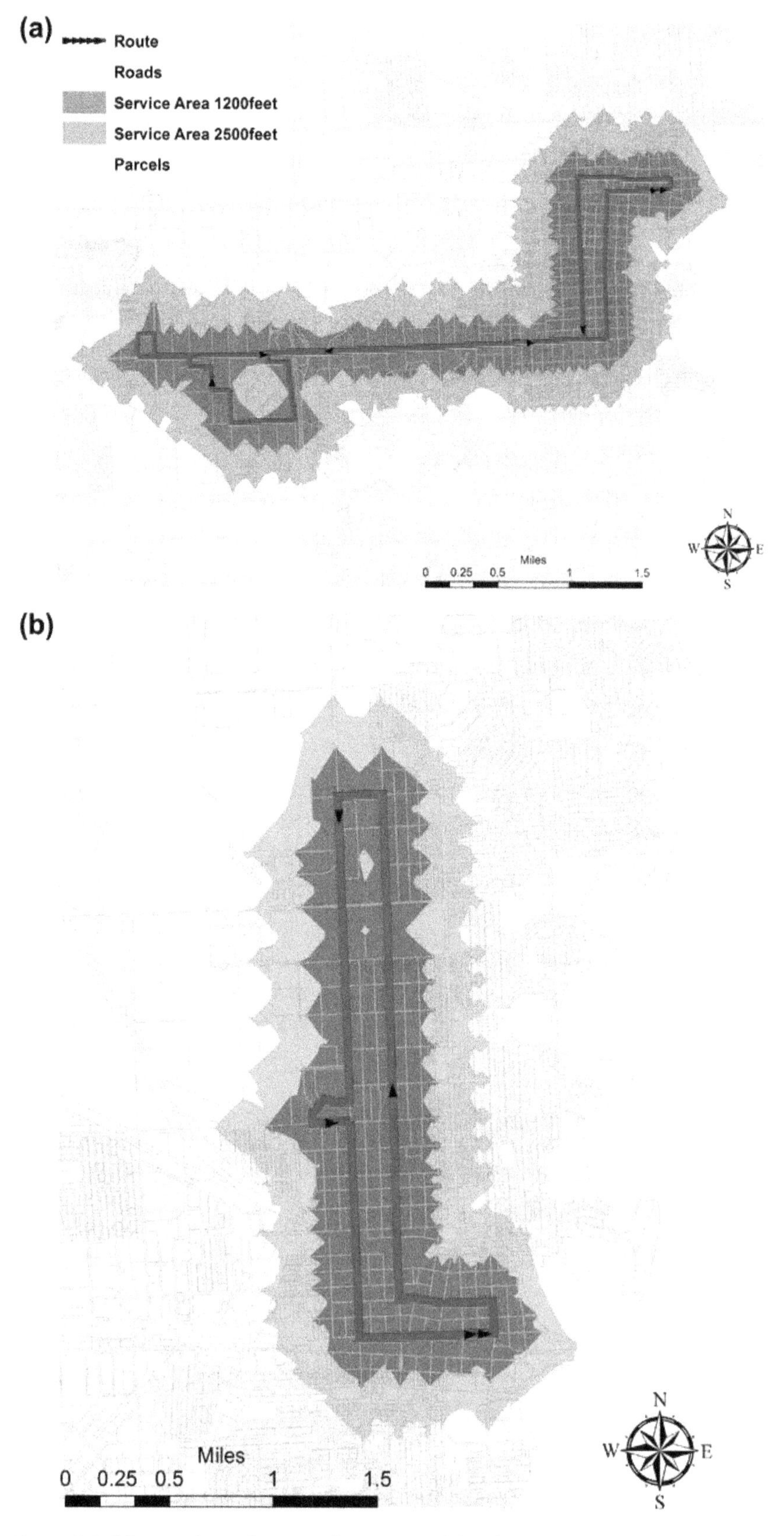

Figure 3. Public transit service areas for route 15 (a) and 13A (b).

Table 1. Walking distance and accessibility

Walking distance (feet)	Route	Service population (person)	Average accessibility	Grocery stores reachable
1,200	11	11,868	0.078581	Hornbacher's North
	13A	14,577	0.129564	Sun Mart North
	14	14,389	0.153888	N/A
	15	16,706	0.144107	Wal-Mart 13th Ave./ Target
	16	13,613	0.091632	Cash wise
	23	10,577	0.048672	Wal-Mart 55th Ave.
	25	15,884	0.055923	Hornbacher's South/ Sun Mart South
2,500	11	21,117	0.148918	Hornbacher's North
	13A	21,915	0.22615	Sun Mart North/ Hornbacher's North
	14	21,917	0.211664	N/A
	15	28,480	0.206012	Wal-Mart /Target
	16	19,933	0.158101	Cash wise
	23	14,351	0.089228	Wal-Mart 55th Ave.
	25	23,129	0.067771	Hornbacher's South/ Sun Mart South

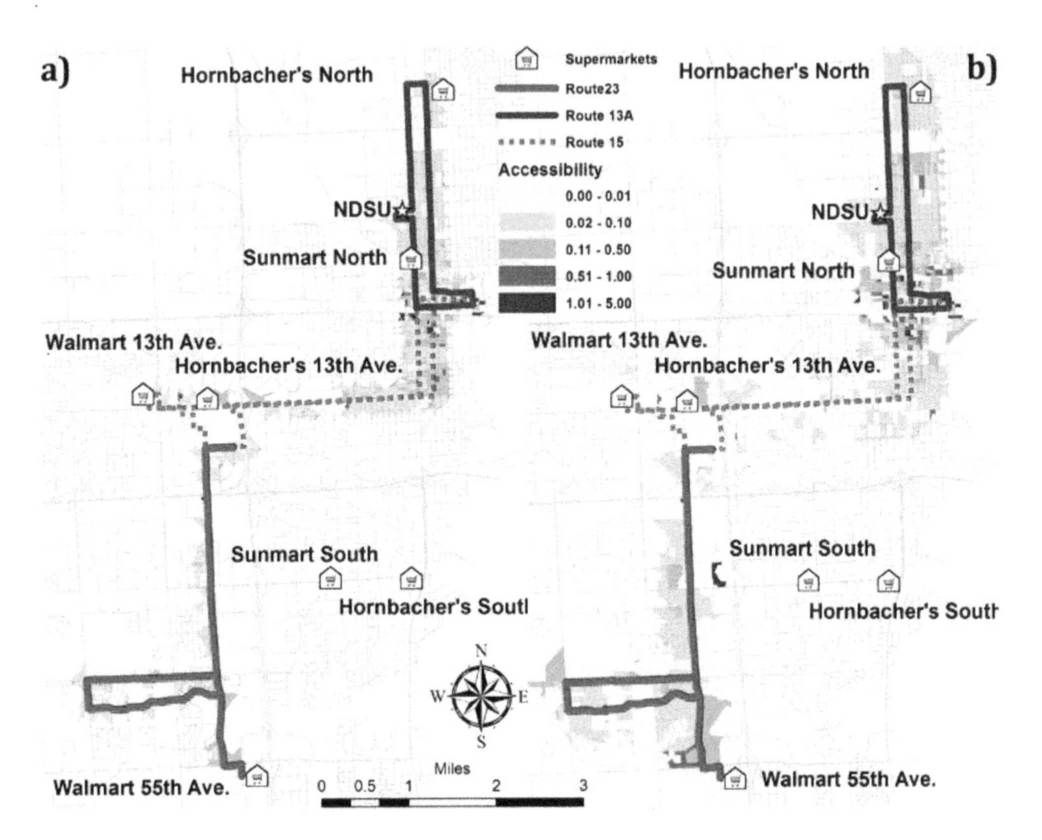

Figure 4. Accessibility of Wal-Mart 55th Ave. with route 23: (a) 1,200-feet walking distance and (b) 2,500-feet walking distance.

Table 2. Route connections and travel time from the university to the grocery stores in town

Walking distance (feet)	Destination	Connections	Travel distance (miles)	Boarding (min, $R_{ik} + R_{kj}$)	Transfer time (min, Q_k)	Walking time (min, W_{oi})	Travel time (min)	Cost ($) for traveling
1,200	Sun Mart North	13A	0.7740	3.0960	0	5	8.0960	2.70
	Hornbacher's North	13A-11	4.3940	17.5760	5	5	27.5760	9.19
	Hornbacher's South	13A-14-25	4.4920	17.9680	10	5	32.9680	10.99
	Hornbacher's 13th Ave.	13A-15	4.7640	19.0560	5	5	29.0560	9.69
	Sun Mart South	13A-14-25	5.4190	21.6760	10	5	36.6760	12.23
	13th Ave.	13A-25	5.6040	22.4160	5	5	32.4160	10.81
	Wal-Mart 55th Ave.	13A-25-23	9.4450	37.7800	10	5	52.7800	17.59
	Cash Wise	13A-16	4.6832	18.7327	5	5	28.7327	9.58
2,500	Sun Mart North	13A	0.7740	3.0960	0	10	13.0960	4.37
	Hornbacher's North	13A-11	4.3940	17.5760	0	10	27.5760	9.19
	Hornbacher's South	13A-25	4.4920	17.9680	10	10	37.9680	12.66
	Hornbacher's 13th Ave.	13A-15	4.7640	19.0560	5	10	34.0560	11.35
	Sun Mart South	13A-25	5.4190	21.6760	10	10	41.6760	13.89
	Wal-Mart 13th Ave.	13A-25	5.6040	22.4160	5	10	37.4160	12.47
	Wal-Mart 55th Ave.	13A-25-23	9.4450	37.7800	10	10	57.7800	19.26
	Cash wise	13A-16	4.6832	18.7327	5	10	33.7327	11.24

Note: Cost is computed by the labor expense of $10 per hour for traveling for a student.

is the only route to reach a grocery store without stopover. In order to save transfer time and increase ease of ride for grocery shopping, Route 13A can be extended to Hornbacher's North and provide connection point to the route 11.

5.2. Discussion

When the study assumes that a student is not making any money, the choice of taking the public transit for grocery shopping at a cost of $0 seems most favorable (Figure 6). This trend continues till the student is willing to pay $4 per hour, which is equilibrium cost for $0.50 vehicle mile cost and free ride for the public transit considering time value. At this threshold, the student is indifferent to choosing the public transit or driving a car at a cost of $0.50 per mile. If the driving cost per mile is $0.80, this threshold is shifted to a cost of $5. The equilibrium cost increases to $6 if the driving cost increases to $1.00 per mile due to fuel price or maintenance cost. In other words, when the vehicle driving cost per mile increases to $1.00 per mile and a student earns $10 per hour, the student is more likely to drive for grocery shopping at Wal-Mart 13th Ave.

Due to the importance of the connectivity of the public transit to places in the city, the mobility and transit service plan play an important role in grocery shopping. Along 13th Ave. South, Wal-Mart, Hornbacher's, and Cash Wise are locations that provide fresh foods. For example, if the city transit system adds a weekend route from NDSU to the West Acres Shopping Center mall area, the total travel

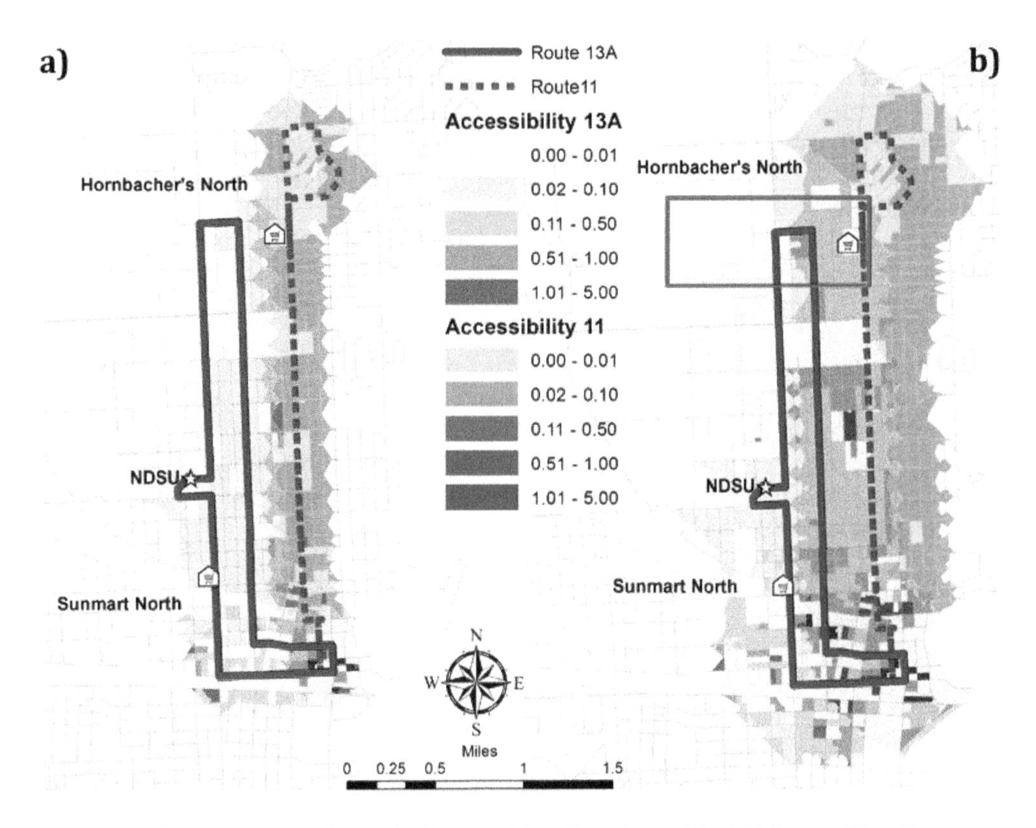

Figure 5. Major routes to reach Hornbacher's and Sun Mart North: (a) 1,200-feet walking distance and (b) 2,500-feet walking distance.

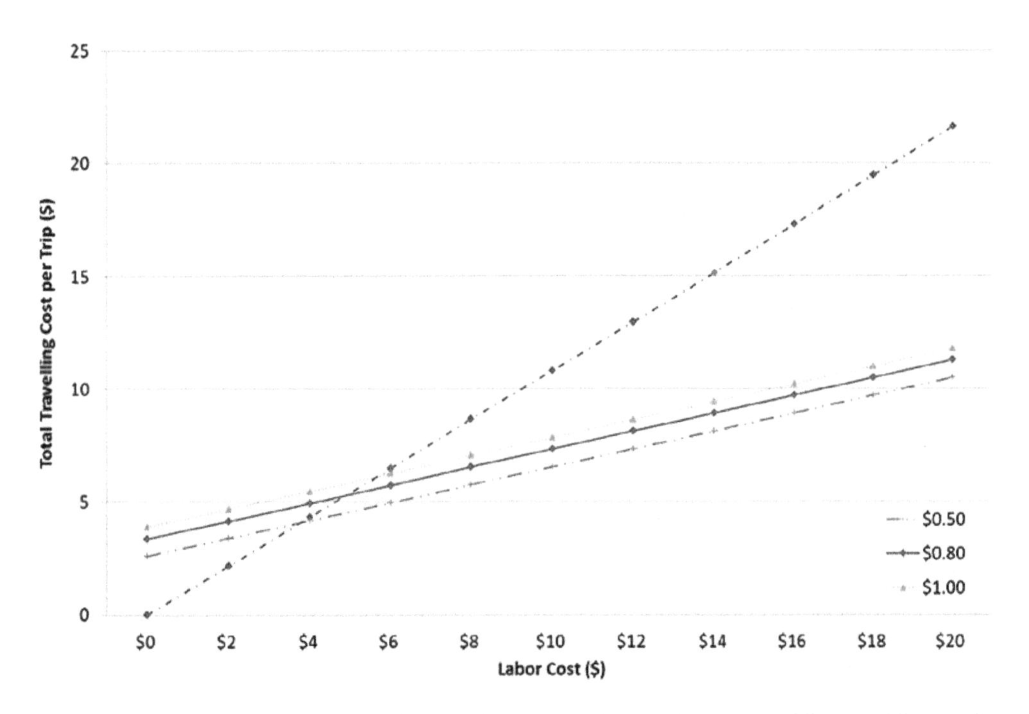

Figure 6. Sensitivity analyses for a trip to Wal-Mart 13th Ave. that compares riding a public transit system with driving a vehicle.

time will be 22 min at a bus speed of 15 miles/h (11 min for 30 miles/h) and it increases the mobility and convenience for the student by providing them opportunity to reach fresh foods from Wal-Mart In addition to the extension of the routes, Hornbacher's 13th Ave. can be added to the proposed transit service route with an additional bus stop.

6. Conclusion and future research

This study shows that MAT BUS in the city of Fargo has adequate routes in the areas where the average annual household income is lower. All grocery stores in the Fargo area are accessible by bus within 8–53 min of one-way travel time for North Dakota State University students.

According to this study, if the pedestrian is willing to travel for 10 min to reach the transit stop, a large area can be considered within the accessible distance of a transit route. This study further shows that Sun Mart North is the closest grocery store to NDSU considering travel time, whereas Hornbacher's North has the highest accessibility (Route 13A). This can be explained by the large residential population surrounding Hornbacher's North. Wal-Mart 55th Ave. is the farthest grocery store from NDSU, and it also is the least accessible since it is located in the newly developing zone.

Considering the cold weather in the city of Fargo, it is practical to use five minutes as the walking distance and the service areas for 1,200 feet (365 m) instead of using 2,500 feet. Therefore, it is important to study the shelter locations. Future research could be conducted to understand how many shelters are available and the accessibility of those shelters.

This research presents an important model and the analysis for the accessibility problem with the help of GIS. This study can be applied to public transit planning and the division of public affairs of the city and the metropolitan organizations. This work can guide the studies to calculate the accessibility of different zones in the new development. Also in urban planning this can help to provide insights into the transit routes and placement of grocery stores. This research can be replicated for other transit agencies which are serving lower income groups.

A survey is proposed for future research to gather inputs from the diverse population. This survey will collect inputs about the willingness of walking to the transit stop, the walking distance, and the time taken for walking. This survey will enhance data about inclination to waiting for a bus at a stop and the time for travel in a bus. The survey data could be used in the GIS study to get the actual distance traveled by individuals to reach the bus. It is important to compare and connect the technical measures and people's perspectives. This proposed survey will help future studies to fulfill this requirement.

Acknowledgement
We would like to thank Jeremy Mattson, Research Fellow at Upper Great Plains Transportation Institute for his intensive review and helpful comments to improve this paper.

Funding
The authors received no direct funding for this research.

Author details
Nimish Dharmadhikari[1]
E-mail: nimish.dharmadhikari@ndsu.edu
ORCID ID: http://orcid.org/0000-0001-8228-2512
EunSu Lee[2]
E-mail: eunsu.lee@ndsu.edu
ORCID ID: http://orcid.org/0000-0003-1840-9368
[1] Indian Nations Council of Governments, Tulsa, OK 74103 USA.
[2] Upper Great Plains Transportation Institute, North Dakota State University, Fargo, ND 58102, USA.

References

Biba, S., Curtin, K. M., & Manca, G. (2010). A new method for determining the population with walking access to transit. *International Journal of Geographical Information Science, 24*347–364. http://dx.doi.org/10.1080/13658810802646679

Burns, C. M., & Inglis, A. D. (2007). Measuring food access in Melbourne: Access to healthy and fast foods by car, bus and foot in an urban municipality in Melbourne. *Health & Place, 13*, 877–885.

Curl, A., Nelson, J. D., & Anable, J. (2011). Does accessibility planning address what matters? A review of current practice and practitioner perspectives. *Research in Transportation Business and Management, 2*, 3–11. http://dx.doi.org/10.1016/j.rtbm.2011.07.001

Currie, G. (2010). Quantifying spatial gaps in public transport supply based on social needs. *Journal of Transport Geography, 18*, 31–41. http://dx.doi.org/10.1016/j.jtrangeo.2008.12.002

Curtin, K. M., & Biba, S. (2011). The transit route arc-node service maximization problem. *European Journal of Operational Research, 208*, 46–56. http://dx.doi.org/10.1016/j.ejor.2010.07.026

Gates, T. J., Noyce, D. A., Bill, A. R., & Van Ee, N. (2006). *Recommended walking speeds for pedestrian clearance timing based on pedestrian characteristics* (pp. 1–21). Washington, DC: TRB 2006 Annual Meeting.

Geurs, K. T., & van Wee, B. (2004). Accessibility evaluation of land-use and transport strategies: Review and research directions. *Journal of Transport Geography, 12*, 127–140. http://dx.doi.org/10.1016/j.jtrangeo.2003.10.005

Jones, A.P., Haynes, R., Sauerzapf, V., Crawford, S. M., Zhao, H., & Forman, D. (2008). Travel times to health care and survival from cancers in Northern England. *European Journal of Cancer, 44*, 269–274. http://dx.doi.org/10.1016/j.ejca.2007.07.028

Kwan, M. P. (1998). Space-time and integral measures of individual accessibility: A comparative analysis using a point-based framework. *Geographical Analysis, 30*, 191–216.

Langford, M., Fry, R., & Higgs, G. (2011). Measuring transit system accessibility using a modified two-step floating catchment technique. *International Journal of Geographical Information Science, 26*, 1–22.

Litman, T. (2015). *Evaluating accessibility for transportation planning—Measuring people's ability to reach desired goods and activities*. Victoria: Victoria Transport Policy Institute.

Lovett, A., Haynes, R., Sunnenberg, G., & Gale, S. (2002). Car travel time and accessibility by bus to general practitioner services: A study using patient registers and GIS. *Social Science & Medicine, 55*, 97–111.

Mavoa, S., Witten, K., McCreanor, T., & O'Sullivan, D. (2012). GIS based destination accessibility via public transit and walking in Auckland, New Zealand. *Journal of Transport Geography, 20*, 15–22. http://dx.doi.org/10.1016/j.jtrangeo.2011.10.001

Miller, H. J. (1999). Measuring space-time accessibility benefits within transportation networks: Basic theory and computational procedures. *Geographical Analysis, 31*(1), 1–26.

Salon, D. (2009). Neighborhoods, cars, and commuting in New York City: A discrete choice approach. *Transportation Research Part A, 43*, 180–196.

Shannon, T., Giles-Corti, B., Pikora, T., Bulsara, M., Shilton, T., & Bull, F. (2006). Active commuting in a university setting: Assessing commuting habits and potential for modal change. *Transport Policy, 13*, 240–253. http://dx.doi.org/10.1016/j.tranpol.2005.11.002

Tomer, A. (2011). *Transit access and zero-vehicle households*. Washington, DC: Brookings.

Tomer, A., Kneebone, E., Puentes, R., & Berube, A. (2011). *Missed opportunity: Transit and jobs in metropolitan America*. Washington, DC: Brookings.

U. Oxford. (2012). *Oxford dictionaries*. Retrieved from http://oxforddictionaries.com/definition/accessible

USDA. (2009). *Food desert locator*. Retrieved May 21, 2012, from http://www.ers.usda.gov/data/fooddesert/about.html

Wachs, M., & Kumagai, T. G. (1973). Physical accessibility as social indicator. *Socio-Economic Planning Sciences, 7*, 437–456). http://dx.doi.org/10.1016/0038-0121(73)90041-4

Zhao, F., Chow, L. F., Li, M. T., Ubaka, I., & Gan, A. (2003). Forecasting transit walk accessibility: Regression model alternative to buffer method. In *Transportation Research Record: Journal of the Transportation Research Board, No. 1835* (pp. 34–41). Washington, DC: Transportation Research Board of the National Academies.

7

Scour due to turbulent wall jets downstream of low-/high-head hydraulic structures

Youssef I. Hafez[1]*

*Corresponding author: Youssef I. Hafez, Royal Commission Yanbu Colleges and Institutes, Yanbu University college, Yanbu, Saudi Arabia

E-mail: mohammedy@rcyci.edu.sa

Reviewing editor: Sanjay Kumar Shukla, Edith Cowan University, Australia

Abstract: To overcome over-prediction of the scour depths by existing methods, a mathematical model is developed based on a work transfer theory. This model predicts the equilibrium scour hole's depth and length due to two-dimensional turbulent wall jets downstream of low-/high-head hydraulic structures. The work transfer theory states that the work done by the attacking jet flow is transferred to work done to remove the volume of the scoured bed material out of the scour hole. This results in an analytical nonlinear equation for predicting the equilibrium scour depth. This unique feature of the nonlinearity of the developed equation shows the mutual dependence of the scour geometry and flow hydrodynamics on each other. Non-circulating wall jet flows and re-circulating jet flows within the scour hole are both considered. A separate equation is developed for predicting the length of the scour hole. Field data at the Nile Grand Barrages in Egypt and the Shimen Arch Dam in China are used to validate the developed model. The developed equation was checked against laboratory data for scour downstream of a spillway for which a complete data-set exists. For scour downstream of grade control structures, an equation was derived from the low-head hydraulic structure equation and tested against laboratory data.

Subjects: Hydraulic Engineering; Water Engineering; Water Science

Keywords: low-/high-head hydraulic structures; barrages; grade control structures; apron; sluice gates; spillways; turbulent wall jets; work transfer theory; scour hole; scour hole depth; scour hole length; Egypt's Grand Barrages; China's Shimen Arch Dam

ABOUT THE AUTHOR

Youssef I. Hafez graduated in 1984 from Cairo University, Egypt, Civil Engineering Department. Soon after, he joined the Nile Research Institute, Egypt. In 1990, he got his Master of Science at Colorado State University, USA, in Civil Engineering, Hydraulics. In 1995, he got his PhD form Colorado State University, USA, in turbulence modeling in closed and open channels using the Finite Element numerical method. In 2000, he became an associate professor at the Nile Research Institute. In 2004 and till now, Youssef has joined the Royal Commission Colleges and Institutes in Yanbu in Saudi Arabia. This paper presents a novel mathematical modeling approach for scour hole predictions downstream of barrages and low-head hydraulic structures. The mathematical model adopts work transfer theory that will open new doors to many analytical developments in the field of scour predictions at hydraulic structures (weirs, spillways, abutments, bridge piers, plunge pool scour, and grade control structures).

PUBLIC INTEREST STATEMENT

It is well known that downstream of hydraulic structures with apron such as low-dams, barrages, spillways, and weirs local scour occurs due to the nature of high velocity of the outflow water jet from the structure gates. These high velocities can destroy the bed's protective layer and erode the unprotected bed downstream of the hydraulic structure. The existence of significant scour holes downstream of a hydraulic structure causes undermining of the bed material below the foundation with a consequent collapse of the hydraulic structure which may be worth millions of US Dollars. The present study provides estimates to the expected maximum scour depth for the safe design and operating purposes.

1. Introduction

It is well known that downstream of low-/high-head hydraulic structures with apron such as barrages, low-head dams, spillways, grade control structures, and weirs local scour occurs due to the nature of high velocity of the outflow water jet from the structure gates (Figure 1). The water jet is controlled by the head difference on the hydraulic structure, gate opening dimensions (width and height), stilling basin characteristics, and the downstream water level. Typical schemes of economic design in which the total gates' width is considerably less than the river width and sometimes gate operations concentrate the whole river discharge to a few gates located close together and accordingly, the resulting narrow flow width induces high outlet jet velocities. These high velocities can destroy the bed's protective layer and erode the unprotected bed downstream of the hydraulic structure. The existence of significant scour holes downstream of hydraulic structures causes undermining of the bed material below the foundation with a consequent collapse of the hydraulic structure. The development of large scour holes downstream of three major barrage structures in Egypt (Isna, Naga Hamadi, and Assiut Barrages) necessitated their replacement with new barrages costing hundred million US dollars. Therefore, it is important to estimate the expected maximum scour depth downstream of low-head hydraulic structures for design purposes and for operating or emergency conditions that might endanger the structure.

Scour investigations can be classified mainly into two approaches, namely: experimental and mathematical approaches. In the experimental approach, flume experiments are used to measure the scour's most influencing factors (Chiew & Lim, 1996; Dietz, 1969; Eggenberger, 1943; Hamidifar, Omid, & Nasrabadi, 2011; Lim & Yu, 2002; Schoklitsch, 1932; Shalash, 1959; Shawky, 2008). Then, dimensional analysis is applied to the experimental data to deduce relations between the equilibrium scour hole's depth and length, on the one hand, and the flow hydrodynamic variables and sediment characteristics, on the other hand. It is noted that most flume experiments have been run under constant discharges and nearly uniform bed sediments. Breusers and Raudkivi (1991) indicated that several of the (scour) equations cannot readily be extended to prototype scale. They further believe that even those which are dimensionally consistent lack corroboration from full scale tests, and the variability of the predictions point at considerable inconsistency even at model scale.

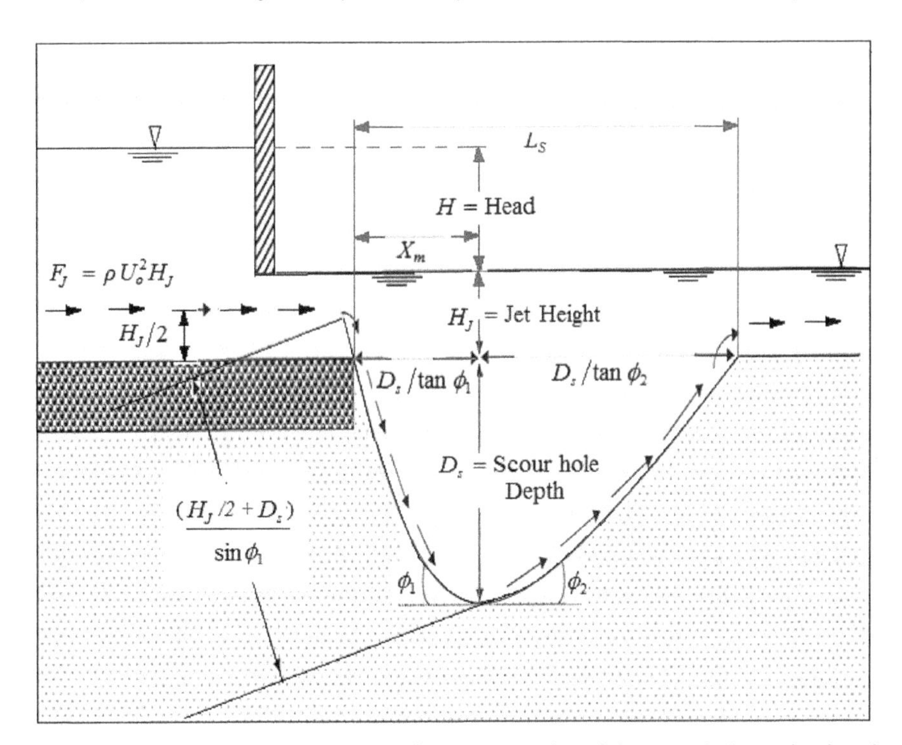

Figure 1. Definition sketch and schematic representation of the scour hole mechanism downstream of low-head hydraulic structures such as barrages.

In the second approach, mathematical models are developed which are solved analytically or numerically. The mathematical treatment to the subject was adopted using often the concept of incipient motion of bed sediment particle (Dey & Sarkar, 2006; Hogg, Huppert, & Dade, 1997; Hopfinger et al., 2004). In this approach, it is assumed that the longitudinal distribution of the excess bed shear stress (local bed shear stress—critical shear stress for sediment particles) is responsible for the motion of the sediment particles out of the scour hole. The shape of the scour hole follows the assumed model for the longitudinal distribution of this excess bed shear stress. Several constants appearing in the mathematical formulations are determined using laboratory data. In the experimental and mathematical approaches, the resulted prediction equations are tested against laboratory flume data often with uniform bed sediments.

1.1. Experimental methods

Schoklitsch (1932) proposed an equation based on model tests for the underflow case with short horizontal sill (which is relevant to the case of the Nile Barrages) as

$$D_s = 0.378\, H_1^{0.5}\, q^{0.35} + 2.15\, a \tag{1}$$

where D_s is the equilibrium or maximum scour depth below the original river bed (m), H_1 is the difference between upstream water level and sill level (m), q is the flow discharge per unit width (m^2/s), and a is the level of the downstream riverbed below the sill level (m). This equation is dimensionally inconsistent.

Eggenberger (1943) performed tests with combined flow over a weir and flow under the weir acting as a sluice gate. If the overflow is zero, the scour resulting from the submerged horizontal jet is

$$D_s + y_o = 7.255\, H^{0.5}\, q^{0.6}\, d_{90}^{-0.4} \tag{2}$$

where y_o is the downstream flow depth (m), H is the head on the structure (m), and D_{90} is sediment size for which 90% of the bed sediment is finer, expressed in mm throughout all the cited equations in this section.

Shalash (1959) included the effect of the apron length with a horizontal end sill as

$$D_s + y_o = 9.65\, H^{0.5}\, q^{0.6}\, d_{90}^{-0.4}\, (L_{min}/L)^{0.6} \tag{3}$$

where L is the apron length and $L_{min} = 1.5\, H$.

Among the few reported formulas for fine sediment scour formulas is that by Dietz (1969) as

$$\frac{D_s}{y_o} = \frac{U_{max} - U_c}{U_c} \tag{4}$$

where U_{max} is the maximum jet velocity and U_c is the critical mean velocity from Shields' curve. Breusers and Raudkivi (1991) reported that Equation (4) predicts very large equilibrium scour depths as, for example, when $U_{max} = 6\, U_c$, the equilibrium scour depth is $5y_o$. This value is used throughout this study for lack of the information needed in determining values of U_c.

Chiew and Lim (1996) developed empirical equations by considering the densimetric Froude number (F_o) as the main characteristic parameter to estimate scour dimensions caused by circular wall jet. The densimetric Froude number is given as

$$F_o = U_o / \sqrt{(\rho_s/\rho - 1)\, g\, d_{50}} \tag{5}$$

where U_o is the jet velocity, ρ_s is the sediment density, ρ is the fluid density, g is the gravitational acceleration, and D_{50} is the median sediment diameter. The scour equations are given as:

$$\frac{D_s}{b_o} = 0.21\,F_o \tag{6}$$

$$\frac{L_s}{b_o} = 4.41\,F_o^{0.75} \tag{7}$$

here b_o is the sluice gate opening and L_s is the scour hole's length.

Lim and Yu (2002) applied regression technique on the database of 161 flume experiments, out of which 61 data-sets were from Nanyang Technological University. Their equation for predicting the maximum scour depth in case of existing apron is

$$\frac{D_s}{b_o} = 1.04\,\sigma_g^{-0.69} F_o^{1.47} \left(\frac{b_o}{d_{50}}\right)^{-0.33} e^{-0.04\beta_1 [L/b_o]^{1.4}} \tag{8}$$

$$\text{and} \quad \beta_1 = \sigma_g^{-0.5}\,F_o^{-0.35} \left[\frac{b_o}{d_{50}}\right]^{0.5} \tag{9}$$

where σ_g is the sediment gradation. They reported that comparison between the calculated and measured scour depths indicates that 76% of the data were within ±20% error band and 91% were within ±30% error band. They indicated that the scour depth would decrease by about 50% if σ_g were increased from 1.2 to 3.13, consistent with the findings of Aderibigbe and Rajaratnam (1998). When there is no apron, Equation (8) becomes

$$\frac{D_s}{b_o} = 1.04\,\sigma_g^{-0.69}\,F_o^{1.47} \left(\frac{b_o}{d_{50}}\right)^{-0.33} \tag{10}$$

Shawky (2008), in his investigation of scour due to high floods, used a scale model with a length scale ratio of 1:32 to study scour at Rossetta Barrage on the Nile River, near Cairo, Egypt (Figure 2). The model was based on Froude number, similarity with the prototype, i.e. assuming only inertial and gravitational forces as the only forces determining the hydraulic scour phenomenon. Due to model space availability and economic factors, only 5 vents (gates) were modeled out of 46 vents which are the total vents of Rossetta Barrage. Shawky (2008) modeled 100 m of the river section upstream the Rossetta Barrage, the stilling basin, and 150 m of the river section downstream of the barrage with the final dimensions of the model as 15 m length and 1.75 m width. He plotted the data of scour areas and length vs. the scour levels and obtained equations with the best fit. These equations could predict the scour area and length downstream of Rossetta Barrage in all cases of discharge (but valid only for Rossetta Barrage). The scour hole's depth or length was not given as an explicit function of the scour influencing factors as in typical scour equations. This approach definitely lacks generality, though it confirmed some useful findings such as the increase of the scour depth due to increase of the jet unit discharge and that the upstream slope of the scour hole is more steeper than the downstream slope.

Hamidifar et al. (2011) examined experimentally the effects of bed roughness on the local scour downstream of an apron. Their results showed that the main characteristics of the scour holes, such as the maximum scour depth and the maximum extension of the scour hole, were much smaller for rough than smooth aprons. A minimum reduction of 60% was obtained for the maximum scour depth with respect to the smooth aprons. A total of 33 experiments were carried out with five different roughness conditions and combined plots of smooth and rough conditions were used to show roughness effects. No explicit scour equations were given and it is noted that the maximum densimetric Froude number reached 13 while Table 1 shows field values of the densimetric Froude number (18th row) higher than 40.

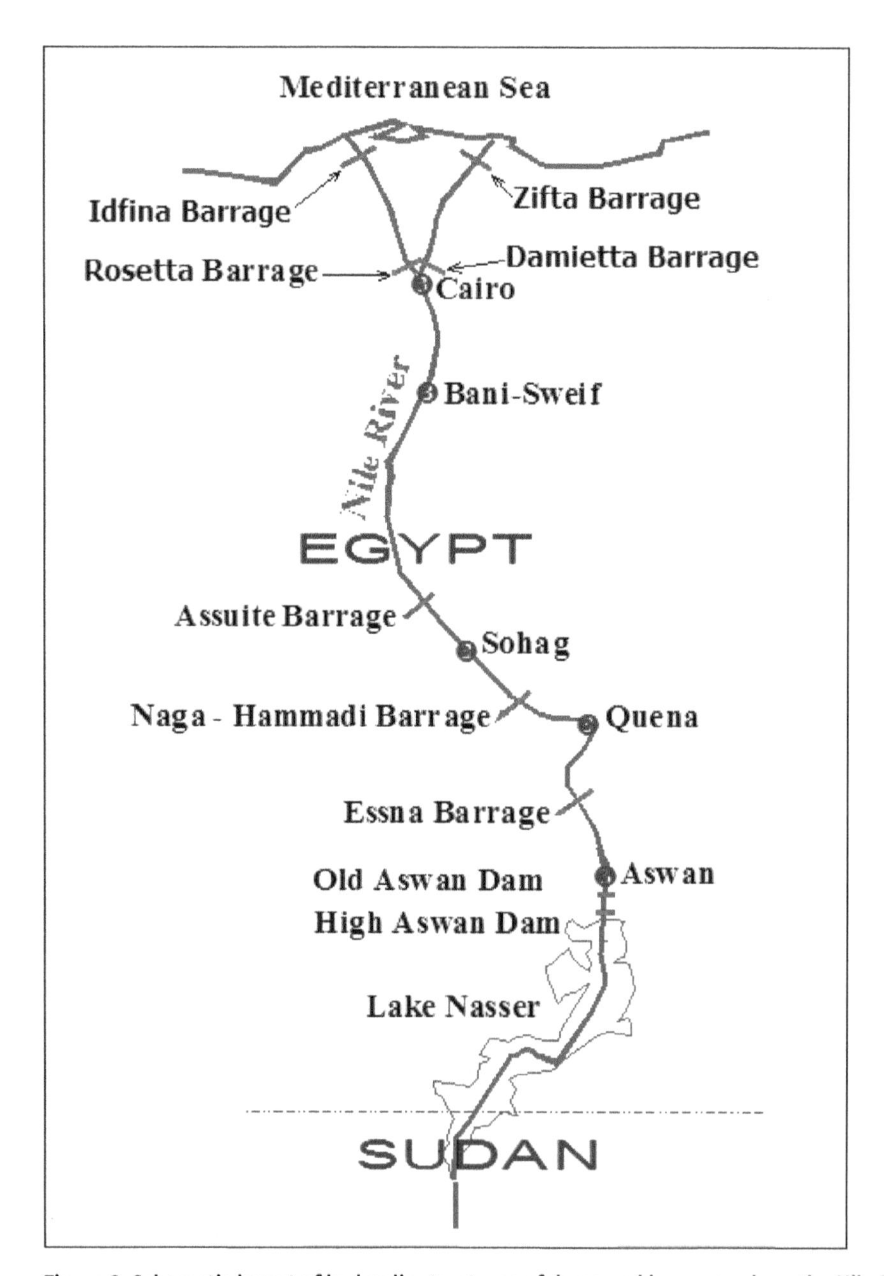

Figure 2. Schematic layout of hydraulic structures of dams and barrages along the Nile River, Egypt.

Bombardelli and Gioia (2005) investigated a closely related problem: the jet-driven turbulent cauldron that plunges into a pool of water and scours a pothole in a cohesionless granular bed. They assumed that the dynamic equilibrium is attained between a localized turbulent flow and a granular bed. They made the following observations on past empirical formulas for predicting the depth of pothole: (1) the formulas often lack dimensional homogeneity; (2) the formulas have often been the product of mangled attempts at dimensional analysis; (3) the formulas have often been predicated on limited experimental data; (4) the formulas have sometimes disregarded important parameters such as the diameter of the grains of the bed; and (5) the formulas do not provide mush insight into the interaction between the granular bed and the turbulent cauldron . The author agrees with these observations except of the importance of the diameter of grains as it influences the rate of scour not the magnitude of the equilibrium scour depth and scour geometry in general.

Table 1. Field data of scour holes at the Nile Grand Barrages, Egypt

Barrages							
Variable	**NEBTG**	**NEBSG**	**ONHB**	**ABV58**	**RBV23**	**DBV24**	**ZB**
H_1 (m)	13.0	13.0	4.9	5.25	5.5	4.5	5.7
a (m)	3.0	3.0	0.0	0.0	0.0	0.5	0.0
L (m)	50.0	50.0	72.0	50.0	65.0	60.0	45.0
H (m)	3.89	3.89	4.5	4.2	3.8	3.8	3.9
y_o (m)	15.11	15.11	5.8	2.75	3.2	4.09	1.8
D_{90} (mm)	1.0	1.0	1.162	0.822	0.92	0.772	1.0
θ	0.4	0.4	0.4	0.4	0.4	0.4	0.4
S_G	2.65	2.65	2.65	2.65	2.65	2.65	2.65
$\tan \varphi_1$	0.11	0.16	0.20	0.20	0.15	0.2	0.2
$\tan \varphi_2$	0.045	0.05	0.10	0.10	0.08	0.1	0.1
α_φ[a]	2.44	3.2	2.0	2.0	1.9	2.0	2.0
$\sin \varphi_1$	0.11	0.158	0.20	0.20	0.2	0.2	0.2
q (m^2/s)	27.77	20.50	36.15	27.24	10.00	10.00	10.64
$H_j = b_o$ (m)	3.0	3.0	3.0	3.0	1.92	1.92	2.0
U_o (m/s)	9.26	6.83	12.05	9.08	5.20	5.20	5.32
F_r[b]	1.71	1.26	2.22	1.67	1.20	1.20	1.20
F_o	72.76	53.71	137.46	78.72	42.68	46.52	41.80
D_s (m)	6.5–8.0	6.0–6.5	10.0	7.5–8.0	3.5	3.25	3.5
L_s (m)	205–213	98	150	100	63	–	–
Aspect ratio[c]	1/25	1/15	1/15	1/12.5	1/18	–	–

Notes: ABV58 = Assiut Barrage Vent No. 58; DBV24 = Damietta Barrage Vent No. 24; NEBSG = New Isna Barrage Sluice Gate; NEBTG = New Isna Barrage Turbine Gate; ONHB = Old Naga Hamadi Barrage; RBV23 = Rossetta Barrage Vent No. 23; and ZB = Zifta Barrage.

[a] $\alpha_\varphi = \frac{\tan \phi_1}{\tan \phi_2}$.

[b] $F_r = U_o / \sqrt{g H_j}$.

[c] Aspect ratio ≈ (D_s/L_s).

It is noted that most scour lab experiments used bed materials (usually sands) with grain sizes comparable to those in field conditions while the flow discharges and flow depths in these experiments are significantly less than their corresponding field values. For example, Schoklitsch (1932) used sediments sizes in the range from 1.5 to 12 mm, Eggenberger (1944) used D_{90} from 1.2 to 7.5 mm, Muller (1947) used D_{50} from 0.43 to 3.67 mm, and Shalash (1959) used D_{50} from 0.7 to 2.65 mm, which are typical sediment sizes found in natural rivers. In lab flumes, usually flow depths are predetermined by the construct of the flume which is in the order of few centimeters and flow discharges by the pump capacity which is in the order of few cubic meters per second. For example, Eggenberger (1944) used unit discharges (q) ranging from 0.006 to 0.024 m^2/s and Shalash (1959) had q values from 0.011 to 0.027 m^2/s. There is no grantee that labs' hydrodynamic and sediment variables are matching their corresponding field values. This, of course, will induce significant scale effects in lab experiments and cause lab-based prediction equations to deviate much when applied to field cases. In lab experiments, small relative flows cause significant scour depths and when scaling these conditions using lab-based equations to field conditions, it results in higher than actually predicted scour depths, thus creating the over-prediction of scour depth problem. Rajaratnam (1981) data indicate upstream slope of the scour holes as 0.5 while field data (Table 1) show values around 0.2. This confirms that scour holes in lab experiments are significantly much steeper than field scour holes. This induces more turbulence structure in lab experiments. Also that field densimetric Froude numbers are more than their lab values points to scale effects.

In agreement with the above statement, despite it is in the context of the closely related bridge scour subject, Jones and Sheppard (2000) stated that "Due to the complexity of the flow and sediment transport associated with local scour processes there are a number of dimensionless groups needed to fully characterize the scour. Many of these groups, such as the ratio of water depth to structure diameter, can be maintained constant between the laboratory model and the prototype structure. However, since there is a lower limit on the sediment particle size before cohesive forces become important, those groups involving sediment size cannot be maintained constant between the model and the prototype. In fact, most laboratory experiments are performed with near prototype scale sediment. If the sediment to structure length scales are not properly accounted for in the predictive equations then problems arise when the equations are applied to situations different from the laboratory conditions on which they are based."

1.2. Mathematical modeling methods

As laboratory experimental investigations are time-consuming and expensive, mathematical/numerical modeling has become very popular lately. Hogg et al. (1997) developed an analytical framework to model the progressive erosion of an initially flat bed of grains by a turbulent wall jet. In their model, the grains are eroded if the shear stress, exerted on the grains at the surface of the bed, exceeds a critical value which is a function of the physical characteristics of the grains (this is why D_{50} or some characteristic particle size appears in all the scour formulae adopting the incipient motion concept). They balanced the mobilizing and resisting moments on the particles at the surface of the sloping bed to obtain critical conditions for incipient particle motion. This enables the prediction of the characteristic dimensions of the steady-state scour pit. They indicated that after the wall jet has been flowing for a sufficiently long period, the boundary attains a steady state, in which the mobilizing forces associated with the jet are insufficient to further erode the boundary. They calculated the profile of the scour pit by considering the shear stress distribution along the surface of the bed. They assumed the following:

(a) The boundary shear stress for the flow of a two-dimensional jet over an erodible boundary is equivalent to the flow of a two-dimensional jet over fixed rough boundary. (b) The boundary profile has negligible influence on the jet flow; or equivalently, that the aspect ratio of the eroded profile is small. However, they admitted that this is a crude approximation because any bed topography will exert an additional drag on the flow, leading to a more rapid attenuation of the boundary shear stress. (c) The mean flow does not separate from the boundary at any downstream location to avoid the need to introduce models of regions in which the flow recirculates. (d) The shear stress along the scour pit boundary varies with Gaussian-like characteristics. They emphasized that this is a somewhat arbitrary model of the shear stress distribution and that any other shape function could be used. However, since the shape of the scour hole was assumed by them similar to the Gaussian-like curve, this in a sense is similar to assuming the bed profile right at the start. Using the aforementioned assumptions, the profile of the boundary is given according to them by integrating the following equation

$$\frac{d\eta}{d\zeta} \equiv \tan\beta = \frac{-\tan\alpha + \left[\tan^2\alpha - (f_b^2 - 1)(f_b^2 - \tan^2\alpha)\right]^{1/2}}{f_b^2 - 1} \tag{11}$$

where η is a non-dimensional vertical distance, ζ is a non-dimensional downstream distance, β is the angle of inclination of the bed, f_b is a non-dimensional bed shear stress, and α is the angle of repose of the granular material of the bed. Equation (11) is solved subjected to the following two boundary conditions: (1) that far from the jet source, erosion is absent and (2) that the volume of the particles is conserved.

They reported that the condition of a fixed volume of particles may not be entirely appropriate for comparison with experiments in which particles are swept downstream into a "sand trap;" however, it is appropriate for geophysical applications. They introduced a relaxation or adjustment process in which the gradient of the upward slopes never exceeds the tangent of the angle of repose. They showed that agreement between the predicted eroded profiles with the experimental data of Rajaratnam (1981) is fairly good except around the downstream locations where the profiles attain a maximum value (dune region). This difference in their opinion is because the model has neglected flow separation over the crest. They obtained the shape of the eroded boundary at intermediate times, before the steady state is attained, by the application of a sediment-volume conservation equation.

It can be seen that some uncertainties exist about Hogg et al.'s (1997) formulation for the boundary shear stress which rests on many empirical constants (up to seven constants were used) and the values of these constants were derived based on laboratory experiments which might differ under field conditions. In addition, scour in field conditions extends to depths of the order of several meters. The one grain size for the bed used in their equation cannot be assumed to represent the grains along the depth of the scour hole. In addition, they adopted a critical shear stress value of 0.05 according to Nielsen (1992). They found from the data of Rajaratnam (1981) that the ratio of the maximum eroded depth, ε_m, to the downstream distance to the position of maximum eroded depth, χ_m, to be constant, σ, where $\sigma = 0.5$. According to field values for tan φ_1 in Table 1 and keeping in mind that $\sigma = \varepsilon_m/\chi_m \approx \tan\varphi_1$, tan φ_1 has a maximum value of about 0.2. In addition, their assumption that the downstream slope of the scour hole is close to the angle of repose which might be true in flume experiments is far from field values for tan φ_2 as seen in Table 1. For example, if the angle of repose is 30° or 35°, then its tangent value is 0.58 or 0.7, respectively, while field values for tan φ_2 in Table 1 have a maximum value of 0.1.

Hopfinger et al. (2004) adopted the model developed by Hogg et al. (1997) but increased the effective shear stress by a constant factor to account for effects of Görtler vortices which were found from observing loose sediment streaks or longitudinal ridges on the upstream-facing sediment slope of the scour hole (that also supports that sediment particles are creeping along the slope of the scour hole). They explained that the contribution of Görtler vortices to bed shear stress is likely to be of the same form as normal turbulent shear stress and is therefore additive, i.e. $\tau_b = \tau_t + \tau_G$ where τ_b is the effective bed shear stress, τ_t is the turbulent shear stress, and τ_G is the shear stress due to Görtler vortices. They showed numerically that Görtler vortices can increase the effective shear stress by an order of magnitude. In addition, they stated that Görtler vortices cause strong up-slope sediment transport and, in turn, strong avalanching which intermittently destabilizes the sediment hill. They emphasized that at least two scouring regimes must be distinguished: a short time regime after which a quasi-steady state is reached, followed by a long time regime, leading to an asymptotic state of virtually no sediment. The quasi-steady-state scour hole's depth, h_s, is given as

$$\frac{h_s}{b_o} = \frac{D_s}{b_o} = B_1 \left(\frac{b_o}{d_{50}}\right)^{-0.11} F_o^{1.1} - B_2 \tag{12}$$

here b_o is sluice gate opening, D_{50} is the mean grain diameter, F_o is the densimetric particle Froude number, and the constants B_1 and B_2 are given as $B_1 = 0.43$ and $B_2 = 0.2$. They stated that because of the narrow range of the experimental conditions, the exponent 1.1 of the densimetric particle Froude number should be taken with some caution. There is also weak dependency of h_s/b_o on b_o/D_{50} but they argued that this is a must for two data points to collapse onto a straight line. They used only five data points to build Equation (12).

Dey and Sarkar (2006) computed the scour profiles downstream of a smooth apron due to submerged wall jets from the threshold condition of the sediments particles on the scour bed which is expressed by the following differential equation:

$$(\Omega^2 - 1)\frac{d\hat{y}}{d\hat{x}} = \mu \pm \left[\mu^2 - (\Omega^2 - 1)((\Omega^2 - \mu^2)\right]^{0.5} \tag{13}$$

where

$$\Omega = \frac{0.0081c^{4\hat{y}/\hat{\delta}} + 0.04(c\hat{y}/\hat{\delta})^2}{[\hat{\tau}_c]_{\beta=0}} \left[\left(2\hat{\delta}\hat{u}_o \frac{d\hat{u}_o}{d\hat{x}} + \hat{u}_o^2 \frac{d\hat{\delta}}{d\hat{x}} \right) \int_o^{\eta} (\psi^2 + \Phi_{11} - \Phi_{22}) d\eta \right]. \tag{14}$$

Here, $\hat{y} = y/b_o$, $\hat{x} = x/b_o$ are the non-dimensional vertical and streamwise distances, respectively, b_o is the sluice gate opening, μ is the Coulomb frictional coefficient of sediment, c is a parameter being a function of $\hat{x}$ and D_{50}, $\hat{u}_o = u_o / U_o$, u_o is local maximum velocity, U_o is the issuing jet velocity, $\hat{\delta}$ is the boundary layer thickness, and $[\hat{\tau}_c]_{\beta=0}$ is the critical bed shear stress on a horizontal bed ($\beta = 0$). In addition, Ψ, Φ_{11}, and Φ_{22} are functional relationships given by Dey and Sarkar (2006). They state that Equation (13) is a first-order differential equation which can be solved numerically by the fourth-order Runge–Kutta method to determine the variation of $\hat{y}$ with $\hat{x}$. It is clear from the complex nature of Equations (13) and (14) that this approach is not practical for the design engineer. In addition, some uncertainty appears about determining the constants needed in this formulation. Uncertainty also appears in defining incipient motion for sediment particles where uniform sediments are assumed with D_{50} representing the bed sediments.

Several numerical models dealing with scour prediction due to turbulent wall jets are summarized in Balachandar and Reddy (2013). A good example is the work of Abedelaziz, Bui, and Rutschmann (2010) who developed a bed load sediment transport module and integrated into FLOW-3D. This model was tested and validated by simulations for turbulent wall jet scour in an open channel flume. Effects of bed slope and material sliding were also taken into account. The hydrodynamic module was based on the solution of the three-dimensional Navier–Stokes equations, the continuity equation, and κ-ε turbulence model (which has several constants not accurately determined for re-circulating flows). The rough logarithmic law of the wall equation was iterated in order to compute the shear velocity and consequently the bed shear stress necessary for bed load computations. The predicted local scour profile fitted well with experimental data; however, the maximum scour depth was slightly underestimated and the slope downstream of the deposition dune was overestimated. In the author's view, such numerical models though highly structured and complex proved somehow successful in fitting the numerical model to certain observed scour data from laboratory flume experiments; however, they cannot be reliable in predicting scour under field conditions. Several constants needed for determining the bed shear stress distribution, as seen before, can have different values under field conditions. In addition, the reliability of these numerical models rests on the reliability of the selected sediment transport formula. How much confidence can be on any sediment transport formula? The need to use a sediment transport formula in any numerical model for scour puts some doubt on this approach due to the fact that there is no reliable sediment transport formula. The presented approach avoids use of sediment transport formulas.

Breusers and Raudkivi (1991) stated that existing equations for predicting the scour depths downstream of low-head structures are limited to laboratory studies using coarse sediments. They report that a lack of verification by field data limits the usefulness of the results. The same authors report that for fine sediments, no general expressions are available for predicting the equilibrium scour depth. Measured field values of scour depths deviate very much from values obtained from existing formulas (Hafez, 2004b) as will be seen in this study. This finding is behind the motivation for developing a novel analytical equation herein for predicting scour downstream of low-head structures. Oliveto and Comuniello (2010) state that "Despite several studies on local scour below low-head spillways have been made, the results appear still inconclusive." In this study, testing is made of the available scour prediction equations along with the developed one in the present study using valuable observed field data at the Grand Nile Barrages in Egypt and the Shimen Arch Dam in China in addition to lab scour data at grade control structures.

2. Mathematical model for scour downstream of low-head hydraulic structures based on work transfer theory

2.1. Case of no flow separation

Following the main lines of the energy balance theory by Hafez (2004a) that was used for modeling and predicting bridge pier scour, an analytical equation for predicting equilibrium scour depth downstream of low-head hydraulic structures is developed herein. However, the concept used here is termed "the work transfer theory" which is found to be more appropriate. Scope of the work here does not include the time development of the profile of scour holes and the three-dimensional aspects of the scour process. The principles of the work transfer theory state that the work done by the attacking fluid flow or jet flow is transferred to the work done in removing the volume of the scoured bed material out of the scour hole. In other words, as the work is equal to the potential energy, the potential energy contained in the attacking fluid flow is converted to a potential energy consumed in removing or transporting the sediment out of the scour hole. The mechanics of energy exchange between the jet flow and the sediment particles are complex and beyond the scope of this work. The basic assumptions or limitations are: (1) two-dimensional steady horizontal wall jets, i.e. the analysis is done in the longitudinal plane that bisects the scour hole as this plane contains the maximum scour depth (Figure 1), (2) granular non-cohesive sand bed porous material or rock bed non-porous type can both be considered, and (3) small aspect ratio of the scour hole dimensions which results in an attached jet to the river bed. The aspect ratio is defined here as the scour hole's depth over the scour hole's length. Small aspect ratios are defined to be less than 1/10. This indicates that no flow separation in the scour hole occurs (however, case of flow separation will be dealt with later), (4) the shape of the scour hole in its longitudinal bisecting plane can be assumed triangular in form with upstream and downstream inclination angels of φ_1 and φ_2, respectively, as seen in Figure 1, (5) the whole scour hole can be considered as a mega porous sediment particle, i.e. one unit or one porous big sediment particle, (6) assuming unlimited bed material along its depth and no armoring, and (7) there is sufficient time for scour formation to reach equilibrium. Kotoulas (1967) found that in case of coarse sand, about 64% of the final scour occurred in the first 20 s, and about 97% of scour depth was attained in 2 h. This leads to assuming in the hypothetical model herein that scour can be considered to occur instantaneously or in a very short time. The first three assumptions are also assumed in the work of Dey and Sarkar (2006), Hogg et al. (1997), and Hopfinger et al. (2004). The triangular shape of the scour hole is evident from many fully developed scour hole profiles as seen in Breusers and Raudkivi (1991) and Hafez (2004b).

The fluid flow force, F_j, or jet flow force per unit width (depth-averaged and assumed acting at the jet flow half-depth) is expressed according to fluid mechanics basics as $F_j = \rho U_o^2 H_j$ where ρ is the fluid (usually water) density, U_o is the jet flow velocity, and H_j is the jet depth. When the bed is initially flat, scour is initiated by the tangential boundary shear stress which results in formation of small scour hole. Once the scour hole is formed with its associated upstream and downstream slopes, the jet flow starts to flow along these slopes as an attached jet and quickly the scour hole grows to its equilibrium profile.

The jet flow force is assumed to flow as an underflow jet that is responsible for eroding the bed material and thus forming the equilibrium scour hole. The jet is assumed to creep along the upstream slope of the scour hole giving its energy to the bed sediment particles. The component of force that is creeping downward along the upstream slope is given as $F_j \cos \phi_1 = \rho U_o^2 H_j \cos \phi_1$. The distance that the jet force is assumed to move along the upstream slope can be assumed as $\frac{(H_j/2+D_s)}{\sin \phi_1}$ (Figure 1) where D_s is the maximum or equilibrium scour depth. Gravity exerts work on this down flow creeping jet and this work or potential energy is the cause of the equilibrium scour hole formation. Thus, the work done (work = force × distance) by gravity on the attacking jet flow force is:

$$\rho U_o^2 H_j \cos \phi_1 \frac{(H_j/2 + D_s)}{\sin \phi_1} = \frac{\rho U_o^2 H_j^2 \cos \phi_1}{\sin \phi_1} \left(\frac{1}{2} + \frac{D_s}{H_j} \right). \tag{15}$$

Now, this work done by the jet flow is converted to work done in raising or moving the weight of the volume of the scoured bed material out of the scour hole to the original bed level before occurrence of scour where it is transported by the flow further downstream. The weight of the volume of the material in the scour hole per unit width (for unit width analysis, the volume is equal to the area of the triangular shape of the scour hole that was filled with sediments) is

$$(\gamma_S - \gamma)(1-\theta)\frac{D_s}{2}\left(\frac{D_s}{\tan\phi_1} + \frac{D_s}{\tan\phi_2}\right). \quad (16)$$

where γ_S is the bed material unit weight, γ is the fluid (water) unit weight, and θ is the bed material or sediment porosity. The weight of the volume of the scoured bed material as given in Equation (16) can be assumed to be concentrated at the scour hole's center of gravity which is at $D_s/3$ for the triangular-shaped scour hole. The work done in removing the scoured material out of the scour hole to the original bed level can be assumed to be the weight force given by Equation (16), times the distance of its center of gravity from the original bed level ($D_s/3$). Thus, the work done in removing the bed material out of the scour hole is

$$(\gamma_S - \gamma)(1-\theta)\frac{D_s}{2}\left(\frac{D_s}{\tan\phi_1} + \frac{D_s}{\tan\phi_2}\right)\frac{D_s}{3} = \rho g(S_G - 1)(1-\theta)\frac{D_s^3}{6}\left(\frac{1}{\tan\phi_1} + \frac{1}{\tan\phi_2}\right). \quad (17)$$

where g is the gravitational acceleration and S_G is the sediment-specific gravity. According to the principles of the work transfer theory mentioned above, the two work done expressions in Equations (15) and (17) are assumed equal at equilibrium conditions. Appendix A explains that Equations (15) and (17) represent the work done during the entire scouring process.

Equality of the right sides of Equations (15) and (17), and after rearranging, yields the following analytically based equation for predicting the scour depth downstream of low-head structures for the case of no flow separation as:

$$\left(\frac{D_s}{H_j}\right)^3 = \frac{6\cos\phi_1}{(S_G - 1)(1-\theta)}\frac{1}{\left(\frac{1}{\tan\phi_1} + \frac{1}{\tan\phi_2}\right)\sin\phi_1}\frac{U_o^2}{gH_j}\left(\frac{1}{2} + \frac{D_s}{H_j}\right) \quad (18)$$

Equation (18), which is dimensionless in form, expresses the general mathematical model for modeling and predicting the maximum or equilibrium scour depth due to turbulent wall jets downstream of low-head hydraulic structures, especially at barrages under the assumptions made above. It is noted that the unknown scour depth exists in both sides of Equation (18) which indicates that the equation is nonlinear in form, thus reflecting the complexity of the phenomenon, the interaction between the variables, and the mutual dependence of the scour geometry and flow hydrodynamics on each other. Also, data are needed when applying Equation (18) about the upstream and downstream slopes of the scour hole which are not known before scour occurrence.

Equation (18) can be simplified and expressed in a more attractive form if the following additional assumptions are made: (1) $\tan\varphi_1 = \alpha_\varphi \tan\varphi_2$ where α_φ is the ratio of the upstream to downstream side slope of the scour hole and (2) φ_1 is a small angle for which it can be assumed that: $\tan\varphi_1 \approx \sin\varphi_1$ and $\cos\phi_1 \approx 1.0$ as seen in Table 1. With reference to point (2) above, it is noted from the data of scour holes at Egyptian Barrages in Table 1 that if $\tan\varphi_1 = 0.2$, then $\sin\varphi_1 = 0.196$, i.e. $\tan\varphi_1 \approx \sin\varphi_1$ and $\cos\phi_1 = 0.981 \approx 1.0$. In fact, the assumption $\tan\varphi_1 \approx \sin\varphi_1$ is valid for angles less than 15° where tan (15°) is ≈ 0.268. The simplifications made in (2) when inserted into Equation (18) yield:

$$\left(\frac{D_s}{H_j}\right)^3 = \frac{6}{(S_G - 1)(1-\theta)}\frac{1}{\left(\frac{\tan\phi_1}{\tan\phi_1} + \frac{\tan\phi_1}{\tan\phi_2}\right)}\frac{U_o^2}{gH_j}\left(\frac{1}{2} + \frac{D_s}{H_j}\right) \quad (19)$$

Using $\tan \varphi_1 = \alpha_{\varphi,} \tan \varphi_2$ in Equation (19) yields

$$\left(\frac{D_s}{H_j}\right)^3 = \frac{6}{(S_G - 1)(1-\theta)} \frac{1}{(1+\alpha_\phi)} \frac{U_o^2}{gH_j} \left(\frac{1}{2} + \frac{D_s}{H_j}\right) \quad (20)$$

Now assuming that $\tan \varphi_1 = 2 \tan \varphi_2$ or simply that $\alpha_\varphi = 2$ (which can be seen from the field data of the scour holes downstream of the Egyptian Barrages in Table 1) in Equation (20) and after simplification yields:

$$\left(\frac{D_s}{H_j}\right)^3 = \frac{2}{(S_G - 1)(1-\theta)} \frac{U_o^2}{gH_j} \left(\frac{1}{2} + \frac{D_s}{H_j}\right) \quad (21)$$

Equation (21) has the advantage over Equation (18) in that no information is needed about the shape of the scour hole through its slopes. The jet flow depth H_j is assumed to be usually as the sluice gate opening, b_o, i.e. $H_j = b_o$ if no data are available. Moreover, Equation (21) can be made more tractable by casting it in terms of the unit width discharge if expressing the unit width discharge of the jet flow as $q = U_o H_j$ and substituting $S_G = 2.65$, $\theta = 0.4$ and $g = 9.81$ m/s^2 to obtain the following simplified equation:

$$D_s^3 = 0.206\, q^2 \left(\frac{1}{2} + \frac{D_s}{b_o}\right) \quad (22)$$

Care must be taken when using Equation (22), which is dimensional, as D_s will be in m, q in m^2/s, and b_o in m. Either of Equations (21) or (22) can be used to obtain the maximum scour depth. Equations (21) or (22) can be solved easily by successive iterations or trials where in the first iteration, D_s is set to zero in the right-hand side of the equation to obtain the first value for D_s in the left-hand side. In the second iteration, the first calculated value of D_s is substituted back in the right-hand side of the equation and accordingly, a new value for D_s is obtained from the left-hand side. This process is continued until the difference between the two values for D_s from the left- and right-hand sides is nearly equal or within a specified tolerance value (assumed here 0.001 m). A simple FORTRAN code was developed to implement this iterative solution process. However, solving Equation (22) iteratively can be easily done with a pocket scientific calculator or any spreadsheet program as well.

In case of unavailability of data about q and U_o, the maximum jet velocity as an under flow could be calculated from the following formula (Hafez, 2004b; Hopfinger et al., 2004)

$$U_o = \sqrt{2gH} \quad (23)$$

where H is the head on the hydraulic structure, i.e. difference between the upstream and downstream water levels. With the jet velocity calculated from Equation (23) and with the jet depth H_j or gate opening b_o known, the maximum unit depth discharge can be calculated and used in the calculations. This jet could be an underflow jet that in the absence of a weir, downstream of the barrage keeps its momentum till it meets the unprotected bed.

The length of the scour hole along its longitudinal axis can be derived once the scour depth was obtained as follows. From Figure 1, the length of the scour hole, L_s, can be stated as

$$L_s = \left(\frac{D_s}{\tan \varphi_1} + \frac{D_s}{\tan \varphi_2}\right) = \frac{(1+\alpha_\phi)}{\tan \varphi_1} D_s \quad (24)$$

Using the assumption stated earlier that $\alpha_\varphi = 2$ or $\tan \varphi_1 = 2 \tan \varphi_2$ in Equation (24) and after simplification yields

$$L_s = \frac{3\,D_s}{\tan\,\varphi_1} \tag{25}$$

Equation (25) predicts the length of the scour hole in the main flow direction in terms of the (predicted) depth of scour and the slope of the upstream face of the scour hole, tan φ_1. In case of unavailability of data about tan φ_1, it can be assumed from 0.10 to 0.2 (Table 1) with a conservative value of 0.1 or assuming ϕ_2 equal to the angle of repose of bed sediments and calculate ϕ_1 from tan φ_1 = 2 tan φ_2.

2.2. Case of flow separation

Scour holes with large aspect ratios (greater than 1/10) are often associated with flow separation. In this case, the jet flow hits the downstream slope of the scour hole, bends down along the downstream slope of the hole, and circulates inside the hole. It transfers its work or potential energy to the sediment particles on the downstream face. The sediment particles fall down in the hole along with the creeping jet and circulate inside the whole till they move up leaving the hole. Owing to the work or energy gained from the down jet flow, the sediment particles are able to move upward out of the scour hole. It is assumed that a portion of the work or energy in the jet flow is transferred to jet flow circulating inside the scour hole through a coefficient η^2 (where $\eta < 1$). Since the jet flow force is proportional to the square of the velocity, η^2 is therefore used. The jet flow force per unit width is thus assumed equal to $\rho\,\eta^2\,U_o^2\,H_j$ and the distance of motion along the downstream slope is $\frac{(H_j/2+D_s)}{\sin\,\phi_2}$. Now the work done by gravity on the attacking jet flow in case of flow separation and circulation is given as

$$\frac{\rho\,\eta^2\,U_o^2\,H_j^2}{\sin\,\phi_2}\left(\frac{1}{2}+\frac{D_s}{H_j}\right) \tag{26}$$

The work done in removing the bed material out of the scour hole is still given by Equation (17). Equality of Equations (26) and (17) gives (after rearranging and assuming ϕ_2 is small, i.e. sin $\phi_2 \approx$ tan ϕ_2):

$$\left(\frac{D_s}{H_j}\right)^3 = \frac{6}{(S_G-1)(1-\theta)}\;\frac{1}{(1+\frac{1}{\alpha_\phi})}\;\frac{\eta^2\,U_o^2}{g\,H_j}\left(\frac{1}{2}+\frac{D_s}{H_j}\right) \tag{27}$$

Examining Equation (27) for the case of flow separation and Equation (20) for the case of no flow separation reveals that the difference lies in the denominator term that has α_φ and η. Both equations can be given in one model equation as

$$\left(\frac{D_s}{H_j}\right)^3 = \frac{6\,C_\phi}{(S_G-1)(1-\theta)}\;\frac{U_o^2}{g\,H_j}\left(\frac{1}{2}+\frac{D_s}{H_j}\right) \tag{28}$$

where for the case of no flow separation:

$$C_\varphi = \frac{1}{(1+\alpha_\varphi)}; \tag{29}$$

and for the case of flow separation,

$$C_\varphi = \frac{\eta^2}{(1+\frac{1}{\alpha_\varphi})}. \tag{30}$$

In addition, $\eta \approx 1$ in case of non-circulating flows while $\eta < 1$ in case of circulating flows. Equation (28) is the general mathematical model for prediction of the scour hole's equilibrium depth whether there is flow separation or not. When $\alpha_\varphi = 2$, the ratio of the scour depths of flow separation to that of no flow separation becomes 1.26. This means that when assuming all variables are the same and $\alpha_\varphi = 2$, the scour depth in case of flow separation increases by 26% due to increase in turbulence activities. For $\alpha_\varphi = 3$ (steeper upstream slope), the percentage increase is 44%. The steeper the upstream slope of the scour hole, the higher is the flow separation. Flow separation is more likely to occur in flume experiments. This might explain why flume-based scour equations tend to over predict equilibrium scour depths when applied to field cases as will be shown later.

3. Applications of the scour formulas for predictions of scour hole's depth

The Grand Nile Barrages in Egypt are among the most important hydraulic structures in the Egyptian water resources system as they are spreading across the 1,000 km stretch of the Nile River in Egypt (Figure 2). The Nile Barrages serve in regulating the Nile flow discharges and water levels in addition to some barrages producing hydropower. Upstream these barrages, off-take canals withdraw water to irrigate most of the agricultural lands in Egypt. The Egyptian Dams and Barrages system starts at the southern border with the High Aswan Dam (HAD) which was built in 1964. Six kilometers downstream from HAD, the Old Aswan Dam (AD) is located which was built between 1899 and 1902. Distances along the Nile River in Egypt are measured starting from Old AD, i.e. it is considered km zero. The Nile Barrages include: Old Isna Barrage at km 166.65, the New Isna Barrage at km 167.85, Naga Hamadi Barrage at km 359.45, Assiut Barrage at km 544.75, Damietta and Rosetta Barrages at km 952.92, Zifta Barrage on Damietta branch at km 1046.7, and Idfina Barrage on Rossetta branch at km 1159.0. Construction has recently been completed for new barrage at Naga Hamadi and has started for a new Assiut low dam. These barrages are very precious hydraulic structures worth millions of US dollars and are therefore very vital to the country's national economy.

Before the construction of HAD in 1964, just upstream Old Aswan dam, all barrages' gates during the flood season were completely opened with the barrage acting as a bridge section with constriction scour only expected. In the rising period of the flood, scour occurred while in the falling period, filling of the scoured holes occurred by the flood sediment-laden water. Therefore, it is thought that the current observed scour holes are due to conditions after HAD, i.e. after 1964, because the water is almost clear of sediments. These conditions are resulting from high-velocity jets either from high-head difference on the barrage or high-unit width discharge due to gate operation schemes. Indeed, there might have been scour developed before the events cited herein, but filling of these scour holes by dumping stones and sand sacks had been a common practice to minimize scour effects. Therefore, it could have been assumed that the river bed downstream of the barrages was restored almost to its nearly pre-scour levels.

In the following sections, the existing and newly developed equations are applied to the Grand Nile Barrages in Egypt. Details of the hydraulic data for this work are found in Hafez (2004b) and summarized in Table 1. For lack of reported data, it is assumed that $D_{50} = 0.7D_{90}$. In Hafez (2004b), scour predictions were made using a similar equation to Equation (18) while substituting the measured upstream and downstream slopes from the field data. In Hafez (2004b), the constant 1/2 added to D_s/H_j in the right-hand side bracket in Equation (18) was assumed erroneously equal to 1.0. It should be noted that field scour data downstream of low-head hydraulic structures are rare. The case of the Shimen Arch Dam, China, is one of these rare cases where a complete data-set exists.

3.1. Local scour prediction downstream of the New Isna Barrage

To replace the historical Old Isna Barrage, the New Isna Barrage in Egypt was built in 1994 at a distance of 168.2 km from Old Aswan Dam. This barrage consists of a powerhouse, flood sluiceway, and navigational lock. The powerhouse has 6 gates that are each 12-m wide and the flood sluiceway has 11 gates that are also each 12-m wide. The floor level elevation is at 66.0 m above the mean sea level (all elevations herein are above the mean sea level).

Scour holes have been formed downstream of the turbine gates and of the flood sluiceway where some measured profiles are shown in Figures 3 and 4. For example, field surveys revealed that the scour depths measured between January 2002 and January 2003 are about 6.5, 8.0, 6.5, 5.0, 4.5, and 3.0 m downstream of the six turbine gates, respectively (Hafez, 2004b). It is noted that at the first three turbine gates, scour is at maximum. This indicates that the first gates were opened more often than the rest. Therefore, conditions of maximum unit discharge prevail at these first three gates with

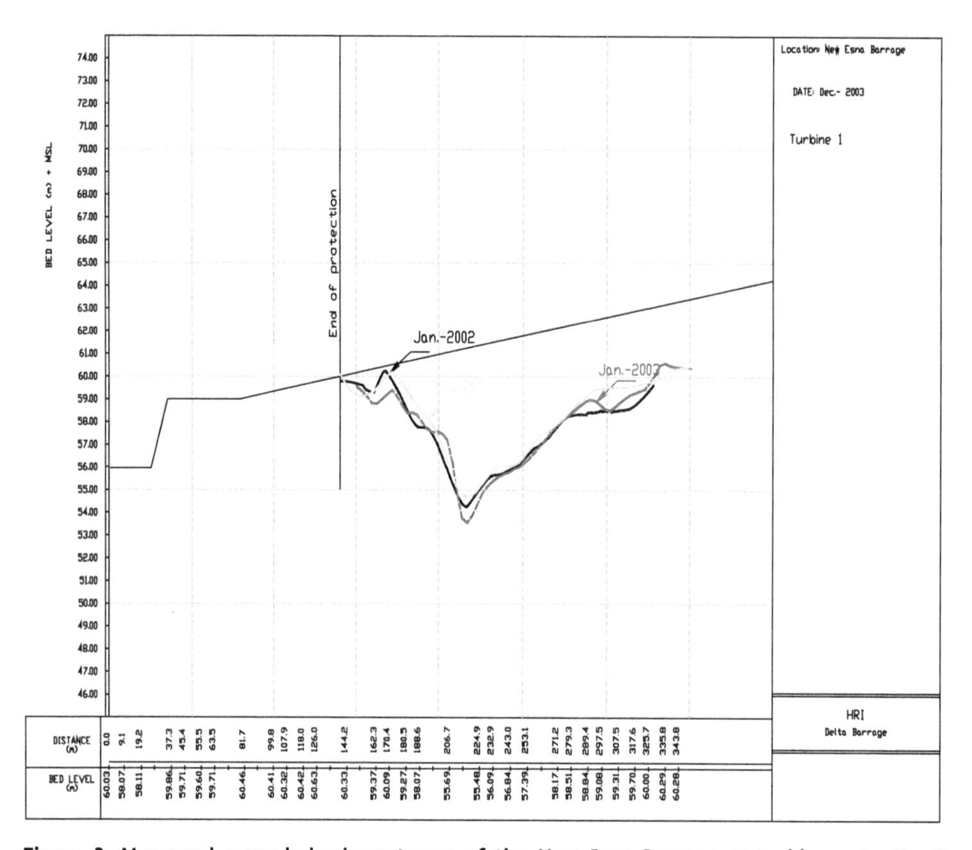

Figure 3. Measured scour hole downstream of the New Isna Barrage at turbine gate No. 1, (Hafez, 2004b).

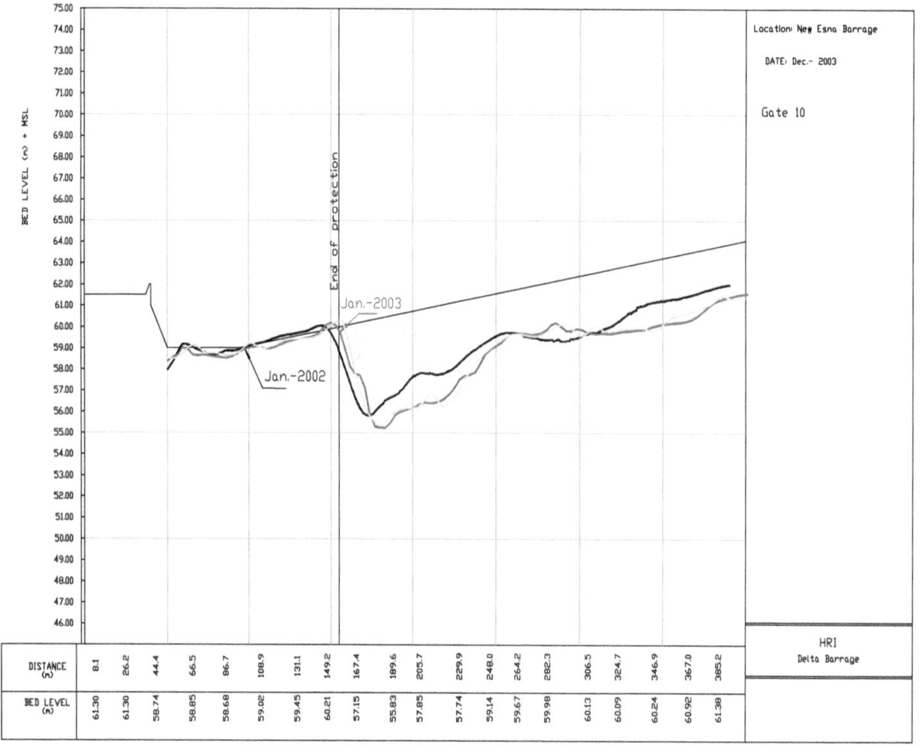

Figure 4. Measured scour hole downstream of the New Isna Barrage at sluiceway No. 10 (Hafez, 2004b).

scour depths of 6.5, 8.0, and 6.5 m. The scour hole with depth 8.0 m at the second gate was symmetrical around the gate axis. Scour is less at the other gates maybe due to partial opening of gates there. Therefore, scour depths ranging from 6.5 to 8.0 m are considered representative of conditions of maximum unit discharge downstream of these turbine gates.

Just right downstream of the 11 flood sluiceway gates, the scour depths are about 3.5, 4.0, 5.0, 6.0, 6.5, 6.0, 5.0, 5.0, 5.5, 4.8, and 3.0 m, respectively (Hafez, 2004b). Again, it is noted that scour is at its maximum at sluiceway gate numbers 4, 5, and 6 with scour depths of 6.0, 6.5, and 6.0 m, respectively. The same pattern of similarity noticed at the turbine gates appears at the sluiceway gates with maximum value of 6.5 m in the middle, indicating opening of the middle gates more often than the rest. Therefore, scour depths ranging from 6.0 to 6.5 m are considered to be produced by conditions of maximum unit discharge downstream of the sluiceway gates. It was observed that these scour holes are located behind the bed protective cover of the downstream part of the barrage.

The maximum flow allowed to the turbines is 2,000 m^3/s which for a total accumulated width of 72 m yields unit width discharge of 27.77 m^2/s. This jet unit discharge is assumed to be responsible for the maximum scour depths ranging from 6.5 to 8.0 m observed downstream of turbine gates No. 1, 2, and 3. The maximum observed flow passing the New Isna Barrage is assumed to be as 234 million m^3/d = 2,708 m^3/s (Hafez, 2004b). With this flow passing only in the flood sluiceway gates along its total width of 132 m (i.e. assuming shutting down the powerhouse), the unit width discharge is 20.5 m^2/s. This jet unit discharge is assumed to be responsible for the maximum scour depths ranging from 6.0 to 6.5 m which are observed downstream of flood sluiceway gates No. 4, 5, and 6. The jet depth is assumed to be as 3.0 m for both cases. The ratio of observed upstream and downstream slopes is seen in Table 1 as 2.44 and 3.2 which differs from the assumed value of 2. Dumping of stones which has been a common practice to stop scour progression may be the cause of higher slope ratios. However, for other barrages, the assumed slope ratio of 2 seems to hold.

The predicted scour depths along New Isna Barrage Turbine gates (NEBTG) are shown in Table 2. Equation (8) is not used because of lack of information in the current field data about the sediment gradation σ_g and according to Equation (8), the influence of σ_g cannot be neglected. It is clear from the table that Equation (22) developed in the present study gives computed scour depth of 7.94 m compared to the measured scour depths between 6.5 and 8.0 m and that the rest of the equations are highly overestimating the scour depth. Equation (1) provides close prediction but not as accurate as Equation (22).

The predicted scour depths along New Isna Barrage sluiceway gates (NEBSG) are also shown in Table 2. Again, Equation (22) yields the best prediction of 6.01 m compared to observed scour depths ranging between 6.0 and 6.5 m followed by Equation (1) while the rest of the equations still yield unrealistic scour depths. It is noted that Equation (1) is not sensitive to the varying flow conditions in the two cases discussed herein, while Equation (22) shows sensitivity to the varying flow conditions.

3.2. Local scour prediction downstream of the Old Naga Hamadi Barrage

The Old Naga Hamadi Barrage was built in 1932 with100 gates, each having a width of 6.0 m. The maximum design head is 4.5 m while the downstream floor level elevation is 58.0 m. Scour holes are severe to the point of initiating the construction of a new barrage. The scour holes (see Figure 5) reached up to 10.0-m depth with upstream slope of 0.20 and downstream slope of 0.10 confirming the assumption previously made that $\alpha_\varphi = 2$. The case that expected to produce this sort of scour is the one for which the outlet jet velocity from the barrage is at its maximum. For a maximum upstream water level of 65.4 m (Hafez, 2004b) and floor level elevation of 58.0 m, the head on the jet is 7.4 m. Note that the head on the jet is not necessarily equal to the head on the barrage. Using Equation (23), the maximum jet velocity becomes 12.05 m/s and for a jet depth of 3.0 m, the resulting maximum unit width discharge becomes 36.15 m^2/s. The measured downstream water level was

Table 2. Predicted scour depths downstream of the Nile Barrages, Egypt

Case	Equation (1) (m)	Equation (2) (m)	Equation (3) (m)	Equation (4) (m)	Equation (6) (m)	Equation (12) (m)	Present study, Equation (22)	Measured scour
NEBTG	10.81	90.03	23.43	75.55	45.84	56.83	7.94	6.50-8.00
NEBSG	10.37	72.52	17.01	75.55	33.84	40.53	6.01	6.00-6.50
ONHB	2.94	118.95	34.30	29.00	55.35	71.25	10.15	10.00
ABV58	2.75	114.05	42.08	13.75	49.59	60.69	7.80	7.50
RBV23	1.99	55.01	14.78	16.00	17.21	20.89	3.68	3.50
DBV24	2.87	60.05	16.69	20.45	19.43	23.28	3.68	3.25
ZB	2.07	57.40	21.35	9.00	17.56	21.37	3.83	3.50

60.72 m in 4/7/1997 and with floor level of 58.0 m; the downstream water depth was 2.8 m, which is close to the assumed jet depth of 3.0 m.

Table 2 reports the scour depth predictions at the Old Naga Hamadi Barrage (ONHB). It is clear that Equation (22) yields the most realistic value. Equation (22) predicts scour depth of 10.15 m compared to a measured scour depth of 10.0 m. Equation (1) highly under predicts the scour depth as 2.94 m while the rest of the equations highly over predict the scour depth (29.0–118.95 m).

3.3. Local scour prediction downstream of Assiut Barrage

Assiut Barrage was built in 1938 at km 544.75 from AD with 111 gates, each gate having a width of 5.0 m. The barrage maximum head difference is 4.2 m, floor level elevation is 43.75 m, and sill level elevation is 43.25 m. Large scour holes (Hafez, 2004b) are formed downstream of the barrage, especially in front of vent No. 58 as seen in Figures 6–8.

Figure 8 reveals two stages of local scour in front of Vent No. 58. The first occurred between January 2002 and January 2003 with a scour depth of about 4.0 m (from level 42.5 to 38.5 m) and the second occurred between January 2003 and December 2003 with a scour depth of about 3.5 m (from level 38.5 to level 35.0 m). The total scour depth between January 2002 and December 2003 is therefore 7.5 m. This large scour hole whose centerline lies downstream of gate No. 58 has dimensions of: length of 110 m, width of 65 m, and depth of 7.5 m, as seen from Figures 7 and 8. The analysis herein is implemented only on the scour hole at gate No. 58 axis with the idea that this scour hole represents a typical fully developed scour hole.

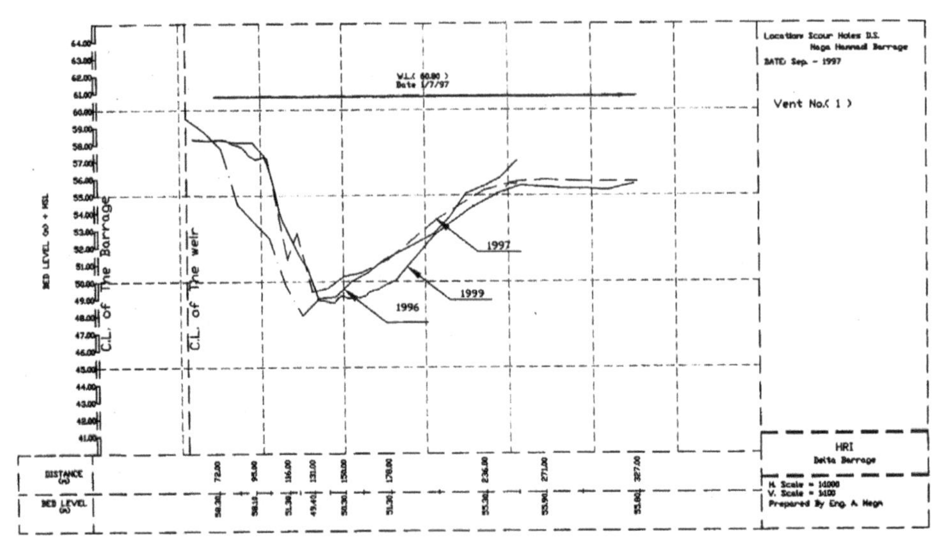

Figure 5. Measured scour hole downstream of the Old Naga Hamadi Barrage at gate No. 1 (Hafez, 2004b).

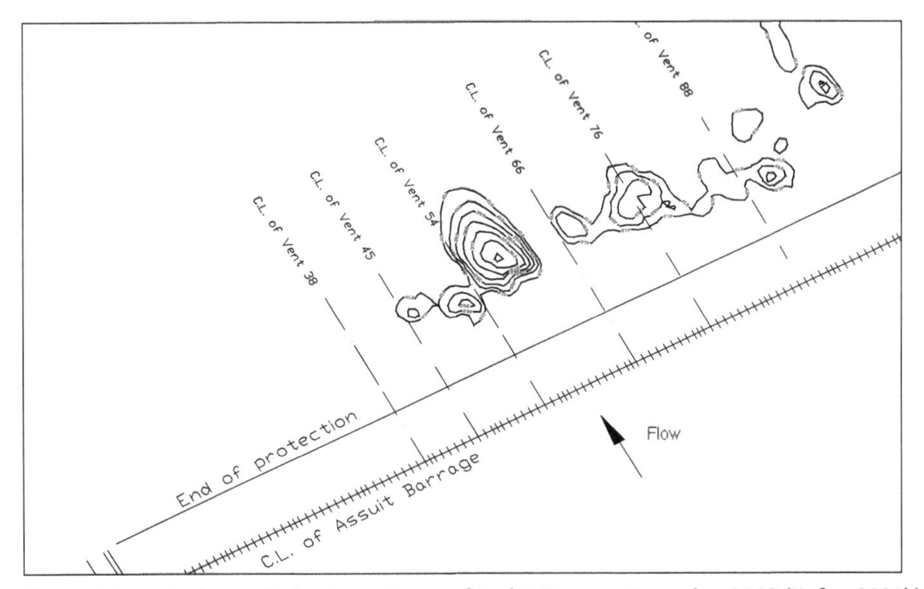

Figure 6. Layout of scour holes downstream of Assiut Barrage, December 2003 (Hafez, 2004b).

The water depth of the discharge jet is assumed to be 3.0 m (measured downstream water level of 45.0 m—floor elevation level of 42.0 m) while a head difference of 4.2 m was measured in December 2003. The upstream water level was 49.0 m while the downstream water level was 44.8 m, giving also a head of 4.2 m. For an upstream water level of 49.0 m and floor level of 43.75 m, the upstream depth (H_1) in Equation (1) becomes 5.25 m. The bed sediment D_{90} downstream of Assiut Barrage was estimated as 0.822 mm. The apron length needed when applying Equation (3) is assumed to be 50.0 m while it is assumed that the initial riverbed level is equal to the apron level (a = 0.0 in Equation (1)). From Figure 8, tan φ_1 and tan φ_2 can be calculated as 0.2 and 0.1, respectively, confirming the assumption previously made that $\alpha_\varphi = 2$.

It is required to predict the maximum equilibrium scour depth of Assiut Barrage downstream of Vent No. 58. It is expected that conditions which produce this maximum scour are a combination of

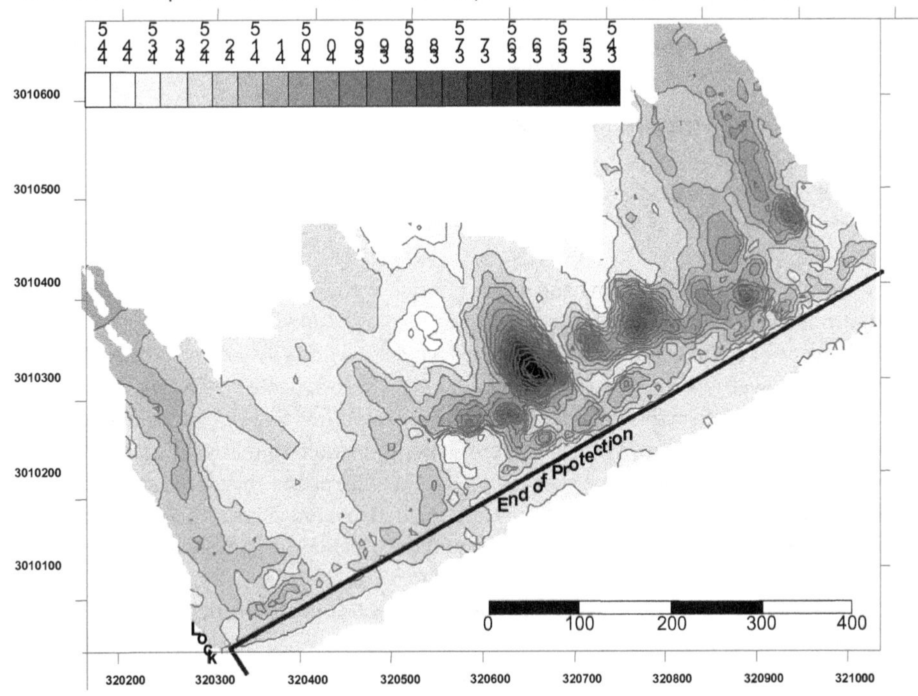

Figure 7. Bed topography of scour holes downstream of Assiut Barrage, December 2003 (Hafez, 2004b).

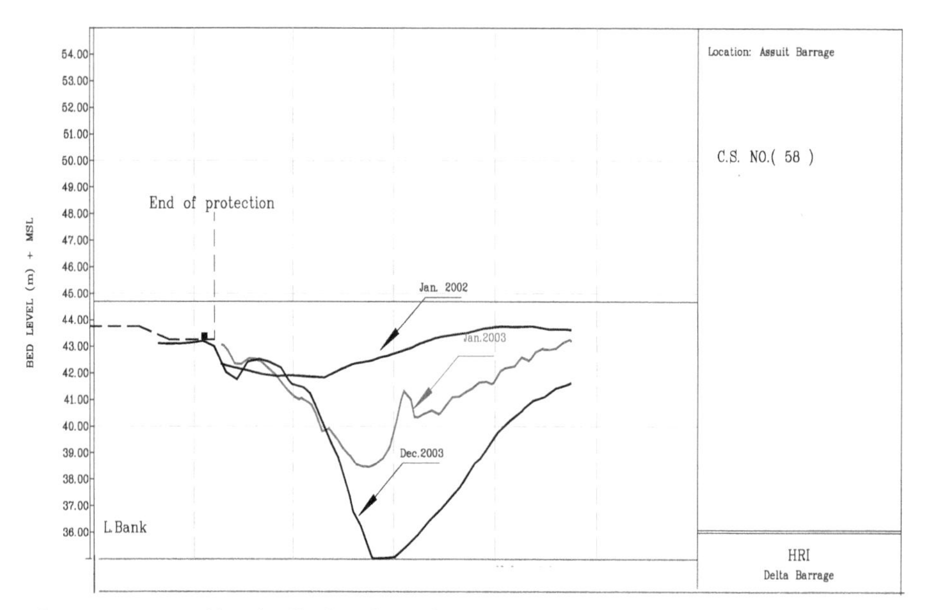

Figure 8. Measured longitudinal profiles of the scour hole downstream of Assiut Barrage at Vent No. 58 (Hafez, 2004b).

maximum head and maximum unit width discharge which together produce a jet with very high velocity enough to erode the bed and cause significant scour in a relatively short time of almost 2 years.

With head (*H*) equal to 4.2 m, Equation (12) yields a velocity of 9.08 m/s and with depth of 3.0 m, the maximum expected unit width discharge is therefore 27.24 m²/s. Table 2 shows the scour prediction at Assiut Barrage Vent NO. 58 (ABV58) resulting from this unit width discharge of 27.24 m²/s. It is clear that the scour depth prediction equation by Equation (22) gives the nearest value of 7.80 m compared to measured scour depth of 7.5 m. Again, Equation (1) highly under predicts the scour depth as 2.75 m while the rest of the equations highly over predict the scour depth (13.75–114.05 m).

3.4. Local scour prediction downstream of Delta Barrage on Rossetta Branch

Delta Barrage on Rossetta Branch was built in 1939 with 46 gates, each having 8.0-m width. The maximum head on the barrage is 3.8 m and the floor level elevation under the barrage is 11.0 m which is sloping down to a floor level of 9.5 m. A weir exists downstream of the barrage to reduce the head and jet velocity. Shawky (2008) reports field measurement of the scour hole at vent number 23 which looks like a well-defined scour hole with an upstream slope of about 0.15 while the downstream slope is about 0.08, yielding $\alpha_{\varphi} \approx 1.9$. The measured field scour depth reported by Shawky (2008) was 3.5 m and the length of the scour hole is 63 m.

It is expected that either the maximum head on the barrage or the maximum unit width discharge through the barrage will produce maximum local scour immediately downstream of the barrage. For a head on the barrage of 3.8 m, Equation (23) yields velocity of 8.63 m/s. Assuming (El Kateb, 1982) that the downstream weir will reduce this velocity by 40%, the expected jet velocity becomes 5.18 m/s. Also, the water depth in this vena contracta region is assumed to be 0.6 of the downstream water depth, i.e. $0.6 \times 3.2 = 1.92$ m. With this information, the resulting unit width discharge becomes 10.0 m²/s. Table 2 proves again the success of Equation (22) as the calculated scour depth at Rossetta Barrage Vent 23 (RBV23) was 3.68 m compared to a measured value of 3.5 m. Equation (1) underestimated the scour depth and the rest of the equations yield unrealistically high values.

3.5. Local scour prediction downstream of Delta Barrage on Damietta Branch

Delta Barrage on Damietta Branch was built in 1939 with 34 gates, each having 8.0-m width. The maximum head on the barrage is 3.8 m and the floor level elevation under the barrage is 12.0 m which is sloping down to a floor level of 10.0 m. A weir is constructed downstream of the barrage to reduce the head and jet velocity. A scour hole (Hafez, 2004b) with depth of 2.0 m exits behind the floor apron downstream of gate No. 18 and another scour hole of depth 3.25 m downstream of gate No. 24.

Following the same lines as in the last section, the resulting jet velocity is 5.18 m/s and with 1.92 m jet depth, the unit depth discharge is 10.0 m²/s. The upstream and downstream slopes of the scour holes are observed as 0.2 and 0.1, respectively, confirming the assumption that $\alpha_{\varphi} = 2$. Table 2 shows the success of Equation (22) at Damietta Barrage (DBV24) compared to the rest of the equations. Equation (22) yields a scour depth of 3.68 m compared to a measured scour depth of 3.25 m. It is noted that Equation (1) closely predicted the scour depth while the rest of the equations over predicted the scour depth by order of magnitudes (16.69–60.05 m).

3.6. Local scour prediction downstream of Zifta Barrage on Damietta Branch

Zifta Barrage (km 1046.7 from AD) on Damietta branch was built in 1903 with 50 gates, each having 5.0-m width. The maximum head on the barrage is 4.0 m and the floor level elevation is 3.5 m. For a head of 4.0 m, the resulting jet velocity from Equation (23) is 8.86 m/s and with 40% reduction due to the weir, the effective jet velocity becomes 5.42 m/s. The jet depth is assumed to be 2.0 (downstream water level of 5.5—floor level 3.5 m). Therefore, the unit width discharge is 10.64 m²/s. For the scour hole along gate No.1 axis, the upstream and downstream slopes are 0.2 and 0.1, respectively, which again confirm the assumption that $\alpha_{\varphi} = 2$.

Using the above data, the maximum predicted scour depth by Equation (22) is 3.83 m. Observed scour depths along the axes of the barrage gates are equal to or less than 3.5 m. Again, Equation (1) under predicts the scour depth as 2.07 m while the rest of the equations highly over predict the scour depth (9.00–57.40 m).

3.7. Scour downstream of the Shimen Arch Dam, China

Lim and Yu (2002) described the case of the Shimen Arch Dam, China. They reported: "The Shimen Arch Dam was built in China along the Bao River in 1973." The structure has a 20 m apron downstream of the six sluice gates (each 7-m wide and 8-m high). In 1978, the apron protection was extended by 30 m to give a total apron length of 50 m. In the period of 14–25 August 1981, the river was inundated with a one in 300 years flood flow and the water released from the six gates was as high as 4,840 m³/s. Scouring occurred downstream of the sluice gates and the maximum scour depth recorded was about 13.6 m, below the original bed level of the downstream channel. The mean efflux jet velocity at the entrance to the apron floor was estimated to be about 20.8–25.4 m/s, based on the head of water upstream of the sluice. The mean efflux flow depth was estimated to be 4.03 m, based on the flood discharge and the apron width. The downstream water depth was about 53 m. The bed material downstream of the dam consists of layers of cipolin or quartzite. Each layer is less than 10-m thick. The high-speed flow destroyed the rocky layers into loose pieces of rock blocks. Some of these rocks were flushed out by the flow and formed the scour hole during the flood. According to survey on the bed material in the scour hole, it was found that the rock blocks had a volume varying between 24 and 26 m³ and assuming it to be spherical would give the bed material size a value of 3.62 m in diameter.

Lim and Yu (2002) used the above data along with assuming a uniform bed material with $\sigma_G = 1.2$ and $D_{50} = 3.62$ m in their equation (Equation 8), and obtained computed maximum scour depths of 12.9–17.0 m for the velocity range of 20.8–25.4 m/s at the entrance to the apron. Their computed mean scour depth of 14.95 m compares favorably with the measured maximum scour depth of 13.6 m. However, such close matching by them between the computed and measured scour depths would have not been possible unless knowledge exists about bed rock pieces with a diameter of 3.62 m which is very difficult to be predicted prior to any given flood event. For any given high floods, it would have not been possible to make an estimate of the newly formed rock spheres due to very high floods.

In order to apply the present study-developed scour equation which is based on the work transfer theory, care must be given to the type of the rocky bed material at the Shimen Dam. It can be assumed that the porosity of the rocky bed material is almost zero ($\theta = 0$ for non-porous materials) and that the specific gravity of quartzite is 2.8. These new values of the porosity and specific gravity are inserted into Equation (21) and after casting it in the same form as Equation (22) yields

$$D_s^3 = 0.1133\,q^2\left(\frac{1}{2}+\frac{D_s}{H_j}\right) \tag{31}$$

For the lower velocity of U_o = 20.8 m/s and jet depth of 4.03 m, the unit flow discharge becomes 83.82 m²/s which is substituted in Equation (31) and solved for D_s to yield a predicted scour depth of 14.97 m. For the higher velocity of U_o = 25.4 m/s and jet flow of 4.03 m, the resulting unit flow discharge becomes 102.36 m²/s and after substituting this value in Equation (31), a value of D_s = 18.09 m can be computed. It is observed that the predicted scour depth (14.97 or 18.09 m) is larger than the measured one (13.6 m); however, filling of the scour hole during the falling period of the flood is likely to cause the maximum scour depth to be larger than the reported value of 13.6 m or that the scour did not reach equilibrium conditions. Keeping in mind that only information about the flood hydrodynamics is used in predicting the scour depth of this high flood, the predictions seem to be very impressive.

4. Application of the developed equation for prediction of scour hole's length

Table 3 shows the results of applying Equation (25) to predict the length of the scour hole in the main flow direction. To use Equation (25), input values for the scour hole's depth and the slope of its upstream face are needed. Equation (22) is used to provide the scour hole's depth while the field observed value of the upstream slope of the scour hole is adopted. From topographic survey maps, the measured scour hole's length can be determined easily; yet, in some cases, its precise value is difficult to be obtained and a range of its value can be settled for.

It is noted that although Equation (25) is very simple, it yields results that are comparable with the observed field values. The predicted values are slightly higher than the measured values which are due to the difference between reality and the theoretical approach used herein. Large differences (Isna and Assiut Barrage cases) can be attributed to the fact that some scour holes have not reached their equilibrium state which is assumed in the theoretical approach herein. The scour length predictions by Chiew and Lim (1996) seem to be of an order of magnitude higher than the measured lengths.

5. Scour downstream of spillway structures due to horizontal jets

In order to test the present approach's scour equation against laboratory data, detailed data are needed. Due to scale effects, the assumptions made earlier that $\varphi_1 = 2\tan\varphi_2$ or simply that $\alpha_\varphi = 2$, $\tan\varphi_1 \approx \sin\varphi_1$ and $\cos\phi_1 \approx 1.0$ (which can be seen from the field data of the scour holes downstream of the Egyptian Barrages in Table 1) cannot be assumed to hold in laboratory data. Therefore, detailed data about the scour hole geometry in terms of its upstream and downstream side slope angles of the scour hole are needed which are rare to find in the case of scour downstream of barrages. Fortunately, such data exist for a nearly similar case which is scour downstream of spillways by Dargahi (2003) as seen in Table 4. Though this is a case of high-head hydraulic structures, the present approach-developed equations still are valid as long as the flow jet downstream of spillway is horizontal with velocity U_o and depth H_j.

Dargahi (2003) experiments were conducted in a 22-m-long, 1.5-m-wide, and 0.65-m-deep flume. An overflow spillway of 0.205-m crest height was placed at 16.5 m from the inlet. Two different uniformly graded bed materials were used, one fine sand with D_{50} = 0.36 mm and the other medium size gravel with D_{50} = 4.9 mm. Two series of experiments were carried out, one with a smooth plate at the toe of the spillway and the other with additional roughness on the plate. The flow discharge was varied from 20 to 100 l/s and the operating head, h_o, varied from 38 to 96.2 mm. Experiment S20R in Table 4 means the bed material is sand (*S*), the flow discharge is 20 l/s, and the protection plate has added roughness (*R*).

Table 3. Predicted length of the scour hole (to nearest meter) downstream of the Nile Barrages, Egypt

Location	Input data		Predicted scour hole-length, L_s Chiew and Lim (1996) Equation (7) (m)	Predicted scour hole-length, L_s present study Equation (25) (m)	Measured scour hole-length, L_s (m)
	Predicted scour depth (D_s) from Equation (22) (m)	Measured slope of the upstream face of the scour hole, tan φ_1			
NEBTG	7.94	0.11	330	217	(205–213)
NEBSG	6.01	0.16	262	113	98
ONHB	10.15	0.2	531	152	150
ABV58	7.80	0.2	350	117	110
RBV23	3.68	0.15	141	74	63

Dargahi (2003) experiments show that the scour geometry varied considerably as the bed material was changed from sand to gravel. In gravel tests, the scour cavity became smaller and the slope angels were reduced by 10–20%. A common feature in all tests was that 40% of the final scour depth was reached after about 20 min or 4% of the test duration which confirms the assumption that scour occurs almost simultaneously. It was found for sand tests that the upstream slope was steeper than the downstream slope as was assumed here also in the present method.

As laboratory experiments are characterized by steep slopes of the scour hole, the equation developed for scour with flow separation will be used here. Equation (27) (without assuming ϕ_2 as small) becomes

$$\left(\frac{D_s}{H_j}\right)^3 = \frac{6}{(S_G-1)(1-\theta)} \frac{1}{\left(\frac{1}{\tan\phi_1}+\frac{1}{\tan\phi_2}\right)\sin\phi_2} \frac{U_o^2}{gH_j}\left(\frac{1}{2}+\frac{D_s}{H_j}\right) \tag{32}$$

Table 4. Predicted scour depths downstream of a Spillway, data of Dargahi (2003)

Test	Q (l/s)	h_o (mm)	φ_1 (degrees)	φ_2 (degrees)	Measured D_s (m)	Calculated D_s (m), Equation (32)	Relative error (%)
S20	20	38.3	27	20	0.2	0.205	2.3
S40	40	56.4	24	23	>0.26	0.239	−8.8
S60	60	73	19	17	>0.26	0.274	5.2
G20	20	38.5	11	19	0.07	0.163	57
G40	40	57	10	18	0.11	0.198	44.5
G60	60	71	13	22	0.16	0.23	30.3
G80	80	83.2	13	21	0.2	0.254	21.1
G100	100	95.8	16	22	>0.26	0.288	9.8
S20R	20	38	21	13	0.1	0.207	51.7
S60R	60	72.8	23	17	>0.26	0.286	9.2
G20R	20	37	8	7	0.03	0.188	84
G60R	60	70.6	14	18	0.13	0.246	47.1
G100R	100	96.2	17	14	0.19	0.323	41.2

It is assumed the jet flow depth along the protection plate is 0.8; the operating head, h_o, i.e. $H_j = 0.8h_o$. Writing the energy equation (neglecting losses) between the midpoint of the operating head and the midpoint in the jet flow depth along the protection plate yields the jet flow velocity, given approximately by

$$U_o = \sqrt{2g(0.205 + h_o(1. - 0.8)/2.0)} \tag{33}$$

With U_o calculated from Equation (33), ϕ_1 and ϕ_2 known from the measured scour profiles, $H_j = 0.8\,h_o$, $S_G = 2.65$, and $\theta = 0.4$, Equation (33) can be used to calculate the scour depths as seen in Table 4.

It should be noted that the maximum scour depths could not be recorded for tests S4, S60, G100, and S60R because the bed sediment thickness was 0.26 m. The calculated scour depths as seen from Table 4 are in good agreement with the laboratory scour data of Dargahi (2003). When the scour depths are very small for gravel beds, such as at G20 and G20R, the deviations are more which is expected. The comparisons with laboratory data herein clearly demonstrate the effectiveness of the present approach in dealing with both laboratory and field cases and its immunity to scale effects.

6. Scour downstream of grade control structures

Scour downstream of grade control structures is a scour phenomenon closely related to wall jet scour that is investigated herein. Laboratory data, Bormann and Julien (1991), for scour downstream of grade control structures are utilized in further testing of the present approach-developed scour equation. These data have the advantage that a large outdoor flume at Colorado State University, USA, was used which reduces model scale effects. In addition, the upstream slope of the scour hole fluctuates around a value of 0.2 in most runs which is close to the assumption made in this study, i.e. having small upstream scour hole slope. Equation (27), after casting it in terms of the unit width discharge by expressing the unit width discharge of the jet flow as $q = U_o H_j$ and substituting $S_G = 2.65$, $\theta = 0.4$, $g = 9.81$ m/s^2, and $\alpha_\varphi = 2$, is modified by adding the height of the drop structure to the work done by the jet flow which yields

$$D_s^3 = 0.412\,\eta^2 q^2 \left(\frac{1}{2} + \frac{D_p}{b_o} + \frac{D_s}{b_o}\right) \tag{34}$$

where D_p is the height of the drop structure. Table 5 shows the number of iterations needed for convergence, the measured input data of: unit discharge, jet flow depth, and drop height, and the predicted scour depths using Equation (34) for three values of η of 1.0, 0.75, and 0.70. It should be noted that the jet flow in grade control structure is not horizontal as assumed in the cases considered in this study but follows the face angle of the structure. The data used here, from Run 1 to 36 the face angle was 45° while for Run 80 to 88 the face angle was ≈ 18.0°. The value of $\eta = 0.7$ provides the closest agreement to the laboratory data of Bormann and Julien (1991) for grade control structures. It is interesting to note that for a jet with an angle of 45°, its velocity horizontal component is U_o cos 45° ≈ 0.707 U_o. Therefore, the grade control sloping jet, the data could be converted to horizontal jet by multiplying the velocities (or q) with 0.7 which is the equivalent to using a value of $\eta = 0.7$.

Figure 9 shows the calculated scour depths vs. the measured ones where most data lie above the line of perfect agreement (only 4 points among 37 points are below the line). Figure 10 shows plot of the ratio of the calculated scour depth over the measured one. From Figure 10, it can be deduced that 25 points lie within 0.99–1.5. In other words, 25/37 or nearly 67% of the predicted scour depths have an error less than +50%. Despite the fact that the jet flow angle was inclined in the data of Bormann and Julien (1991) while the analysis here is for horizontal jets, however, very good agreement between the data and the predictions can be assumed.

7. General observations

(1) It should be noted that the mathematical modeling approach developed herein from the principles of the work transfer theory is some sort of a global or general incipient motion concept but in terms of work (or potential energies) rather than in terms of forces or moments. In the classical incipient motion, balance of forces (Dey & Sarkar, 2006; Hopfinger et al., 2004) or moments (Hogg et al., 1997) is applied to a typical sediment bed particle usually D_{50}. But, below the original river bed at depths 3.0 m and deeper, the D_{50} value may change from the D_{50} value in the river bed. In the work transfer theory, balance of work is used instead of balance of forces or moments. The whole material in the scour hole is considered as a one big porous particle (mega sediment particle or one unit having a triangular shape). When the jet flow exerts work that is equal to the work needed to move this mega sediment particle out of the scour hole, the equilibrium geometry of the scour hole is attained. This produces a scour equation that is void of the sediment sizes as the sediment unit weight (submerged weight) is more important. This explains theoretically the observation that some scour equations do not include sediment sizes. Simply, what the work transfer theory is stating is that at a certain flow hydrodynamic condition (velocity or depth) and certain sediment properties (sediment-specific gravity and porosity), such conditions produce certain work or energy which can erode the bed material (assuming unlimited bed material along its depth and no armoring) to an extent that the exerted work is exactly the work needed to lift or carry the sediment particles out of the scour hole. From the laws of mechanics, the work done by a group of forces is equal to the work done by the resultant of these forces. This last statement is utilized to deal with the scour hole as one mega particle having resultant weight force acting at the center of mass of the scour hole.

(2) The effect of bed load can be taken in a more direct manner by considering the loss of work or energy due to exerting work in moving the bed load. The work done in moving the bed load inside the scour hole can be given as:

$$W_{bl} = \rho_s q_b U_b \left(\frac{D_s}{\sin \varphi_1} \right) \tag{35}$$

where W_{bl} is the work per unit width exerted by the fluid jet flow in moving the bed load along the sloping length of the scour hole, q_b is the bed load discharge, and U_b is the bed load velocity. This work is subtracted from the work done by the fluid flow which was given in Equation (15) as:

$$W_{net} = \frac{\rho U_o^2 H_j^2 \cos \phi_1}{\sin \phi_1} \left(\frac{1}{2} + \frac{D_s}{H_j} \right) - W_{bl} \tag{36}$$

Here, W_{net} is the net work done by the attacking jet flow considering the energy lost in moving the bed load. This will result in less work available for removing the bed material out of the scour hole and the scour depth should decrease due to the bed load. It can be assumed that the bed load motion takes the form of dunes with scour hole profile of Figure 1 passing equally through the crests and troughs so that the same area (volume) as given by Equation (16) is still valid. A similar treatment can be done for loss of energy due to suspended load if it is present.

(3) Examining the field data in Tables 1 and 2 reveals the existence of the similarity principal in the cases of scour due to jet flows issuing from gate opening. For example, at the New Isna Barrage turbine gates No. 1 and 3, the same scour depth of 6.5 m for both gates was observed under the same conditions of maximum unit discharge, jet depth, and same bed sediment properties. At the sluiceway, nearly equal scour depths of 6.0–6.5 m appear at gates No. 4, 5, and 6 where hydrodynamic conditions are nearly similar. Even similarity exists between scour at the turbine and sluiceway gates. When the unit discharge was 27.77 m²/s at the turbine gates, scour reached 6.5 m and when the unit discharge was 20.5 m²/s, scour reached 6.0 m for the same jet flow depth of 3.0 m and nearly same sediment properties. A more evident

Table 5. Predicted scour depths for grade control structures data of Bormann and Julien (1991) using Equation (38)

Run No.	No. of iterations	q m²/s	H_j (m)	D_p (m)	D_s (m) Equation 32 η = 1.0	D_s (m) Equation 32 η = 0.75	D_s (m) Equation 32 η = 0.70	D_s (m) measured
Runs-1-2-3	7	2.25	0.94	0.15	1.735	1.35	1.271	1.02-1.12
Runs-4-5	7	2.22	0.57	0.15	2.076	1.597	1.500	1.40-1.46
Runs-6-7-8	6	1.72	0.88	0.15	1.403	1.095	1.033	1.01-1.07
Runs-9-10-11-12	7	1.71	0.48	0.15	1.752	1.349	1.268	1.07-1.28
Run-13	6	1.81	1.19	0.25	1.357	1.069	1.010	0.72
Run-14	6	1.78	0.75	0.25	1.561	1.217	1.148	0.98
Run-15	7	1.78	0.46	0.25	1.886	1.457	1.370	1.30
Run-16	4	1.16	2.4	0.25	0.805	0.648	0.616	0.55
Run-17	7	2.32	0.55	0.25	2.232	1.72	1.617	1.32
Run-18	6	1.93	1.17	0.05	1.383	1.082	1.021	0.66
Run-19	7	2.27	0.79	0.05	1.828	1.41	1.326	1.08
Run-20	8	2.32	0.54	0.05	2.170	1.66	1.557	1.21
Run-21	7	1.94	0.52	0.05	1.865	1.428	1.341	0.97
Run-22	7	1.99	0.53	0.05	1.895	1.451	1.362	1.26
Run-23	6	2.47	1.13	0.23	1.792	1.4	1.321	0.96
Run-24	7	2.32	0.58	0.23	2.176	1.678	1.578	1.06
Run-25	7	2.32	0.55	0.23	2.224	1.713	1.611	1.39
Run-26	6	1.42	1.04	0.23	1.149	0.906	0.857	0.70
Run-27	7	1.46	0.43	0.23	1.614	1.248	1.175	0.89
Run-28	7	1.46	0.4	0.23	1.662	1.284	1.208	1.10
Run-29	5	0.61	0.69	0.23	0.647	0.514	0.487	0.27
Run-30	6	0.58	0.34	0.23	0.784	0.615	0.581	0.29
Run-31	6	0.6	0.2	0.23	0.994	0.772	0.727	0.56
Run-32	6	0.59	0.19	0.23	1	0.776	0.731	0.62
Run-33	5	0.34	0.25	0.23	0.558	0.44	0.416	0.1
Run-34	6	0.34	0.14	0.23	0.697	0.545	0.514	0.15
Run-35	6	0.33	0.12	0.25	0.73	0.57	0.537	0.39
Run-36	6	0.33	0.12	0.23	0.723	0.564	0.532	0.93
Run-80	6	2.44	1.13	0.23	1.773	1.386	1.308	0.59
Run-81	6	1.47	1.1	0.23	1.163	0.918	0.868	0.30
Run-82	7	1.42	0.42	0.23	1.589	1.229	1.157	0.71
Run-83	5	0.59	0.62	0.23	0.65	0.516	0.488	0.15
Run-84	6	0.61	0.24	0.23	0.936	0.729	0.687	0.57
Run-85	6	0.55	0.19	0.23	0.94	0.73	0.688	1.52
Run-86	6	0.33	0.16	0.23	0.644	0.505	0.476	0.48
Run-87	6	0.32	0.13	0.23	0.682	0.533	0.502	0.56
Run-88	6	0.29	0.11	0.23	0.67	0.523	0.493	0.97
Shimen-Dam-L*	10	83.82	4.03	0	20.82	15.827	14.83	13.6
Shimen-Dam-H*	11	102.36	4.03	0	25.22	19.138	17.92	13.6

*Equation (37) is used for the Shiemen Arch Dam, China.

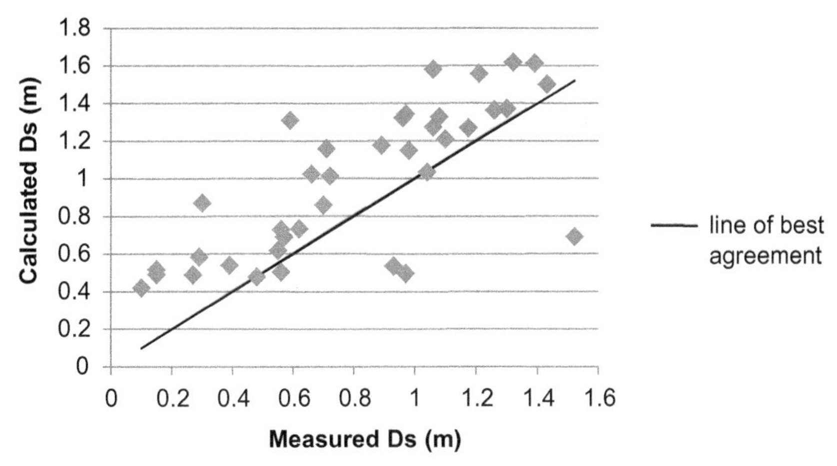

Figure 9. Calculated vs. measured scour depths (m) for grade control structures, measured data of Bormann and Julien (1991).

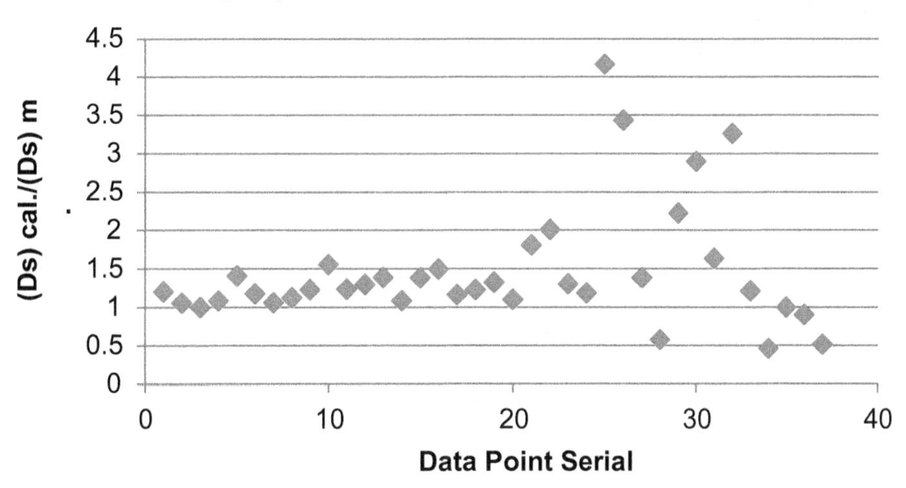

Figure 10. Ratio of calculated over measured scour depths for grade control structures, measured data of Bormann and Julien (1991).

case is the similarity at three different locations at Rossetta, Damiatta, and Zifta Barrages. The unit discharge at these three barrages is around 10.0 m²/s, the jet flow depth is about 2.0 m, D_{90} is from 0.772 to 1.0 mm, and the maximum scour depth varies in a narrow range between 3.25 and 3.5 m while the scour hole's shape is the same with $\alpha_\varphi \approx 2$. In summary, similar hydrodynamic and sediment conditions at different barrage locations produce similar scour pattern (scour depth and scour hole's shape) which confirms the similarity principle.

(4) The theoretical findings in the present study were validated via field data; however, some other studies and laboratory flume data can give more support. For example, Hogg et al. (1997) report that $x_m/x_c = 0.28$ from their model in addition to the experimental results of Rajaratnam (1981) where x_m is the location of the maximum depth of scour and x_c is distance to the crest of the sediment hill. However, plot of the data between x_m/x_c and a flow parameter indicates that all the data points are above the line $x_m/x_c = 0.28$. Now the ratio x_m/x_c can be calculated in the present study as follows. From Figure 1, the distances x_m and x_c are assumed equal to $\frac{D_s}{\tan\varphi_1}$ and L_s, respectively. Therefore,

$$\frac{x_m}{x_c} = \frac{\frac{D_s}{\tan\varphi_1}}{\frac{D_s}{\tan\varphi_1} + \frac{D_s}{\tan\varphi_2}} = \frac{1}{1 + \frac{\tan\varphi_1}{\tan\varphi_2}} = \frac{1}{1 + \alpha_\varphi} \tag{37}$$

It is assumed herein that x_c is nearly equal to the length of the scour hole. For $\alpha_\varphi = 2$, Equation (37) yields $x_m/x_c = 0.33$. The line $x_m/x_c = 0.33$ fits better than the line $x_m/x_c = 0.28$ by Hogg et al. (1997) through the data of Rajaratnam (1981). This agreement confirms not only the similarity in shape of the scour hole between the laboratory data and the theoretical findings here but also the assumption that $\alpha_\varphi = 2$.

(5) The equation for the case of flow separation can also be tested against the barrage field data. Equation (27) for the case of flow separation can be expressed in terms of the unit discharge and jet flow depth in addition to assuming $\alpha_\varphi = 2$, SG = 2.65, $\theta = 0.4$, and $g = 9.81$ m/s^2 which yields

$$D_s^3 = 0.412\ \eta^2\ q^2 \left(\frac{1}{2} + \frac{D_s}{b_o}\right) \tag{38}$$

Equation (38) is similar in form to Equation (22); however, the coefficient η needs to be assumed properly. Table 6 shows the predicted scour depths downstream of the Nile Barrages using different values for η such as 1.0, 0.75, and 0.70. It is clear from Table 6 that for $\eta = 1$, the predicted scour depths are relatively higher than with the other two η values of 0.75 and 0.70. For these two values, the predicted scour depths are in excellent agreement with the field data.

(6) The case of the Shimen Arch Dam, China, is considered here using a version of Equation (27) (flow separation) in a form similar to Equation (31). In that case, using porosity value θ = zero in addition to assuming $\alpha_\varphi = 2$, SG = 2.65, and $g = 9.81$ m/s^2 yields

$$D_s^3 = 0.2266\ \eta^2\ q^2 \left(\frac{1}{2} + \frac{D_s}{b_o}\right) \tag{39}$$

The last two entries in Table 5 show the predicted scour depths using Equation (39) which considers jet flow re-circulation or flow separation. The two values of scour depth prediction (14.83 and 17.92 m), which correspond to the two reported estimated velocity values of 20.8–25.4 m/s, respectively, are acceptable predictions to the measured scour depth of 13.6 m giving the complexity of the problem.

8. Conclusions and recommendations

The developed equations for predicting scour hole's depth and length due to two-dimensional turbulent wall jets in the present study from the mathematical model that is based on the work transfer theory proved to be very reliable for wall jet scour under field conditions. It yields predictions of the scour hole's depth and length very close to the observed values and realistic in their order of magnitude when applied to the cases of the Grand Egyptian Barrages contrary to existing scour formulas found in the literature. Close match between the predicted and measured scour depths occurs also for the case of the 300-year flood at the Shimen Arch Dam, China. The success of predicting scour hole's depth occurred in both cases of non-circulating and re-circulating jet flow. The developed equation for predicating the length of the scour hole showed outstanding performance when compared to field measurements. The close matching of the predicted and measured scour depth and length supports the validity of the work transfer theory and its underlying assumptions such as considering the whole scour hole as mega sediment particle or one unit, the triangular shape of the scour hole section, and that the upstream slope is twice the downstream one.

In addition, scour in case of re-circulating currents in the scour hole can be expressed by the present mathematical model. Comparison of the developed equation against laboratory scour data for scour downstream of a spillway where the data are almost complete clearly demonstrates the effectiveness of the present method. The developed scour equation in this case predicted very well the scour depths measured downstream of grade control structures. The developed equation is based on incipient motion concept in terms of energies rather than in terms of forces or moments, avoids the inclusion of representative bed sediment sizes such as D_{50} but rather includes the bed sediment

Table 6. Predicted scour depths considering flow separation downstream of the Nile Barrages, Egypt

Case	D_s Equation (38) (m) η = 1.0	D_s Equation (38) (m) η = 0.75	D_s Equation (38) (m) η = 0.70	D_s measured (m)
NEBTG	10.97	8.38	7.86	6.50–8.00
NEBSG	8.26	6.34	5.95	6.00–6.50
ONHB	14.09	10.73	10.05	10.00
ABV58	10.77	8.23	7.72	7.50
RBV23	5.05	3.88	3.65	3.50
DBV24	5.05	3.88	3.65	3.25
ZB	5.27	4.04	3.80	3.50

submerged unit weight and porosity as the sediment's most influencing factors in wall jet scour. In agreement with Melville and Lim (2013), the jet Froude number is found to be the flow hydrodynamic influencing variable while past scour formulas considered the denismetric Froude number. Inclusion of porosity in the scour depth equation enabled successful prediction of scour at a rocky non-porous or non-granular bed material at the Shimen Arch Dam in China.

Based on the above findings, further testing of the analytically developed equation to other cases of low-head hydraulic structures is recommended, especially the ratio α_φ which is the ratio of the scour hole upstream to downstream slopes. This ratio can be related to the hydrodynamic forces affecting the scour hole formation and also some soil parameters that reflect soil resistivity to erosion. The generalized nature of the work transfer theory might enable developing expressions for scour at abutments (groins or dykes) and scour due to free falling jets at plunging pools, but these will be the subject of future publications.

Acknowledgments
The author would like to present special thanks and deepest appreciations to the editors and reviewers of this Journal for their valuable comments, suggestions, discussions, and feedback. The author expresses also his deepest appreciation to the staff of the Hydraulic Research Institute (HRI), Egypt, for providing him with the valuable data-sets about the Egyptian Barrages while he was working there in the period 2003–2004. Thanks also go to Mr. Haider A. Chishti, Yanbu Industrial college library services for providing several valuable references. Many Thanks go to Eng. Chandrakant Shitole and Mr. Rajamail Gurumoorthi from Royal Commission Yanbu Colleges and Institutes for their help in some graphical work.

Funding
The author received no direct funding for this research.

Author details
Youssef I. Hafez[1]
E-mail: mohammedy@rcyci.edu.sa
ORCID ID: http://orcid.org/0000-0002-9503-8348
[1] Royal Commission Yanbu Colleges and Institutes, Yanbu University college, Yanbu, Saudi Arabia.

References

Abedelaziz, S., Bui, M. D., & Rutschmann, P. (2010). Numerical simulation of scour development due to submerged horizontal jet, River Flow. In A. Dittrich, K. Koll, J. Aberle, & P. Geisenhainer. ISBN 978-3-93923-000-7.

Aderibigbe, F., & Rajaratnam, N. (1998). Generalized study of erosion by circular horizontal turbulent jets. *Journal of Hydraulic Research, 36*, 613–635.

Balachandar, R., & Reddy, P. (Eds.). (2013). *Scour caused by wall jets, sediment transport* (pp. 177–210). ISBN 980-953-307-557-5.

Bombardelli, F. A., & Gioia, G. (2005). Towards a theoretical model for scour phenomena. In G. Parker & M. Garcia (Eds.), *Proceedings of the 4th IAHR Symposium on River, Coastal and Estuarine Morphodynamics, RCEM 2005* (Vol 2, pp. 931–936). Urbana, IL: Taylor & Francis.

Bormann, N. E., & Julien, P. Y. (1991). Scour downstream of grade-control structures. *Journal of Hydraulic Engineering, 117*, 579–594.

Breusers, H. N. C., & Raudkivi, A. J. (1991). *Scouring*. Rotterdam: Balkema.

Chiew, Y. M., & Lim, S. Y. (1996). Local scour by a deeply submerged horizontal circular jet. *Journal of Hydraulic Engineering, 122*, 529–532. http://dx.doi.org/10.1061/(ASCE)0733-9429(1996)122:9(529)

Dargahi, B. (2003). Scour development downstream of a spillway. *Journal of Hydraulic Research, 41*, 417–426. http://dx.doi.org/10.1080/00221680309499986

Dey, S., & Sarkar, A. (2006). Response of velocity and turbulence in submerged wall jets to abrupt changes from smooth to rough beds and its application to scour downstream of an apron. *Journal of Fluid Mechanics, 556*, 387–419.

Dietz, J. W. (1969). Kolkbildung im fenem oder leichter sohlmaterialen bei strömenden abfluss [Scour in feneme or lighter Sohlmaterialen in flowing drain]. *Mitt. Theodor Rehbock Flussbaulab, Heft, 155*, 1–119.

Eggenberger, W. (1943). *Kolkbildung bei ueberfall und unterströmen* (Dissertation). ETH Zürich - Versuchsanstalt für Wasserbau, Hydrologie und Glaziologie, Zürich.

Eggenberger, W. (1944). Die Kolkbildung beim reinen Überströmen und bei der Kombination Überströmen-Unterströmen. *Mitteliungen aus Versuchsanstalt für Wassenrbau* (Vol. 5). Zürich: ETH Zürich.

El Kateb, H. (1982). *Irrigation design part II regulators and barrages*. Faculty of Engineering, Cairo University.

Hafez, Y. I. (2004a). A new analytical bridge pier scour equation. In *Proceedings The Eighth International Water Technology Conference, IWTC8* (pp. 587–600). Alexandria.

Hafez, Y. I. (2004b). *Local scour downstream the Nile Barrages* (Unpublished Technical Report No. 37/2004). Cairo: Hydraulics Research Institute Publications.

Hamidifar, H., Omid, M. H., & Nasrabadi, M. (2011). Scour downstream of a rough rigid apron. *World Applied Sciences Journal, 14*, 1169–1178.

Hogg, A. J., Huppert, H. E., & Dade, W. B. (1997). Erosion by planar turbulent wall jets. *Journal of Fluid Mechanics, 338*, 317–340. http://dx.doi.org/10.1017/S0022112097005077

Hopfinger, E. J., Kurniawan, A., Graf, W. H., & Lemmin, U. (1999). Sediment erosion by Görtler vortices: The scour-hole problem. *Journal of Fluid Mechanics, 520*, 327–342. http://dx.doi.org/10.1017/S0022112004001636

Jones, J. S., & Sheppard, D. M. (2000, July 30–August 2). Scour at wide bridge piers. In *Joint Conference on Water Resources Engineering and Water Resources Planning and Management, ASCE* (p. 10). Minneapolis, MN.

Kotoulas, D. (1967). *Das kolkproblem unter berüchsichtigung der faktoren zeit und geschiebemischung im rahmen der wildbachverbauung* (dissertation) , Braunschweig: T.U. Braunschweig.

Lim, S.-Y., & Yu, G. (2002). Erosion below sluice gate. *Civil Engineering Research Bulletin, 15*, 100–101.

Melville, B. W., & Lim, S. Y. (2013). Scour caused by 2D horizontal jets. *Journal of Hydraulic Engineering, 140*, 149–155. doi:10.1061/(ASCE)HY.1943-7900.0000807

Müller, R. (1947). Die Kolkbildung beim reinen Ünterströmen und allgemeinere Behandlung des Kolkproblemes, *Mitteliungen aus Versuchsanstalt für Wasserbau* (Vol. 5). Zürich: ETH Zürich.

Nielsen, P. (1992). *Coastal bottom boundary layers and sediment transport*. Singapore: World Scientific. http://dx.doi.org/10.1142/ASOE

Oliveto, G., & Comuniello, V. (2010). *Local scour progress downstream of low-head stilling basins*. Edinburgh, UK: Heriot-Watt University - School of the Built Environment.

Rajaratnam, N. (1981). Erosion by plane turbulent jets. *Journal of Hydraulic Research, 19*, 339–358. http://dx.doi.org/10.1080/00221688109499508

Schoklitsch, A. (1932). Kolkbildung unter Überfallstrahlen. Die [Kolkbildung unter Ueberfallstrahlen]. *Wassserwirtschaft, 24*, 341-343.

Shalash, M. A. E. (1959). *Die kolkbilung beim ausfluss unter schützen [The scour the outflow Protecting]*. (Diss. T.H). Müchen.

Shawky, Y. (2008). Local scour downstream Rosetta Barrages due to high floods. In *Twelfth International Water Technology Conference, IWTC12* (pp. 1345–1364). Alexandria.

Appendix A

There are several ways of proving Equation (15) as follows.

First method

As mentioned before, it is assumed that steady flow conditions exist (U_o and H_j are constants). In addition, it is assumed that the shape of the scour hole does not change with time, i.e. the upstream and downstream slopes of the scour hole are constant (ϕ_1 and ϕ_2 are constants). The constancy with respect to time of the scour hole can be seen in the experiments of Rajaratnam (1981) and also the theoretically calculated profiles by Hogg et al. (1997). This leaves the scour depth is the only variable changing with time. Writing Equations (15) and (17) at any given time *t* yields:

$$W_{in}(t) = \frac{\rho U_o^2 H_j^2 \cos\phi_1}{\sin\phi_1}\left(\frac{1}{2} + \frac{D_s(t)}{H_j}\right) \tag{A1}$$

$$W_{out}(t) = \rho g (S_G - 1)(1 - \theta)\frac{\{D_s(t)\}^3}{6}\left(\frac{1}{\tan\phi_1} + \frac{1}{\tan\phi_2}\right) \tag{A2}$$

where W_{in} is the work done by the jet flow and W_{out} is the work required to lift the scoured material out of the scour hole, where all the variables are constant with respect to the time, *t*, except the scour depth $D_s(t)$. Now after a very long time at which equilibrium will be established, i.e. at an infinite time, taking the limits of each of Equations (A_1) and (A_2) as $t \to \infty$ and observing that in this case $D_s(t) \to D_s$ yields Equations (15) and (17) and the rest of the procedure follows.

Second method

Considering the time development of the scour hole, it is assumed that at each time interval (for which significant scour occurs), the scour hole has incremental scour depths of ΔDs_1, ΔDs_2, ΔDs_3 ..., ΔDs_n where n is the number of time increments in order to reach scour equilibrium. Again, it is assumed that the shape of the scour hole remains constant during the scouring process. The work done by the jet flow to reach scour hole equilibrium, W_{in}, in this case is expressed as:

$$W_{in} = \frac{\rho U_o^2 H_j^2 \cos \phi_1}{\sin \phi_1} \left(\frac{1}{2} + \sum_n \left(\frac{\Delta D_{s1}}{H_j} + \frac{\Delta D_{s2}}{H_j} + \frac{\Delta D_{s3}}{H_j} + \ldots + \frac{\Delta D_{sn}}{H_j} \right) \right) \tag{A3}$$

Now, as $n \rightarrow \infty$ (or very large), the summation in Equation (A_3) tends to D_s/H_j and Equation (A3) becomes identical to Equation (15) and the rest of the procedure follows. The expression inside the summation sign in Equation (A_3) is similar to integrating the work term over the duration of scour but in a discrete manner.

Third method

As mentioned before, Kotoulas (1967) found that in case of coarse sand, about 64% of the final scour occurred in the first 20 s, and about 97% of scour depth was attained in 2 h. This leads to assuming in the hypothetical model herein that scour can be considered to occur instantaneously or in a very short time for which Equations (15) and (17) can be established and the rest of the procedure follows. In other words, the analysis is done instantaneously in a very short time in which both Equations (15) and (17) can be written. This leads to considering the whole scour hole as a mega sediment porous particle, i.e. one unit or one big sediment porous particle where the jet flow water flow through it.

8

Optimizing rib width to height and rib spacing to deck plate thickness ratios in orthotropic decks

Abdullah Fettahoglu[1]*

*Corresponding author: Abdullah Fettahoglu, Department of Civil Engineering, Bursa Orhangazi University, Yildirim 16310, Turkey
E-mail: abdullahfettahoglu@gmail.com

Reviewing editor: Amir H. Alavi, Michigan State University, USA

Abstract: Orthotropic decks are composed of deck plate, ribs, and cross-beams and are frequently used in industry to span long distances, due to their light structures and load carrying capacities. Trapezoidal ribs are broadly preferred as longitudinal stiffeners in design of orthotropic decks. They supply the required stiffness to the orthotropic deck in traffic direction. Trapezoidal ribs are chosen in industrial applications because of their high torsional and buckling rigidity, less material and welding needs. Rib width, height, spacing, thickness of deck plate are important parameters for designing of orthotropic decks. In the scope of this study, rib width to height and rib spacing to deck plate thickness ratios are assessed by means of the stresses developed under different ratios of these parameters. For this purpose a FE-model of orthotropic bridge is generated, which encompasses the entire bridge geometry and conforms to recommendations given in Eurocode 3 Part 2. Afterwards necessary FE-analyses are performed to reveal the stresses developed under different rib width to height and rib spacing to deck plate thickness ratios. Based on the results obtained in this study, recommendations regarding these ratios are provided for orthotropic steel decks occupying trapezoidal ribs.

Subjects: Computer Aided Design (CAD); Structural Engineering; Transportation Engineering

Keywords: steel bridge; orthotropic deck; trapezoidal rib; Eurocode 3; FEM

ABOUT THE AUTHOR

Abdullah Fettahoglu, born 1979 in Trabzon is an assistant professor in Bursa Orhangazi University in Turkey. He worked in private sector and in universities prior to his current employment. He lectures structural mechanics, steel structures, highways, and railways. His research areas are orthotropic bridges, road pavement materials, highways and railways. His doctoral study and most of his articles are on the design recommendations of orthotropic steel bridges. The findings of this article highlight the dimensional ratios of orthotropic deck in terms of stresses developed in the deck. Therefore, other researchers focused on this subject can produce stress reducing new design solutions for orthotropic decks by means of the results provided by this article.

PUBLIC INTEREST STATEMENT

The three bridges spanning Bosphorus and connecting Asia and Europe together in Istanbul are the most important bridges of Turkey. All of these bridges have orthotropic deck structure and are designed by foreign companies, since the design of this type of structure is not well known to Turkish engineers. In this article, design of this bridge type is investigated using current engineering methodologies. Improving design methods and knowledge sharing related to this subject with international colleagues will be a benefit for Turkey for future projects.

1. Introduction

Construction of orthotropic decks with deck plate, cross-beams and trapezoidal ribs going through the cutouts in cross-beam webs started approximately in 1965 and is still widely used in industry (Jong, 2007). Orthotropic deck structure is a common design, which is used worldwide in fixed, movable, suspension, cable-stayed, girder, etc. bridge types. In Japan, Akashi Kaikyo suspension bridge, Tatara cable-stayed bridge (Honshu Shikoku Bridge Authority, 2005), Trans-Tokyo Bay Crossing steel box-girder bridge (Fujino & Yoshida, 2002), which are among the longest bridges in the world, have orthotropic deck structure. In France, Millau viaduct has a box girder with an orthotropic deck with trapezoidal stiffeners (Virlogeux, 2004). In England, Germany, and Netherlands there are a lot of steel highway bridges having orthotropic decks (Jong, 2007). In USA San Francisco Oakland Bay Bridge, Self Anchored Suspension Span in California and in Italy Strait of Messina Bridge are examples of orthotropic steel bridges. In Turkey, the Golden Horn Bridge, First Bosphorus Bridge and Fatih Sultan Mehmet Bridge are also examples of orthotropic steel bridges (Kennedy, Dorton, & Alexander, 2002). In Troitsky (1987), Huang and Mangus (2008), Hoopah (2004), Korniyiv (2004), and Choi, Kim, Yoo, and Seo (2008), Design Manual of Orthotropic Bridge- 2 (2012) examples of bridges, in which orthotropic deck is used, are given in detail. The spacings of longitudinal stringer and cross-beam are in general 300 mm and 3–5 m, respectively. In addition to orthotropic deck structure, wearing surface lying on deck plate and main girders transmitting load to supports are two important components of orthotropic bridges. While wearing surface might be of asphalt or concrete, main girder might be of a girder, a truss, a cable-stayed or a tied-arch system.

Orthotropic decks resist against corrosion by means of traditional anti-corrosive paintings used in industry. The top of the orthotropic deck is covered by wearing course and individual ribs are sealed with end plates to prevent moisture from entering the interior of the rib (Connor et al., 2012). Deck plate forms the flanges of ribs, cross-beams and main girders, hence leads an integral behavior of whole orthotropic deck and results in fewer material use. The closed ribs became dominant on open ribs in industry, because they have much more torsional, buckling rigidities, distribute wheel loads much better on deck plate, require half amount of welding than open ribs, provide less steel material needed in bridge orthotropic deck and so lighter dead load, which makes them also cost effective against orthotropic decks of open ribs. As a result, they have become an inevitable part of orthotropic decks to span long distances. In Figure 1 types of closed ribs are given as trapezoidal, U-shaped, and V-shaped forms, in which trapezoidal ribs became paramount in time. Experienced cracks in orthotropic bridges revealed that the design of orthotropic decks should be performed with respect to fatigue analysis because of repetitive wheel loads varying in type and magnitude. Therefore, the fatigue strengths of orthotropic deck details are provided by engineering standard, Eurocode 3 Part 1-9 (2003). To calculate stresses developed under wheel loads the solution method chosen shall enclose the entire bridge geometry, which can be achieved today using FEM instead of conventional analytical and numerical methods used in history in the absence of FEM. In addition to the correct

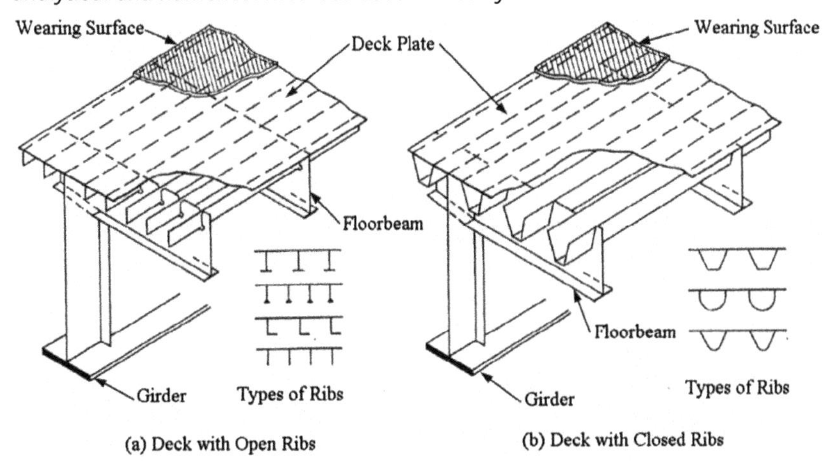

Figure 1. Orthotropic deck with (a) open and (b) closed ribs (American Institute of Steel Construction, 1963; Connor et al., 2012).

analysis and according to the design of orthotropic decks, their fabrication, shipping to construction area, and workmanship shall be done flawlessly and with care to obtain the desired service life. For that reason, all steps until and during the construction of orthotropic decks require necessary quality control measures so as to provide the required service life. Because of their higher initial costs, if orthotropic steel decks are produced under permanent surveillance of quality control measures, they can supply a 100 year service life, which is demonstrated by laboratory studies (Connor et al., 2012). In the scope of this study, the analysis based on conventional techniques of orthotropic bridges is summarized in the next section. Afterwards, FEM applied in this study is introduced in Section 3. In Section 4, the influence of width to height ratio of trapezoidal rib on stress distribution in orthotropic deck is handled. Subsequently, rib spacing to deck plate thickness ratio is evaluated and results are supplied in Section 5. Consequently, conclusions and recommendations for the design of orthotropic decks are given in the last section.

2. Analysis of orthotropic deck using conventional methods

Orthotropic steel decks of bridges are subject to fluctuating wheel loads of different magnitudes. Wheel loads are first dispersed by wearing course and introduced in deck plate. Subsequently, longitudinal stringers transmit wheel loads to cross-beams. Finally wheel loads are transferred from cross-beams over main girders to the supports. Although an orthotropic deck forms an integrated structure to resist against wheel loads, the assumed load transmitting scheme is generally accepted as given in Figures 2–8. In Figure 2, transverse flexural stress develops in deck plate and rib as a result of local deck plate deformation.

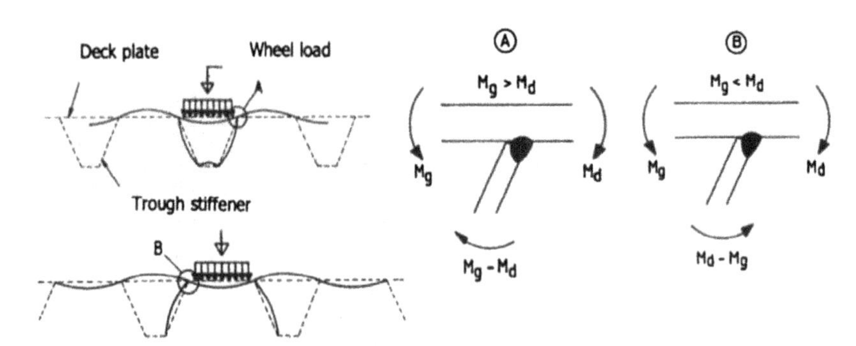

Figure 2. Transverse flexural stress in deck plate and rib (Connor et al., 2012).

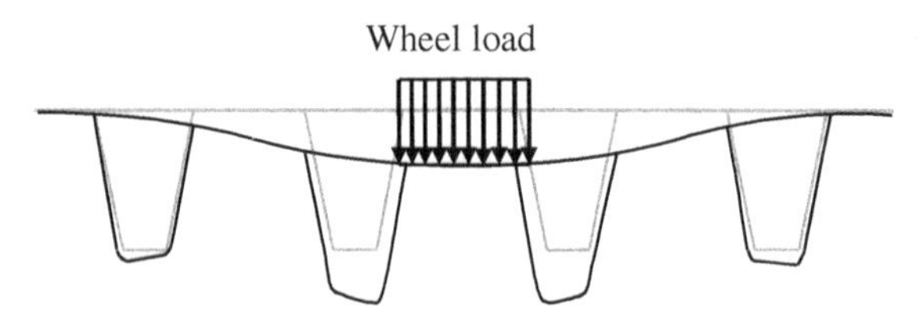

Figure 3. Transverse deck stress from rib differential displacements (Connor et al.,2012).

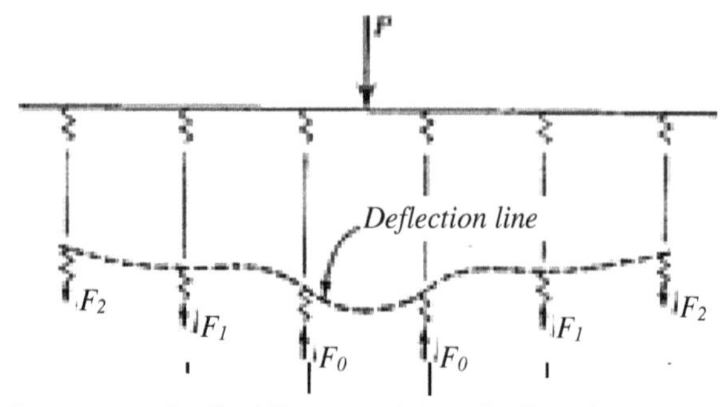

Figure 4. Longitudinal flexure and shear in rib acting as a continuous beam on flexible floor beam supports (Connor et al., 2012).

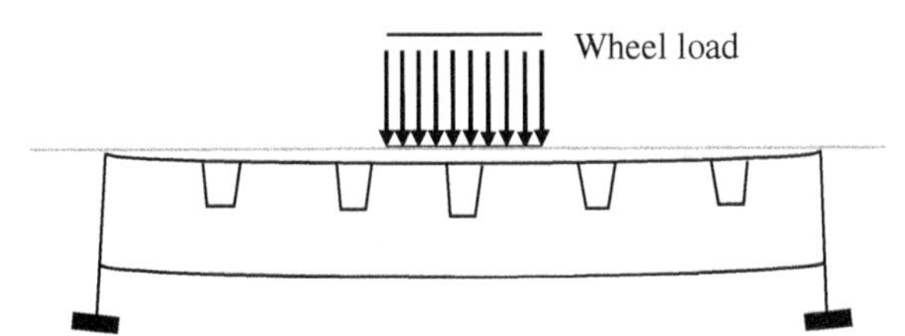

Figure 5. Flexure and shear in cross-beam acting as beam spanning between rigid girders (Connor et al., 2012).

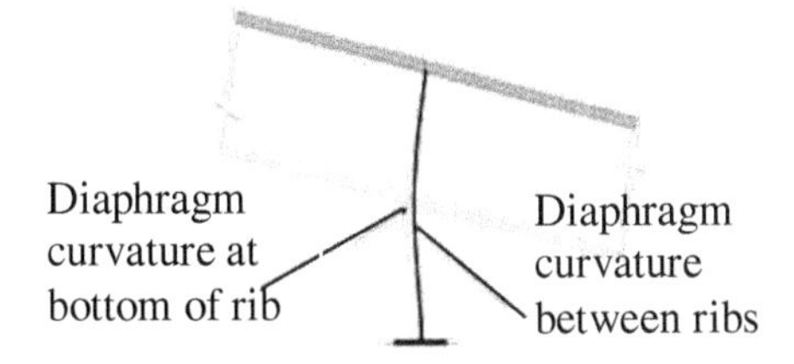

Figure 6. Out-of-plane flexure of cross-beam web at rib due to rib rotation (Connor et al.,2012).

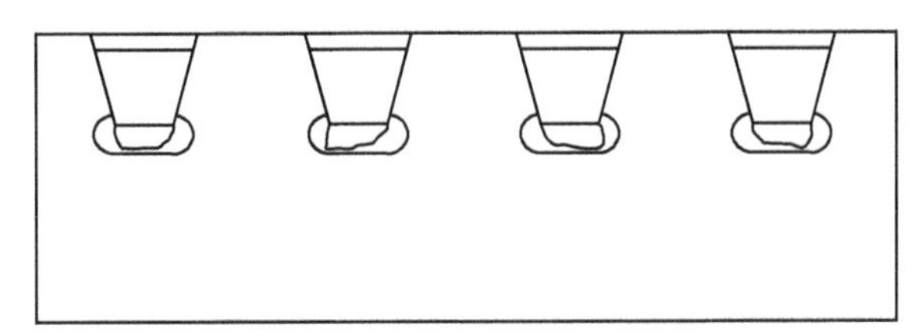

Figure 7. Local flexure of rib wall due to cross-beam cutout (Connor et al., 2012).

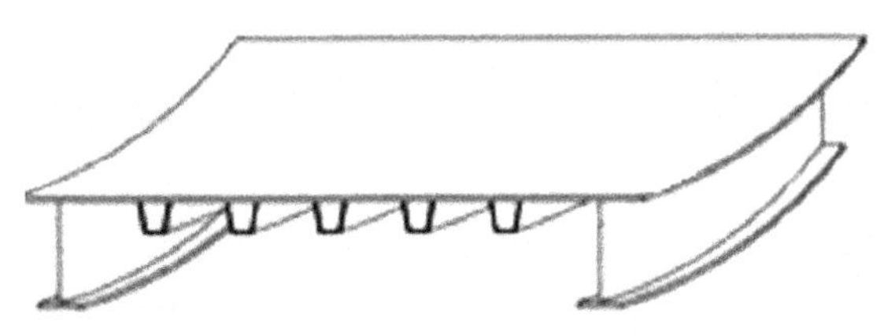

Figure 8. Axial, flexural, and shear stresses from supporting girder deformations (Connor et al., 2012).

In Figure 3, panel deformation yields transverse deck stress from rib differential displacements.

In Figure 4, longitudinal flexure and shear in rib acting as a continuous beam on flexible floorbeam supports results from rib longitudinal flexure.

In Figure 5, cross-beam in plane flexure results in flexure and shear in cross-beam acting as beam spanning between rigid girders.

In Figure 6, out-of-plane flexure of cross-beam web at rib due to rib rotation occurs as a result of cross-beam distortion.

In Figure 7, rib distortion causes Local flexure of rib wall due to cross-beam cutout.

In Figure 8, axial, flexural, and shear stresses develop from supporting girder deformations.

3. Analysis of orthotropic deck using FEM

So as to compare stresses developed under different structural thicknesses, spacings, and spans, all dimensions of the bridge shall be defined as variables in ANSYS (2010). Therefore, an algorithm to provide this condition is written by means of APDL (Ansys Parametric Design Language). Afterwards, thicknesses, spacings, and spans of structural parts, which are of interest, are entered in ANSYS using this algorithm. Stresses developed under different parameter values are given in the subsequent sections. The FE-model of the bridge is generated using SHELL 181, which is illustrated in Figure 9.

The FE-model of Huurman et al. (2002) inspired the researcher to create FE-model of the bridge used in this research (Fettahoglu, 2012; Fettahoglu & Bekiroglu, 2012; Fettahoglu, 2013a, 2013b, 2013c). However, in the FE-model, which is generated using ANSYS (2010) and used in this study, stiffened main girder and pedestrian road are also generated, which are not included in the FE-model of Huurman et al. (2002) (see Figure 10). Because of mesh refinement process the number of nodal unknowns increase excessively and as a result, spans of the bridge used in this research are chosen as short as possible. Figure 11 depicts the perspective front view of the whole orthotropic steel bridge, while Figure 12 shows the wheel loads and their arrangement on the entire bridge geometry. To decrease further the number of nodal unknowns solely the quarter of the bridge shown in Figure 13 is modeled by applying the necessary boundary conditions. As a result, number of elements and nodes in the FE-model of the bridge are 284 010 and 293 491 respectively, when rib width, height and spacing are 300, 275 and 300 mm, respectively. However, element and node numbers vary slightly, when rib width, height, or spacing is changed. Width of pedestrian road and deck plate in transverse direction are 1.1 and 6.3 m, respectively, while width of deck plate changes, when rip spacing changes.

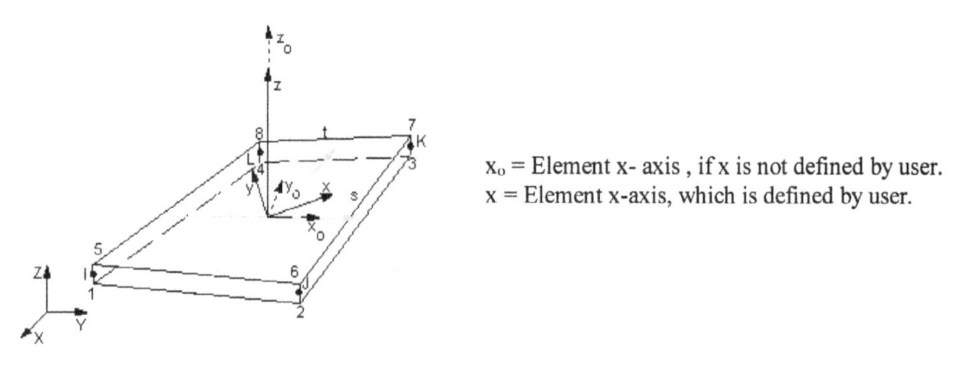

Figure 9. Shell 181 finite element, which is used in this study (ANSYS, 2010).

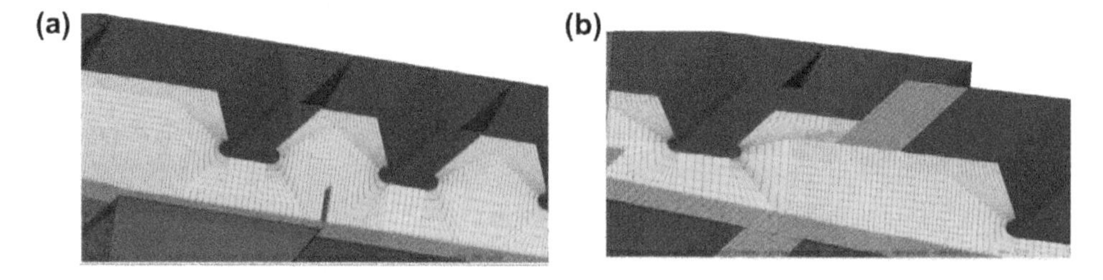

Figure 10. (a) Connection of cross-beam to main girder, (b) connection of deck plate to pedestrian road.

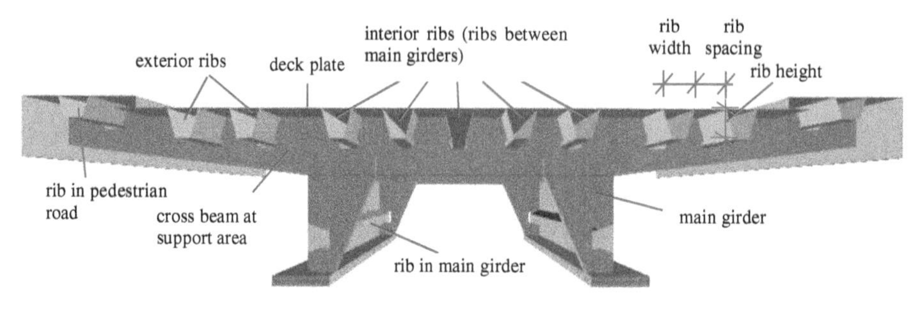

Figure 11. Traditional orthotropic steel bridge as to Eurocode 3 Part 2 (2006).

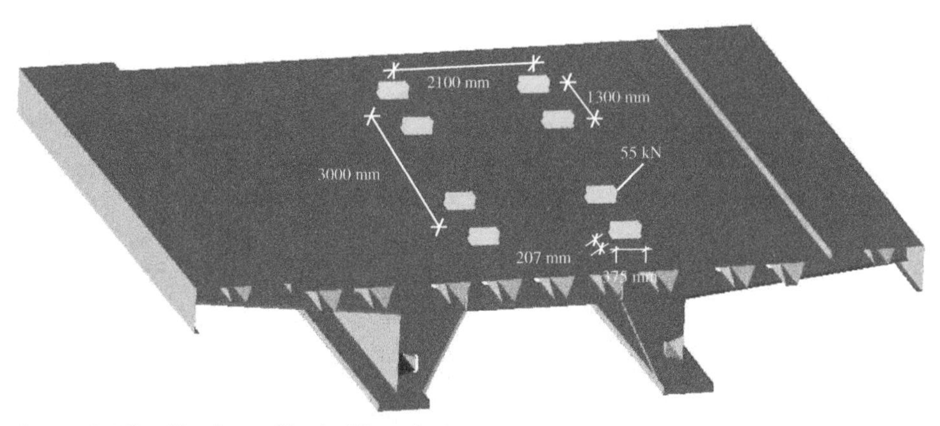

Figure 12. Wheel loads used in the FE-analyses.

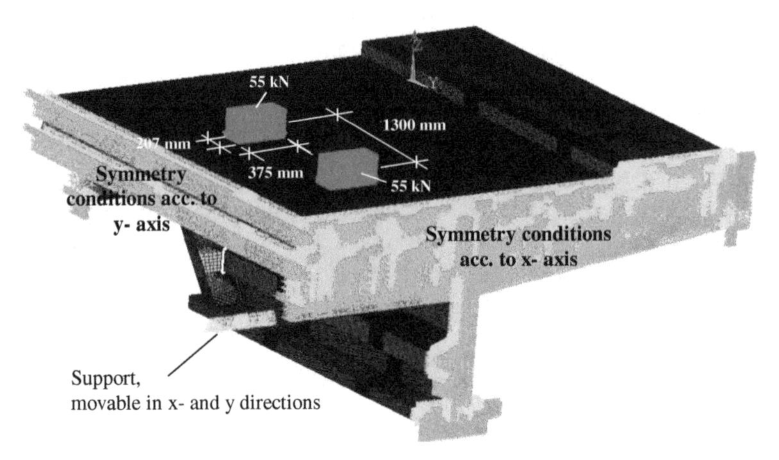

Figure 13. FE-model and boundary conditions of bridge quarter.

The bridge analyzed in this study spans 6 m in longitudinal direction and has stiffened main girders at supports, normal main girders at field (outside support areas), 2 exterior ribs at each side, 5 interior ribs in the middle, 1 rib in each main girder and 1 rib in each pedestrian road. The initial height, width and spacing of the ribs used in orthotropic deck are 275, 300, and 300 mm, respectively. However, one of these parameters is changed in FE-analyses to evaluate its effect on results, while number of ribs and other dimensions are kept constant.

According to Capital 3.2 of Eurocode 3 Part 1-1 (2001) material properties of the selected steel material (S 355H) are given in Table 1. The FE-analyses are based on geometric and material non-linear theories, details of which can be found in Fettahoglu and Bekiroglu (2012).

Wheel loads and areas used in this study on FE-model of the bridge are given in Figure 13.

Table 1. Material properties

Yield strength of steel (*fy*)	355 N/mm^2	Shear module (G)	81,000 N/mm^2
Ultimate strength (*fu*)	510 N/mm^2	Poisson ratio (υ)	0.3
Elasticity module (*E*)	210,000 N/mm^2	Density (ρ_{steel})	78.5 kN/m^3

4. Rib width to height ratio

Rib width to height ratio of trapezoidal longitudinal stiffener is assessed by means of the stresses revealed for different height of ribs, while the width of rib is always kept as 300 mm. Results are compared with each other to interpret the deformation and stress behavior of the bridge for small, moderate, and high values of rib height. Figure 14 shows the parameters used in this study to evaluate the effect of dimensional ratios on stresses developed in orthotropic deck. Here a, b, e and t are rib width, height, spacing, and deck plate thickness, respectively. Results are illustrated mainly in two different ways, namely using contour graphics and tables. Since they both are scalar values and represent all displacement and stress components well, respectively at the point of interest, displacement vector sum and von Mises stress are chosen to interpret the results using contour graphics. Second, tables are used to present the values of vectorial stress components. Table 2 shows the stresses developed in deck plate, cross-beam and rib separately for different rib width to height ratios. In the FE-analyses all dimensions except rib height are always kept constant to be able to understand the effect of rib width to height ratio on deformation and stress behavior clearly. In the scope of this study, five different FE-analyses are performed for rib height values of 150, 200, 275, 300, and 375 mm, while other dimensions of bridge are kept constant.

In terms of deformation behavior of the whole structure max. deformation occurs, when the shortest rib height is used as seen in Figure 15. Max. deformation developed in deck plate decreases, while rib height increases, that can be seen in Figures 15, 17, 19, 21, and 23. So rib height is a parameter to limit the deformations rising in deck plate and also in the wearing surface, which lies directly on deck plate and deforms simultaneously together with deck plate. Figure 25 indicates that variation of deck plate deformations depending on rib height is not linear. Slope of this variation reduces as the rib height increases. After rib height of 300 mm, which is equal to rib width, the change in deformation is so less, which can be neglected.

In Table 2 max. stresses develop as follows,

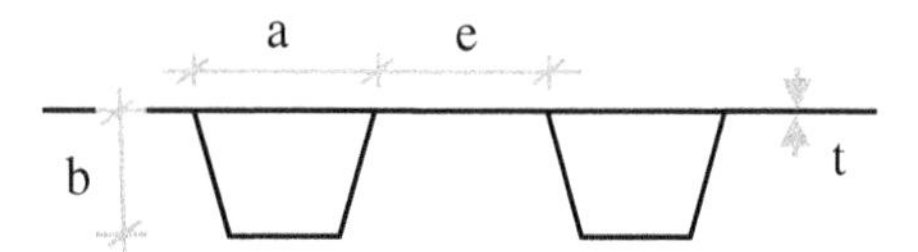

Figure 14. Parametric dimensions, which are of interest.

Table 2. Stresses developed in structural parts for different rib width to height ratios (rib width is 300 mm)

Type of structure	Type of Stress (MPa)	b = 150 mm (a/b = 2.00)		b = 200 mm (a/b = 1 50)		b = 275 mm (a/b = 1.0)9		b = 300 mm (a/b = 1.0)0		b = 375 mm (a/b = 0.8)0	
		max. tens.	max. comp.	max. tens.	max. comp.	max. tens.	max. comp.	max. tens.	max. comp.	max. tens.	max. comp.
Deck plate	σ_x	129.11	121.92	128.74	128.17	135.11	136.05	136.26	137.25	139.59	138.18
	σ_y	113.12	88.80	79.22	80.88	60.02	73.94	58.09	72.42	55.25	69.21
	σ_z	~0									
Cross-beam at field	σ_x	399.00	319.15	244.09	178.07	117.22	85.91	130.50	92.31	184.32	131.78
	σ_y	12.40	9.62	12.85	10.29	12.33	10.41	11.95	10.31	10.52	11.31
	σ_z	58.96	196.85	78.91	207.72	113.51	212.20	124.86	212.67	162.08	215.79
Rib	σ_x	190.02	217.58	127.83	137.08	84.14	80.49	81.32	77.53	81.16	76.98
	σ_y	193.38	343.99	164.39	265.29	145.48	180.98	143.92	156.59	150.53	146.37
	σ_z	294.23	284.85	330.03	238.43	244.22	164.45	216.18	161.68	212.38	169.39

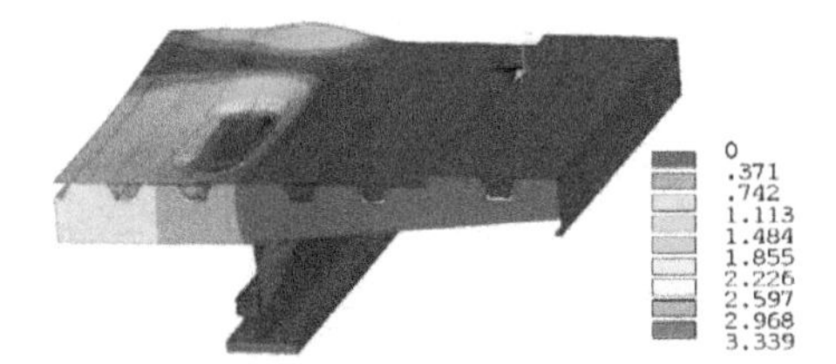

Figure 15. Max. disp. vector sum = 3. 339 mm, when rib height is 150 mm.

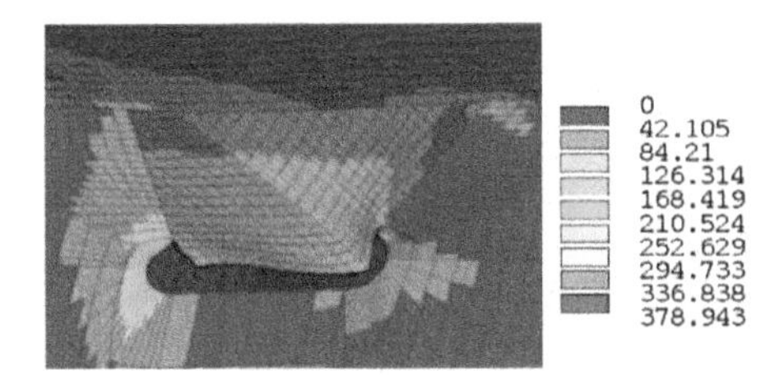

Figure 16. Max. von Mises stress = 378.943 MPa, when rib height is 150 mm.

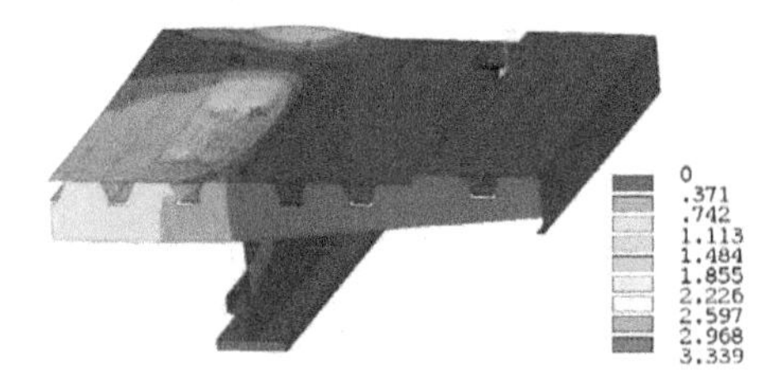

Figure 17. Max. disp. vector sum = 2.398 mm, when rib height is 200 mm.

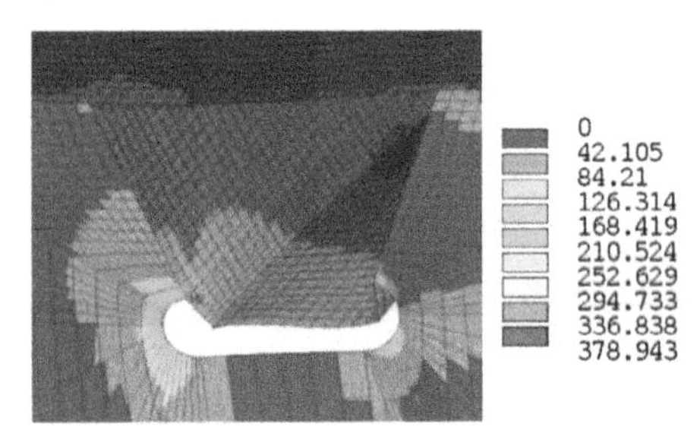

Figure 18. Max. von Mises stress = 367.808 MPa, when rib height is 200 mm.

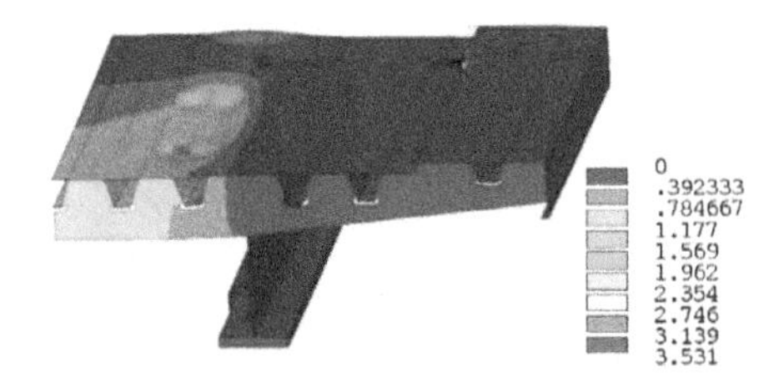

Figure 19. Max. disp. vector sum = 1.729 mm, when rib height is 275 mm.

- In deck plate,
  - σ_x develops direct under wheel load, due to bending of deck plate. Max. comp. σ_x develops at top, while max. tens. σ_x develops at bottom of deck plate.
  - σ_y develops under wheel load due to bending of deck plate. Two exceptions exist out of this situation, when $a/b = 1.5$ and 0.8. At these situations max. tens. σ_y develops at the top of deck plate, where deck plate is connected to normal cross-beam.
  - σ_z is zero in all points.

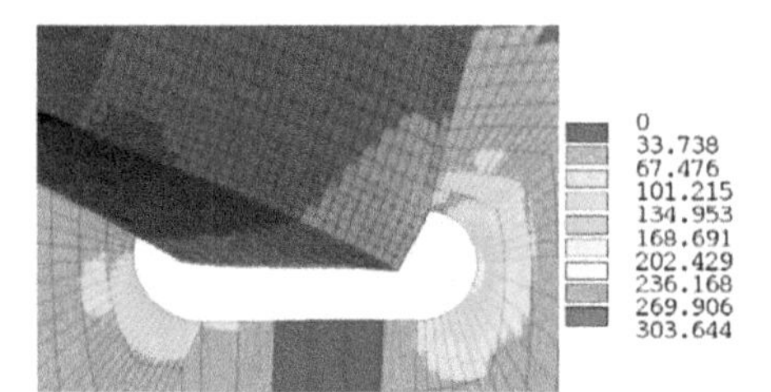

Figure 20. Max. von Mises stress = 255.279 MPa, when rib height is 275 mm.

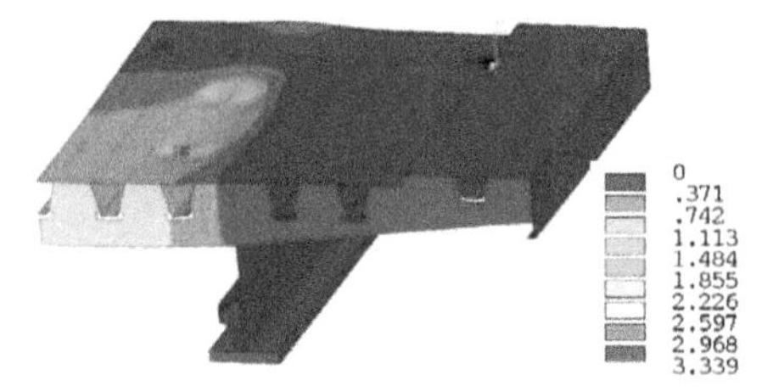

Figure 21. Max. disp. vector sum = 1.677 mm, when rib height is 300 mm.

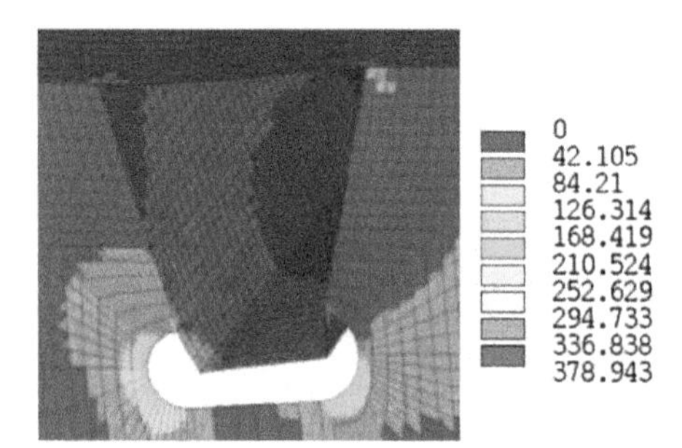

Figure 22. Max. von Mises stress = 224.739 MPa, when rib height is 300 mm.

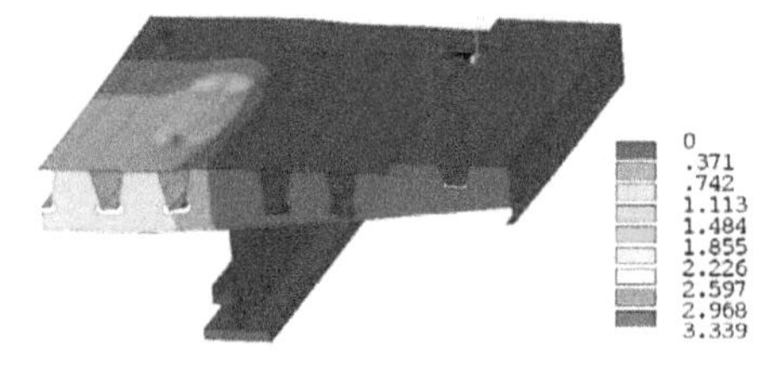

Figure 23. Max. disp. vector sum = 1.641 mm, when rib height is 375 mm.

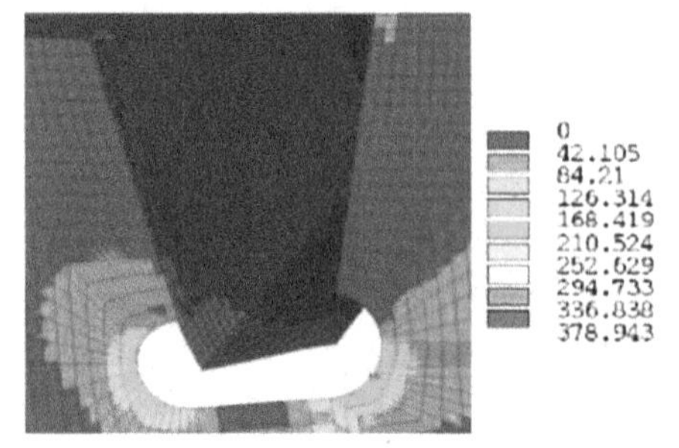

Figure 24. Max. von Mises stress = 229.745 MPa, when rib height is 375 mm.

- In cross- beam at field,
  - For $b = 150$ and 200 mm σ_x develops at rib web to normal cross-beam connection, where cope hole starts. In other points, σ_x develops at normal cross-beam cope hole rounds.

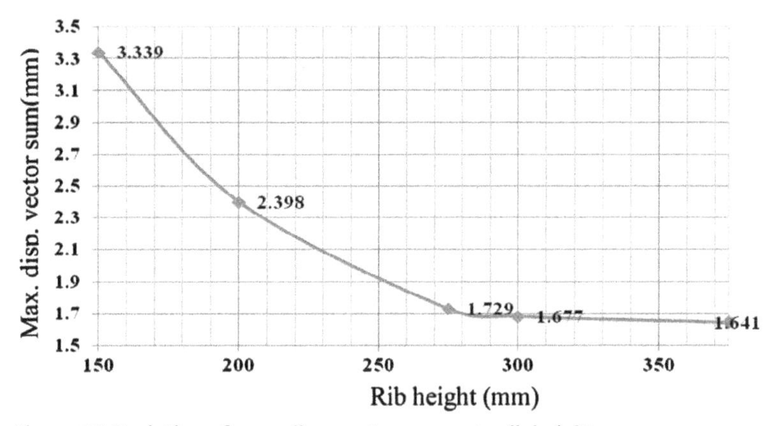

Figure 25. Variation of max. disp. vector sum as to rib height.

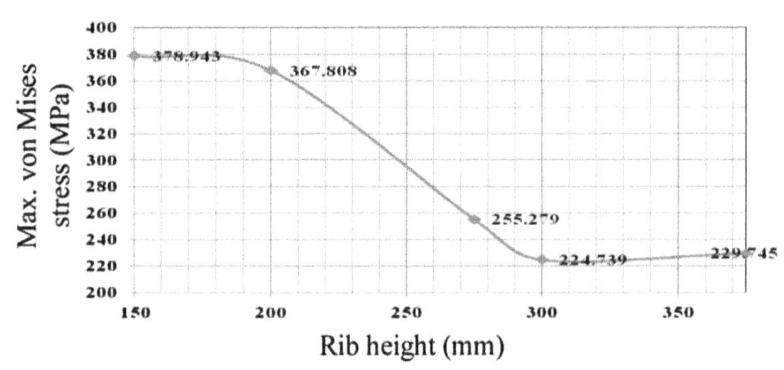

Figure 26. Variation of max. von Mises stress as to rib height.

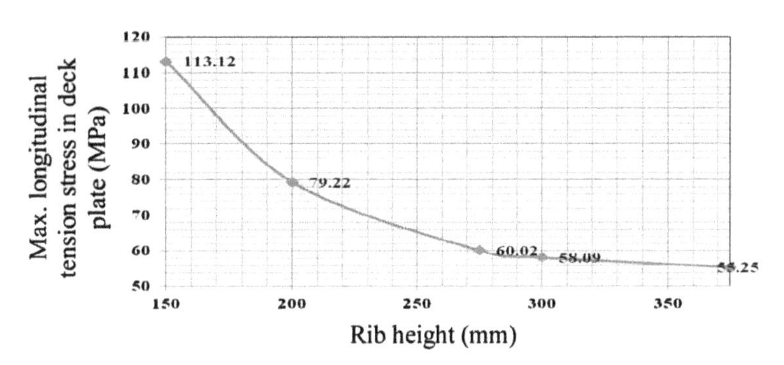

Figure 27. Variation of max. longitudinal tension stress in deck plate as to rib height.

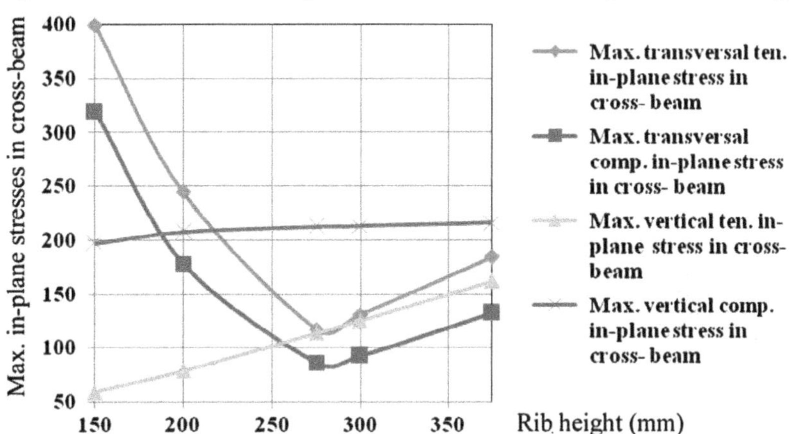

Figure 28. Variation of max. in-plane stresses in cross-beam as to rib height.

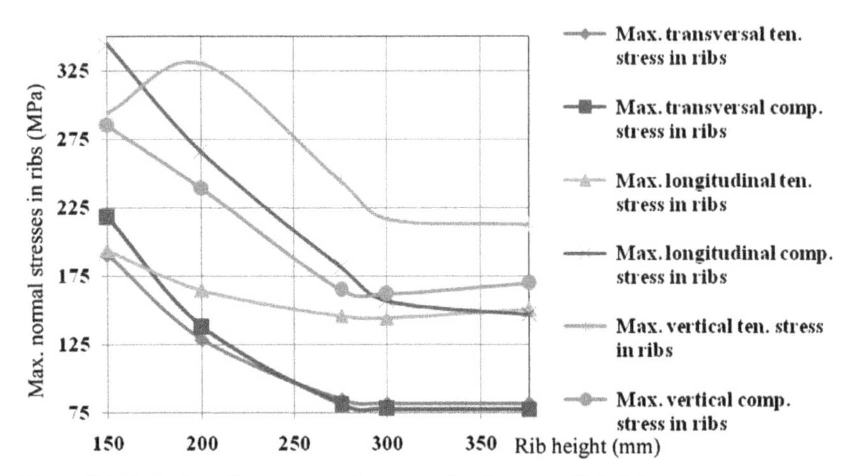

Figure 29. Variation of max. normal stresses in ribs as to rib height.

- For b = 150 and 200 mm σ_y develops at the bottom flange of normal cross-beam. Max. comp. σ_y develops at the top of flange plate, while max. tens. σ_y develops at the bottom of flange plate.
- Max. tens. and comp. σ_z stresses develop at the rounds of cope hole. This cope hole is located in the interior side next to main-girder.

- In ribs,

  - For b = 150 and 200 mm σ_x develops at the rib bottom flange next to main-girder. Max. and min. σ_x stresses are both located in the longitudinal mid of bridge. σ_x develops at the bottom flange of rib, where rib web connects to stiffened cross-beam in other points.
  - For b = 150 and 200 mm max. tens. σ_y develops at the rib bottom flange, longitudinally at the wheel load position. Max. comp. σ_y develops at the rib, cross- beam and cope hole connection point. This rib is located next to main-girder. For b = 275 and 300 mm max. tens. σ_y develops at the rib web closer to stiffened cross-beam. Max. comp. σ_y develops at the rib web to normal cross-beam connection. This rib is located next to main-girder. For b = 375 mm both max. σ_y stresses develop at the rib web closer to stiffened cross-beam longitudinally.

Table 3. Stresses developed in structural parts for different rib spacings (a = 300 mm, t = 12 mm, b = 275 mm)

Type of structure	Type of stress (MPa)	e = 150 mm (e/t = 12.5 and a/e = 2.00)		e = 225 mm (e/t = 18.75 and a/e = 1 33)		e = 300 mm (e/t = 25 and a/e = 1.00)		e = 375 mm (e/t = 31.25 and a/e = 0.80)		e = 450 mm (e/t = 37.5 and a/e = 0.67)	
		max. tens.	max. comp.	max. tens.	max. comp.	max. tens.	max. comp.	max. tens.	max. comp.	max. tens.	max. comp.
Deck plate	σ_x	122.88	116.39	137.44	129.40	135.11	136.05	157.61	159.15	182.05	179.53
	σ_y	65.47	75.88	73.2	89.39	60.02	73.94	107.69	121.01	132.69	140.60
	σ_z	~0									
Cross beam at field	σ_x	39.14	26.08	73.61	53.21	117.22	85.91	148.90	110.33	160.35	130.72
	σ_y	4.93	5.51	9.95	7.88	12.33	10.41	12.70	11.63	12.04	11.78
	σ_z	26.09	81.87	56.60	151.53	113.51	212.20	152.95	221.21	157.21	187.85
Rib	σ_x	27.48	25.00	32.11	32.33	84.14	80.49	108.94	110.66	142.21	140.79
	σ_y	41.98	78.19	75.57	106.43	145.48	180.98	167.70	202.93	137.11	229.88
	σ_z	69.27	60.63	88.30	77.67	244.22	164.45	299.41	218.40	310.11	231.25

Table 4. Stresses developed in structural parts for different deck plate thickness (a = e = 300 mm, b = 275 mm)

Type of structure	Type of stress (MPa)	t = 8 mm (e/t = 37.5)		t = 10 mm (e/t = 30.0)		t = 12 mm (e/t = 25)		t = 14 mm (e/t = 21.43)		t = 16 mm (e/t = 18.75)	
		max. tens.	max. comp.	max. tens.	max. comp.	max. tens.	max. comp.	max. tens.	max. comp.	max. tens.	max. comp.
Deck plate	σ_x	343.22	336.91	208.86	208.80	135.11	136.05	90.38	91.41	69.23	80.50
	σ_y	120.80	138.87	77.79	98.02	60.02	73.94	55.97	58.97	52.30	49.01
	σ_z	~0									
Cross beam at field	σ_x	131.58	90.12	124.04	87.85	117.22	85.91	116.81	84.09	115.91	82.28
	σ_y	12.70	11.08	12.55	10.73	12.33	10.41	12.06	10.11	11.77	9.82
	σ_z	111.94	222.45	113.16	217.83	113.51	212.20	113.24	205.93	112.47	199.28
Rib	σ_x	90.48	86.96	258.53	171.17	84.14	80.49	81.11	77.53	78.25	74.75
	σ_y	158.24	201.94	151.98	191.58	145.48	180.98	139.14	170.56	133.19	160.64
	σ_z	273.26	178.01	87.28	83.63	244.22	164.45	230.55	157.87	217.78	151.51

- For b = 150 mm max. tens. σ_z develops at the inner side of rib web, cross-beam and cope hole connection, while max. comp. σ_z develops at the outer side of rib web at the same place. This rib is located next to main girder. For b = 200 mm max. tens. and comp. σ_z stresses develop at the opposite webs of the rib, which is adjacent to main-girder. For b = 275 and 300 mm max. tens. σ_z develops at the rib web to normal cross-beam connection, while max. comp. σ_z develops at the rib web to stiffened cross-beam connection. This rib is located next to main girder. For b = 375 mm max. tens. and comp. σ_z stresses develop at the opposite webs of the rib, which is adjacent to main girder.

With respect to von Mises stress distribution as rib height increases, the place and value of max. von Mises stress changes. For rib heights of 150, 200, and 275 mm max. von Mises stress rises at the rib web, where it is welded to the web of cross-beam. At this intersection, rib is aligned in traffic direction through cutouts in cross-beam. Therefore, welding between webs of rib and cross-beam is not continuous at the bottom flange of rib, which results in excessive distortion, von Mises stress value and so yielding of material as given in Figures 16 and 18. At rib heights of 275 and 300 mm max. von Mises stress takes much lesser values below yield stress and develops still at the same point, that is the bottom of welding between rib and cross-beam (see Figures 20 and 22). At rib height of 375 mm, which is higher than rib width, max. von Mises stress takes a slight higher value than the value developed at rib height of 300 mm as given in Figures 24 and 26. However max. von Mises stress occurs at the cross-beam cutout edge, when the rib height is higher than rib width. From these results, it is recommended to select rib height equal to rib width to avoid excessive deformations in deck plate, yielding of steel material, and stress concentrations at the intersection point between rib and cross-beam webs under wheel loads used in this study.

Table 2 presents normal stress values as per type of structure and rib width to height ratio. Shear stresses developed in structural parts are so less that they are not included in the assessment of results. Regarding normal stresses, variation of rib height effects longitudinal max tension stress in deck plate, in-plane stresses in cross-beam, and normal stresses in ribs according to the stress values given in Table 2. Dependence of other stress components on rib height is of no importance as seen in Table 2. Increasing rib height results in lesser max. longitudinal tension stress in deck plate as given in Figure 27. Nevertheless, this stress decreases rapidly between rib heights of 150 and 275 mm and the slope of stress decrease becomes very small and can be assumed constant between rib heights of 300 and 375 mm.

Figure 28 illustrates the variation of in-plane stresses in cross-beam. Here, max. vertical compressive in-plane stress increases steadily proportional to rib height. However, this increase has a so small a slope that the change of this stress component can be disregarded. Max. vertical tensional in-plane stress increases also steadily proportional to rib height with an almost constant slope and the change of this slope cannot be disregarded. This stress component takes its min. value (58.96 MPa) at the rib height of 150 mm and its max. value (162.08 MPa) at the rib height of 375 mm. Max. transversal both tensional and compressive in-plane stresses first decrease between rib heights of 150 and 275 mm, then increase between rib heights of 275 and 350 mm.

In trapezoidal ribs variation of all normal stresses is of importance and assessed subsequently as per Figure 29. Max. transversal stresses decrease rapidly between rib heights of 150 and 275 mm. The decrease of this stress component between 275 and 375 mm is almost of no importance. Max. longitudinal tension stress decreases when the rib height is 300 mm and increases between 300 and 375 mm. Max. longitudinal compressive stress decreases rapidly between rib heights of 150 and 300 mm and slightly between rib heights of 300 and 375 mm. Max. vertical tension stress first increases between rib heights of 150 and 200 mm, then decreases rapidly between rib heights of 200 and 300 mm and then a slight decrease in this stress component occurs between rib heights of 300 and 375 mm. Max. vertical compressive stress decreases rapidly between rib heights of 150 and 275 mm and slightly between 275 and 300 mm, but afterwards increases slightly between rib heights of 300 and 375 mm.

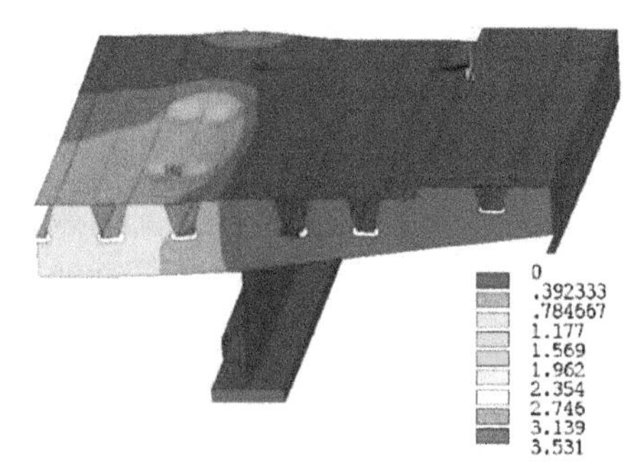

Figure 30. Max. disp. vector sum = 1.729 mm, when t = 12 mm and e = 300 mm.

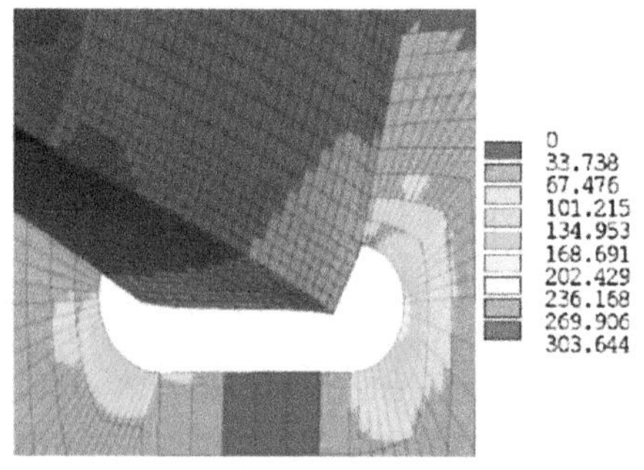

Figure 31. Max. von Mises stress = 255.279 MPa, when t = 12 mm and e = 300 mm.

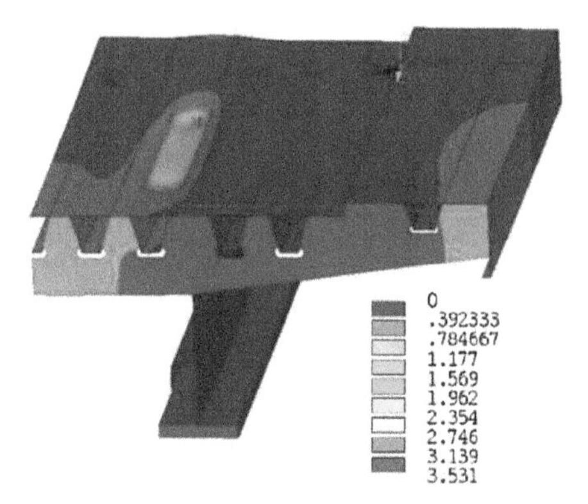

Figure 32. Max. disp. vector sum = 1.284 mm, when t = 12 mm and e = 150 mm.

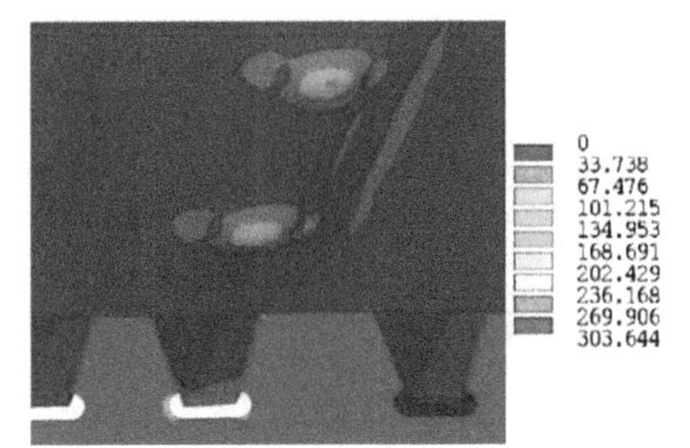

Figure 33. Max. von Mises stress = 106.497 MPa, when t = 12 mm and e = 150 mm.

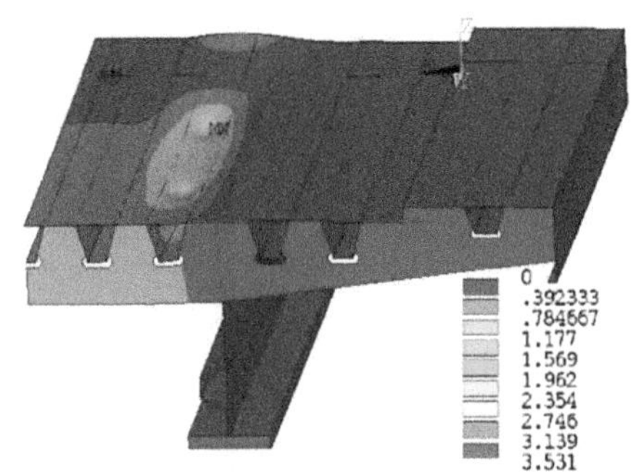

Figure 34. Max. disp. vector sum = 1.796 mm, when t = 12 mm and e = 225 mm.

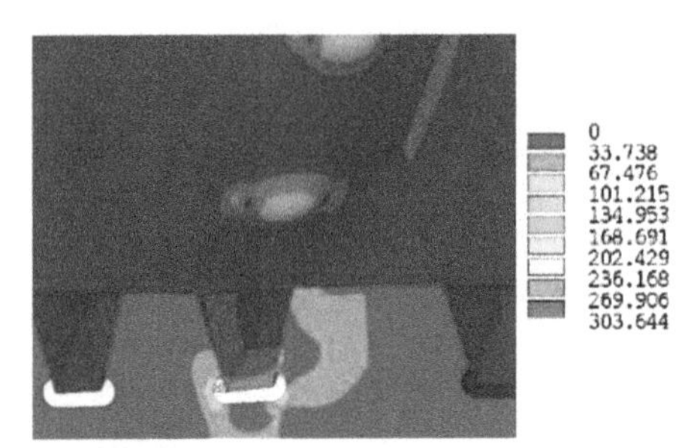

Figure 35. Max. von Mises stress = 144.719 MPa, when t = 12 mm and e = 225 mm.

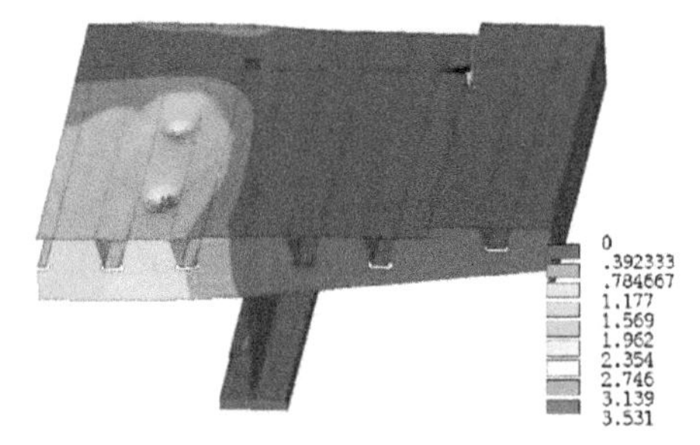

Figure 36. Max. disp. vector sum = 2.793 mm, when t = 12 mm and e = 375 mm.

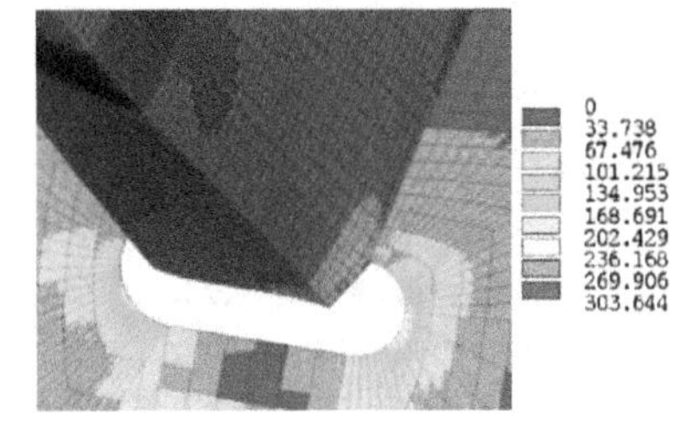

Figure 37. Max. von Mises stress = 295.286 MPa, when t = 12 mm and e = 375 mm.

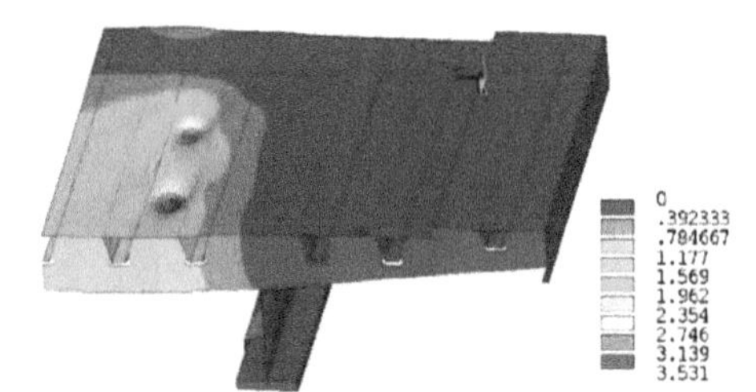

Figure 38. Max. disp. vector sum = 3.531 mm, when t = 12 mm and e = 450 mm.

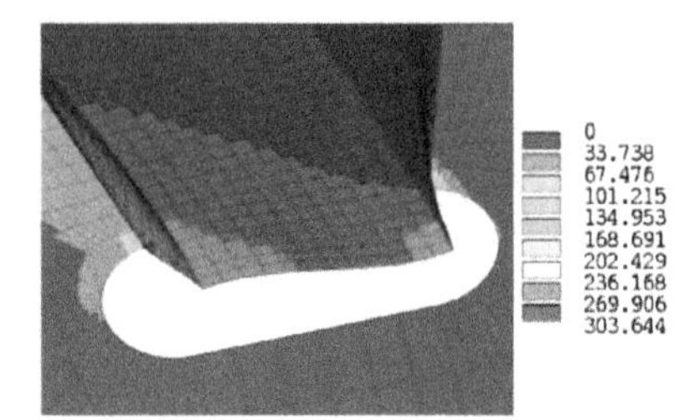

Figure 39. Max. von Mises stress = 303.644 MPa, when *t* = 12 mm and e = 450 mm.

Figure 40. Max. disp. vector sum = 2.665 mm, when t = 8 mm and e = 300 mm.

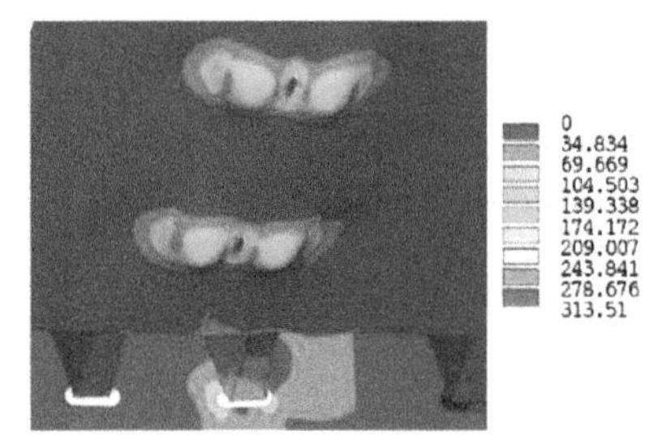

Figure 41. Max. von Mises stress = 313.51 MPa, when *t* = 8 mm and e = 300 mm.

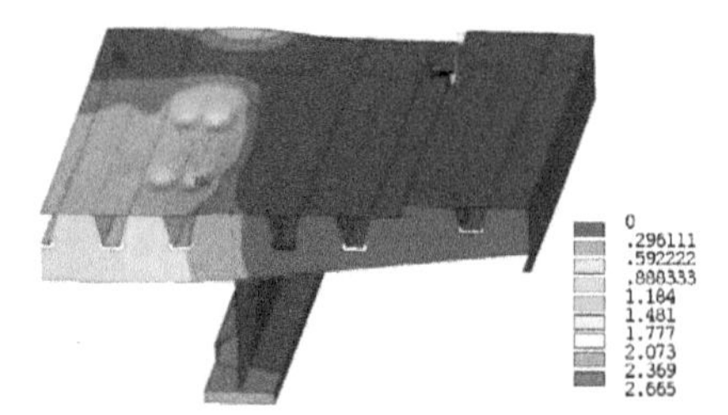

Figure 42. Max. disp. vector sum = 2.04 mm, when *t* = 10 mm and e = 300 mm.

5. Rib spacing to deck plate thickness ratio

Tables 3 and 4 show the stresses developed in deck plate, cross-beam and rib separately for different rib spacing to deck plate thickness ratios. In the FE-analyses of this parameter study rib width and height are always kept constant as 300 and 275 mm, respectively, then variation of rib spacing to deck plate thickness ratio is evaluated separately. Initial values of rib spacing and deck plate

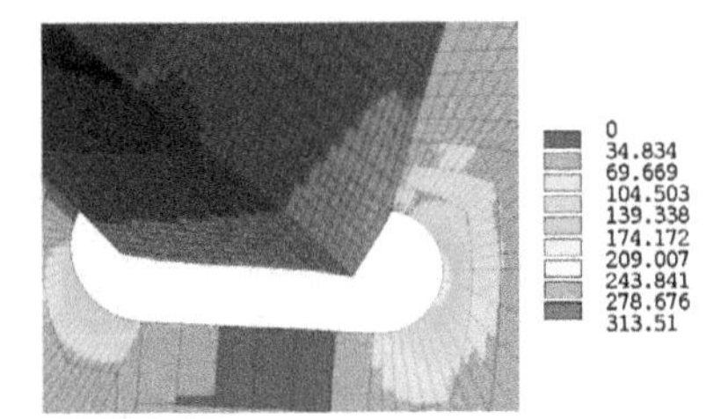

Figure 43. Max. von Mises stress = 269.658 MPa, when t = 10 mm and e = 300 mm.

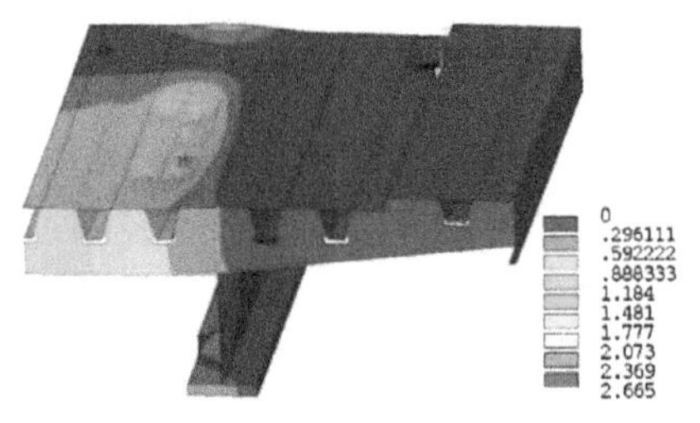

Figure 44. Max. disp. vector sum = 1.591 mm, when t = 14 mm and e = 300 mm.

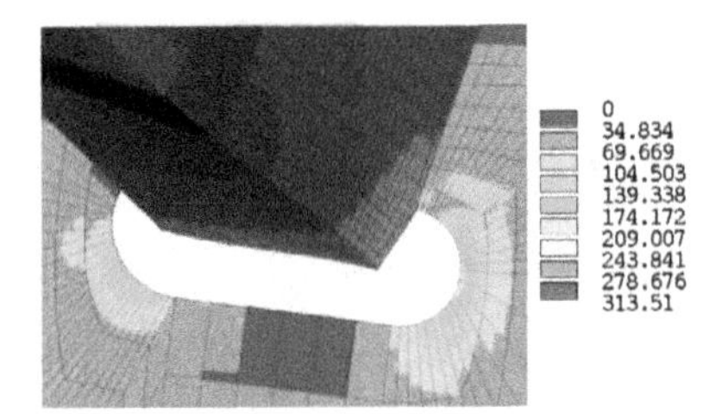

Figure 45. Max. von Mises stress = 241.05 MPa, when t = 14 mm and e = 300 mm.

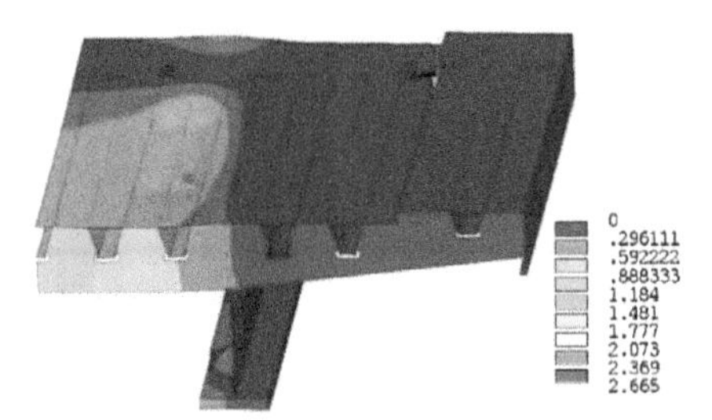

Figure 46. Max. disp. vector sum = 1.511 mm, when t = 16 mm and e = 300 mm.

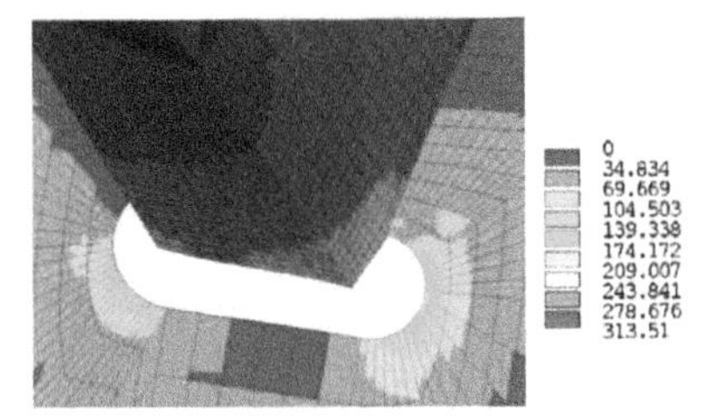

Figure 47. Max. von Mises stress = 227.38 MPa, when t = 16 mm and e = 300 mm.

thickness are 300 and 12 mm, respectively and results of these FE-analyses for initial dimensional values are given in Figures 30 and 31. Subsequently, four different FE-analyses are performed, when rib spacing equals 150, 225, 375, and 450 mm, while other dimensions of bridge are kept constant. Displacements of steel highway bridge depending on rib spacing to deck plate thickness ratio is assessed using contour graphics, which depict the change in max. displacement vector sum. Figures 30, 32, 34, 36, 38, and 48 illustrate the variation of max. displacement vector sum on the whole bridge depending on rib spacing. As a result, max. displacement vector sum develops always in deck

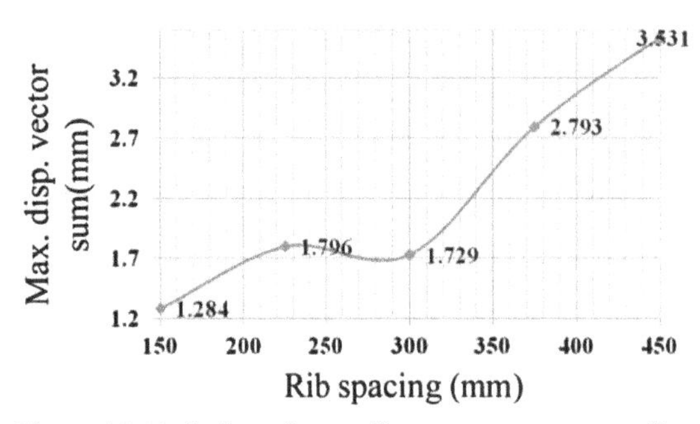

Figure 48. Variation of max. disp. vector sum as to rib spacing.

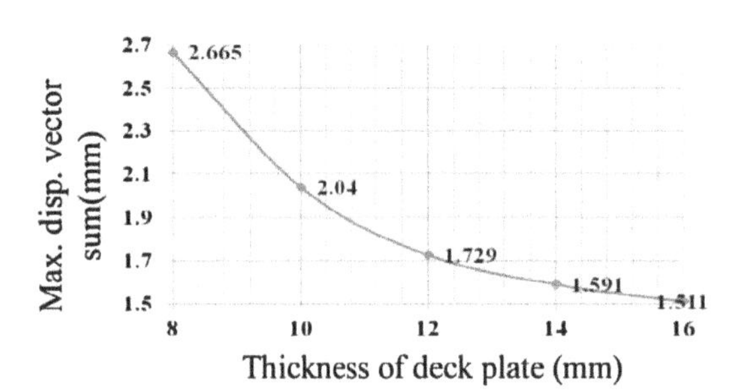

Figure 49. Variation of max. disp. vector sum as to thickness of deck plate.

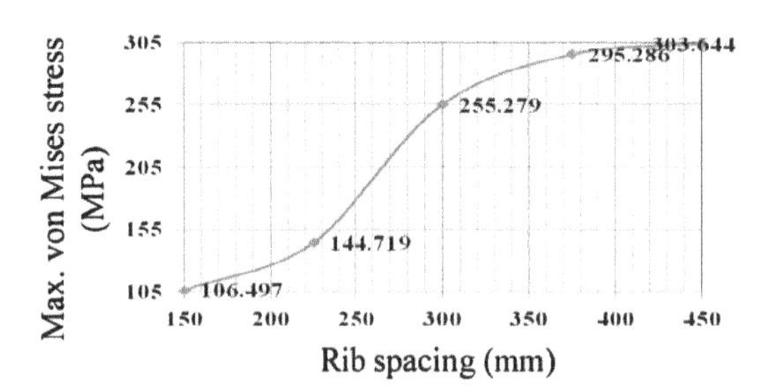

Figure 50. Variation of max. von Mises stress as to rib spacing.

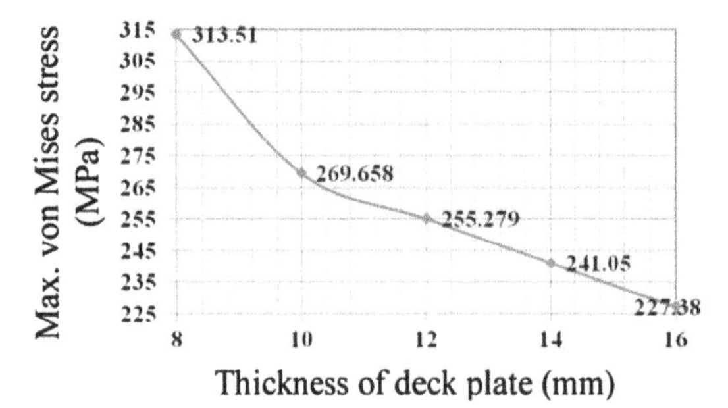

Figure 51. Variation of max. von Mises stress as to thickness of deck plate.

plate and increases proportional to rib spacing. In Figure 48, it is observed that the value at the rib span of 300 mm is out of this trend. The reason of deviation from trend line at this point is the place of wheel loads, which rest on deck plate and rib web at the rib spacing of 300 mm, while wheel loads rest solely on deck plate at the FE-analyses of other rib spacings. While the distance between wheel loads (axle distance of the vehicle) remains always same, the width of bridge changes in transversal direction as rib spacing changes. If wheel loads rested always on rib web and deck plate or only on deck plate, the slope of curve given in Figure 48 would always have a positive slope.

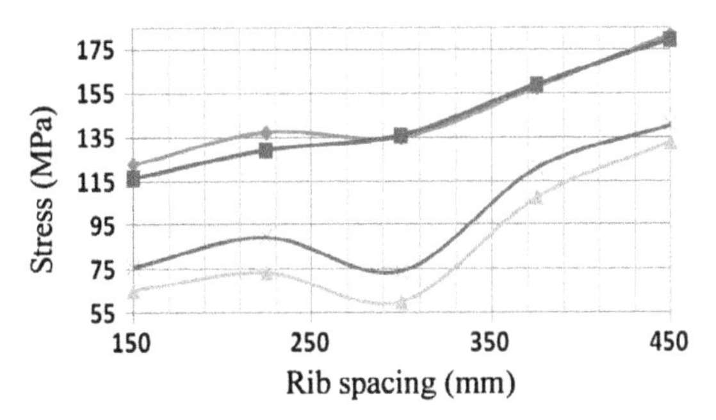

Figure 52. Variation of max. stress values in deck plate as to rib spacing.

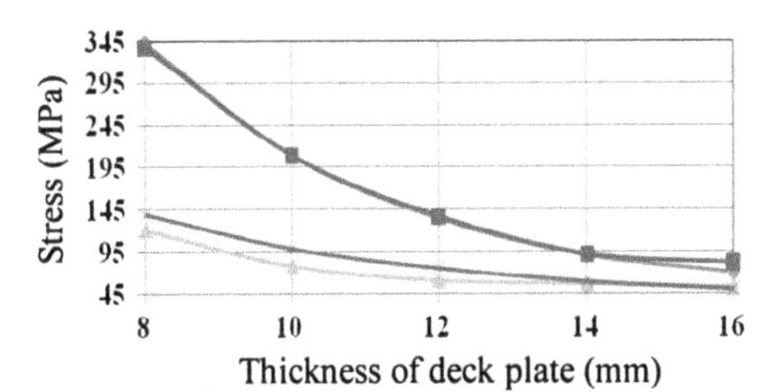

Figure 53. Variation of max. stress values in deck plate as to thickness of deck plate.

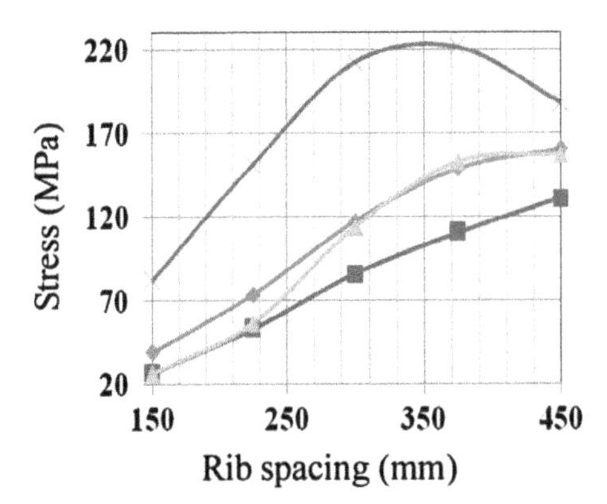

Figure 54. Variation of max. stress values in cross-beam as to rib spacing.

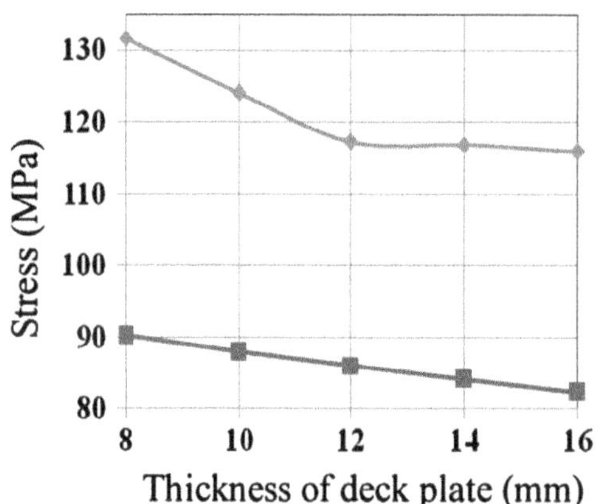

Figure 55. Variation of max. stress values in cross-beam as to thickness of deck plate.

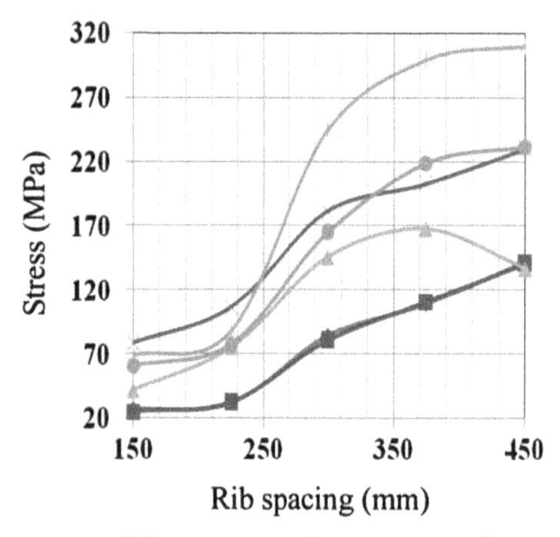

Figure 56. Variation of max. stress values in ribs as to rib spacing.

In the second set of FE-analyses instead of rib spacing, thickness of deck plate is changed to investigate the variations of displacement and stresses. Initial values of a, b, e and t (see Figure 14) are same as 300, 275, 300, and 12 mm, respectively. Afterwards, four different new FE-analyses are done for t = 8, 10, 14, and 16 mm, while other dimensions of bridge are kept constant. Results of these FE-analyses are given below.

From the assessment of Figures 30, 40, 42, 44, 46, and 49 max. displacement vector sum is inversely proportional to the thickness of deck plate and decreases as deck plate thickness increases. Eurocode 3 Part 2 (2006) Annex C1.2.2 recommends $e/t \leq 25$ and $e \leq 300$ mm for design of orthotropic deck. According to this recommendation e = 300 mm and t = 12 mm provide min. conditions to design orthotropic decks. Figure 49 shows that deviations from this min. condition by decreasing deck plate thickness results in localized and rapidly increased displacements in deck plate. This

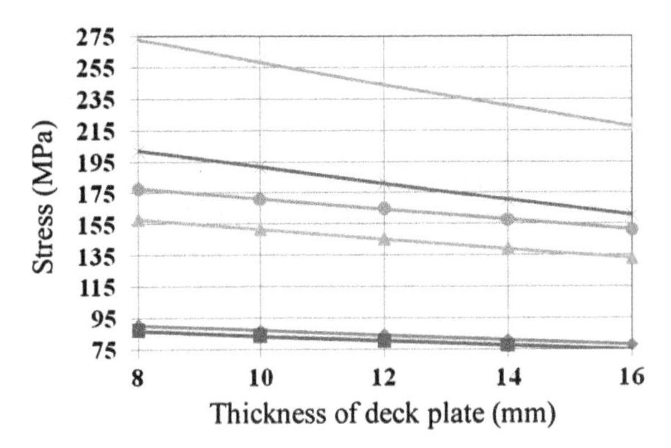

Figure 57. Variation of max. stress values in ribs as to thickness of deck plate.

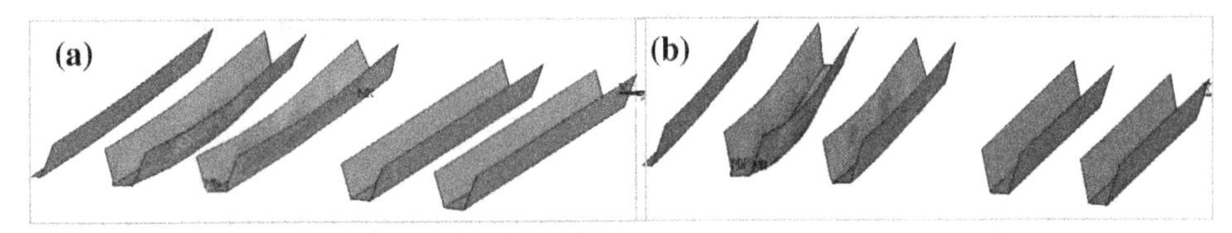

Figure 58. Comparison of deformation behavior and places at which max. longitudinal stress develops (a) at rib spacing of375 mm and (b) at rib spacing of 450 mm.

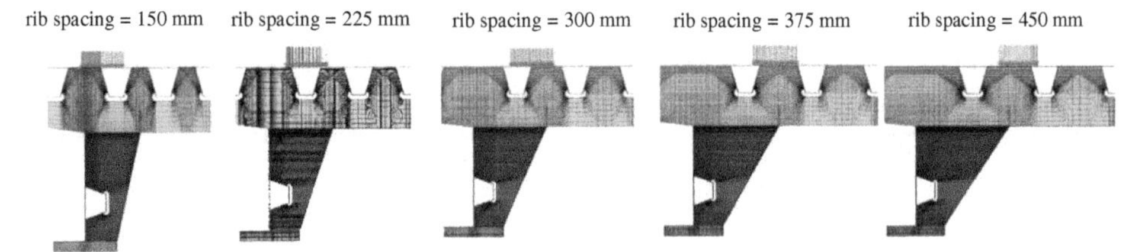

Figure 59. Wheel load positions depending on rib spacing.

result is also valid if deviations from min. condition occur by increasing rib spacing as proven in Figure 48. Finally the numerical results obtained under wheel loads used in this study support recommendations of Eurocode 3 Part 2 (2006) regarding rib spacing to deck plate thickness ratio based on the displacements developed in deck plate in terms of deformation criteria.

Increasing rib spacing yields in increase in von Mises stress as seen in Figure 50. The slope of the curve given in Figure 50 states that max. von Mises stress increases rapidly between rib spacing of 150 and 300 mm, nevertheless it increases with a decreasing slope after rib spacing of 300 mm, which is a point of inflection, where the sign of curve' s curvature changes. The cause of this "S" shape of curve is the changing of stress behavior between rib spacings of 150 and 300 mm. At the rib spacings of 150 and 225 mm max. von Mises stress rises in deck plate and in cross-beam, respectively as given in Figures 33 and 35. Max. von Mises stress develops always in the connection of rib and cross-beam webs as given in Figures 31, 37, and 39 during and after rib span of 300 mm. Accordingly, Figure 50 shall be seen as two curves before and after rib spacing of 300 mm, which is the min. recommendation of Eurocode 3 Part 2 (2006) and satisfies the ratio, $e/t = 25$. Figure 51 reveals that increasing of deck plate thickness causes decreasing of max. von Mises stress. At the deck plate thickness of 8 mm, max. von Mises stress localized direct under wheel loads takes a very high value close to yield stress (see Figure 41). Therefore, deck plate thickness under 10 mm shall not be allowed in orthotropic decks under wheel loads used in this study. Max. von Mises stress develops always in the connection between cross-beam and rib webs as given in Figures 31, 43, 45, and 47 in deck plates having thickness higher than 8 mm. The magnitudes of shear stresses in comparison with normal stresses developed in orthotropic deck are so less, that they are not involved in the assessment of results. The stress components, which vary essentially depending on rib spacing to deck plate thickness ratio are given in Figures 52–57.

In Table 3,

- In deck plate,
    - For e = 150, 225 and 300 mm σ_x develops direct under wheel load, due to bending of deck plate. Max. comp. σ_x develops at top, while max. tens. σ_x develops at bottom of deck plate. For e = 375 mm max. tens. σ_x develops at the bottom of deck plate under wheel load closer to normal cross-beam. Max. comp. σ_x develops at the top of deck plate under wheel load closer to stiffened cross-beam. For e = 450 mm max. tens. σ_x develops at the bottom of deck plate under wheel load closer to stiffened cross-beam. Max. comp. σ_x develops at the top of deck plate under wheel load closer to normal cross-beam.
    - For e = 150, 225 and 300 mm σ_y develops under wheel load due to bending of deck plate. Max. tens. σ_y develops at the bottom of deck plate under wheel load closer to normal cross-beam. For e = 375 and 450 mm max. comp. σ_y develops at the top of deck plate under wheel load closer to stiffened cross-beam.
    - σ_z is always zero in deck plate.
- In cross- beam at field,
    - For e = 150, 225 and 300 σ_x develops at normal cross-beam cope hole rounds. For e = 375 and 450 max. tens. and comp. σ_x arise at the rib web, normal cross beam and cope hole connection points at opposite sides of cope hole.
    - σ_y develops at the bottom flange of normal cross-beam. Max. comp. σ_y develops at the top of flange plate, while max. tens. σ_y develops at the bottom of flange plate.
    - Max. tens. and comp. σ_z stresses develop at the rounds of cope hole. This cope hole is located in the interior side next to main girder.
- In ribs,
    - For e = 150 mm both σ_x stresses develop at the bottom flange of rib adjacent to main girder. They arise in the longitudinal mid of bridge. For e = 225 and 300 mm σ_x develops at the bottom flange of rib, where rib web connects to stiffened cross-beam. For e = 375 mm max. tens. σ_x arises at bottom flange of rib, located longitudinally closer to stiffened cross-beam. Max. comp. σ_x arises at bottom flange of rib, located longitudinally closer to normal crossbeam. For e = 450 mm both σ_x stresses develop at the bottom flange of rib adjacent to main girder. They arise in the longitudinal mid of bridge.
    - For e = 150 mm max. comp. σ_y develops at the bottom flange of rib adjacent to main girder. They arise in the longitudinal mid of bridge. Max. tens. σ_y develops at the bottom flange of rib adjacent to main girder, closer to stiffened crossbeam. For e = 225 mm max. comp. σ_y develops at the bottom flange of rib adjacent to main girder located longitudinally at normal cross-beam location. Max. tens. σ_y develops at the rib web longitudinally closer to stiffened crossbeam. For e = 300 mm max. tens. σ_y develops at the rib web closer to stiffened cross-beam. Max. comp. σ_y develops at the rib web to normal cross-beam connection. This rib is located next to main-girder. For e = 375 mm max. tens. σ_y arises at the rib web, located longitudinally at stiffened cross-beam position. Max. comp. σ_y arises at the rib web, located

longitudinally at normal cross-beam position. For e = 450 mm both σ_y develop at the opposite rib webs of the rib adjacent to main girder. Their position is longitudinally at the normal cross-beam location.

- For e = 150 mm max. tens. σ_z develops at the rib web adjacent to main girder. Max. comp. σ_z develops at the rib web to deck plate connection, longitudinally closer to stiffened cross-beam. For e = 225 mm max. tens. σ_z arises at the rib web longitudinally very close to stiffened cross-beam. Max. comp. σ_z arises at the rib to deck plate connection longitudinally almost between normal and stiffened cross-beams. For e = 300 mm max. tens. σ_z develops at the rib web to normal cross-beam connection, while max. comp. σ_z develops at the rib web to stiffened cross-beam connection. This rib is located next to main girder. For e = 375 mm max. comp. σ_z arises at the rib web, located longitudinally at stiffened cross-beam position. Max. tens. σ_z arises at the rib web, located longitudinally at normal cross-beam position. For e = 450 mm both σ_z develop at the opposite rib webs of the rib adjacent to main girder. Their position is longitudinally at the normal cross-beam location.

In Table 4,

- In deck plate,
  - For *t* = 8 and 10 mm both σ_x arise at deck plate under wheel load, longitudinally closer to normal cross-beam. Max. tens. σ_x arises at top, while max. comp. σ_x arises at the bottom of deck plate. For *t* = 12 mm σ_x develops directly under wheel load, due to bending of deck plate. Max. comp. σ_x develops at top, while max. tens. σ_x develops at the bottom of deck plate. For *t* = 14 mm both σ_x arise at deck plate under wheel load, longitudinally closer to normal cross-beam. Max. tens. σ_x arises at top, while max. comp. σ_x arises at the bottom of deck plate. For *t* = 16 mm max. tens. σ_x arises at the bottom of deck plate, under wheel load, longitudinally closer to stiffened cross-beam, max. comp. σ_x arises at bottom of deck plate, where deck plate connects to normal cross-beam.
  - For *t* = 8 and 10 mm max. tens. σ_y develops at bottom of deck plate, closer to normal cross-beam, max. comp. σ_y develops at the top of deck plate, closer to stiffened cross-beam. For *t* = 12 mm σ_y develops under wheel load due to bending of deck plate. For 14 mm max. comp. σ_y arises at the top of deck plate under wheel load closer to stiffened cross-beam. Max. tens. σ_y arises at the top of deck plate, and deck plate to normal cross-beam connection. For 16 mm max. tens. σ_y arises at the bottom of deck plate, under wheel load, longitudinally closer to stiffened cross-beam, Max. comp. σ_y arises at the bottom of deck plate, where deck plate connects to normal cross-beam.
  - σ_z is zero in all points of deck plate.

- In cross- beam at field,
  - For *t* = 8 and 10 mm max. tens. σ_x arises at rib, normal cross-beam and cope hole connection. Max. comp. σ_x arises at round of normal cross-beam's cope hole. For *t* = 12 mm σ_x develops at normal cross-beam cope hole rounds. For *t* = 14 and 16 mm both σ_x stresses develop in the round of normal cross-beam's cope hole.
  - For *t* = 8 and 10 mm max. tens. σ_y develops at the top of normal cross-beam's bottom flange, while max. comp. σ_y develops at bottom of flange. For *t* = 12 mm σ_y develops at the bottom flange of normal cross-beam. Max. comp. σ_y develops at the top of flange plate, while max. tens. σ_y develops at the bottom of flange plate. For *t* = 14 and 16 mm max. tens. σ_y develops at the top of normal cross-beam's bottom flange, while max. comp. σ_y develops at bottom of flange.
  - For *t* = 8 and 10 mm max. σ_z stresses develop at the round of normal cross-beam cope hole. For *t* = 12 mm max. tens. and comp. σ_z stresses develop at the rounds of cope hole. This cope hole is located in the interior side next to main girder. For *t* = 14 and 16 mm max. σ_z stresses develop at the round of normal cross-beam cope hole.

- In ribs,
  - For t = 8 and 10 mm both σ_x stresses arise at bottom flange of rib adjacent to main girder and longitudinally at the stiffened cross-beam location. For t = 12 mm σ_x develops at the bottom flange of rib, where rib web connects to stiffened cross-beam. For t = 14 and 16 mm both σ_x stresses arise at bottom flange of rib adjacent to main girder and longitudinally at the stiffened cross-beam location.
  - For t = 8 and 10 mm max. tens. σ_y develops at rib web adjacent to main girder, and longitudinally at the stiffened cross-beam location. Max. comp. σ_y develops at the rib web adjacent to main girder and longitudinally at the normal cross-beam location. For t = 12 mm max. tens. σ_y develops at the rib web closer to stiffened cross-beam. Max. comp. σ_y develops at the rib web to normal cross-beam connection. This rib is located next to main-girder. For t = 14 and 16 mm max. tens. σ_y develops at rib web adjacent to main girder, and longitudinally at the stiffened cross-beam location. Max. comp. σ_y develops at the rib web adjacent to main girder and longitudinally at the normal cross-beam location.
  - For t = 8 and 10 mm max. tens. σ_z arises at the rib web adjacent to main girder and longitudinally at the normal cross-beam location. Max. comp. σ_z arises at the rib web adjacent to main girder, and longitudinally at the stiffened cross-beam location. For t = 12 mm max. tens. σ_z develops at the rib web to normal cross-beam connection, while max. comp. σ_z develops at the rib web to stiffened cross-beam connection. This rib is located next to main girder. For t = 14 and 16 mm max. tens. σ_z arises at the rib web adjacent to main girder and longitudinally at the normal cross-beam location. Max. comp. σ_z arises at the rib web adjacent to main girder, and longitudinally at the stiffened cross-beam location.

As seen in Figure 52 and in Figure 53 max. transversal and longitudinal stresses in deck plate increases with the increase of rib spacing and decrease of deck plate thickness.

Variation of deck plate thickness results in slight changes in max. transversal in-plane stresses developed in cross-beam and is inversely proportional with transversal in-plane stresses as given in Figure 55. However increase of rib spacing causes rapidly higher transversal and vertical in-plane stresses in cross beam (see Figure 54).

All stress components developed in ribs are inverse linear proportional to deck plate thickness as given in Figure 57. According to Figure 56, transversal stresses in ribs almost do not change between rib spacings of 150 and 225 mm, however they increase rapidly proportional to rib spacing between rib spacings of 225 and 450 mm. Max. vertical stresses in ribs increase very slightly from rib spacings of 150–225 mm and from rib spacings of 375–450 mm. However vertical stresses in ribs increase enormously from rib spacings of 225 mm to rib spacings of 375 mm. Longitudinal stresses developed in ribs increase also proportional to rib spacing as seen in Figure 56.

There is an exception, max. longitudinal tension stress decreases from rib spacings of 375 mm to rib spacing of 450 mm as seen in Figure 56, since the point and phenomenon of stress behaviors are different at rib spacing of 375 and 450 mm. The cause of this exception is explained in Figure 58. While max. tension and compression longitudinal stresses rise in the rib next to main girder at rib spacing of 375 mm, they develop at the second rib away from the main girder at rib spacing of 450 mm. The main reason of this issue is actually increasing the width of deck plate in transversal direction proportional to rib spacing, as the axle distance between wheel loads as to bridge' s midpoint (symmetry center) remains same (see Figure 59).

6. Conclusions

In this study a FE-model established by researcher is used to evaluate the stresses developing in orthotropic decks. It is investigated the influence of dimensional parametric ratios on the deformations and stresses developed in orthotropic deck. These dimensional parametric ratios are the rib width to height ratio and the rib spacing to deck plate thickness ratio under used wheel loads employed in this study. The results summarized below are the findings of this study,

- Numerical results obtained in this study support recommendations of Eurocode 3 Part 2 (2006) regarding rib spacing to deck plate thickness ratio as $e/t \leq 25$ and $e \leq 300$ mm under wheel loads used in this study. Since same results with Eurocode 3 Part 2 (2006) are obtained in this study, numerical model used is considered satisfactory.
- Another result obtained in this study is deck plate thickness less than 10 mm leads to localized high stress concentrations direct under wheel loads and should not be allowed to be used for designing of orthotropic decks.
- It is concluded from the overall assessment of numerical results that height of trapezoidal rib should be equal to its width with respect to used wheel loads, deck dimensions used in this study, deformations and stresses developed in orthotropic decks. Although it would be appropriate to compare the numerical results with experimental test, author considers the affinity of results between this study and Eurocode 3 Part 2 (2006) can be seen as benchmark of his FE-model. However, a new study of the author, which will take one or two years including comparison of numerical results with experimental tests and which is funded by the government is now in progress.
- To avoid high to moderate stresses or stress concentrations in cross beam or in ribs, dimensional parametric ratios should be as $e/t \leq 225/12 = 18.75$ and $a/e \geq 300/225 = 1.33$ under wheel loads used in this study.
- To allow moderate stress values in orthotropic deck structure dimensional parametric ratios should be as $25 > e/t \geq 18.75$ and $1.33 > a/e \geq 1$.
- If $e/t >> 25$ and $a/e << 1$, stress concentrations at the intersection between rib and cross-beam webs together with yielding of material can occur under wheel loads used in this study.
- Of course, dimensions of orthotropic deck should be determined as per fatigue strength, however instead of repeating fatigue calculations after every FE-analysis, it is preferred in this study using von Mises stress, which is a function of stress phenomena at every structural point and a scalar value. As a result, von Mises stress is used for comparison of results of FE-analyses with each other. Consequently, dimensional parametric ratios investigated in this research are of great importance to define the places and values of max. stresses and so control the stresses developing in orthotropic deck. Author will focus on the effect of different loading schemes such as quasi- static and dynamic loads in his future studies.

Funding
The author received no direct funding for this research.

Author details
Abdullah Fettahoglu[1]
E-mail: abdullahfettahoglu@gmail.com
[1] Department of Civil Engineering, Bursa Orhangazi University, Yildirim 16310, Turkey.

References

American Institute of Steel Construction. (1963). *Design manual for orthotropic steel plate deck bridges*. Chicago, IL: Author.

ANSYS. (2010). *ANSYS: User manuals*.

Choi, D., Kim, Y., Yoo, H., & Seo, J. (2008, August). Orthotropic steel deck bridges in Korea. In *Proceedings of 2008 Orthotropic Bridge Conference, ASCE*. Sacramento, CA.

Connor, R., et al. (2012). *Manual for design, construction, and maintenance of orthotropic steel deck bridges* (Publication No. FHWA-IF-12-027). Washington, DC: US Department of Transportation Federal Highway Administration.

Eurocode 3 Part 1-1. (2001). *Design of steel structures-general structural rules*. Brussel: European Committee for Standardization.

Eurocode 3 Part 1-9. (2003). *Design of steel structures-fatigue*. Brussel: European Committee for Standardization.

Eurocode 3 Part 2. (2006). *Design of steel structures-steel bridges*. Brussel: European Committee for Standardization.

Fettahoglu, A. (2012, October). Effect of deck plate thickness on the structural behaviour of steel orthotropic highway bridges. *Advanced in Civil Engineering*, Ankara.

Fettahoglu, A. (2013a). Arranging thicknesses and spans of orthotropic deck for desired fatigue life and design category. *International Journal of Advances in Engineering & Technology, 6*, 1512–1523.

Fettahoglu, A. (2013b). A FEA study conforming recommendations of DIN FB 103 regarding rib dimensions and cross-beam span. *International Journal of Civil Engineering Research, 4*, 197–204.

Fettahoglu, A. (2013c). Assessment on web slope of trapezoidal rib in orthotropic decks using FEM. *Sigma, Journal of Engineering and Natural Sciences*, Accepted.

Fettahoglu, A., & Bekiroglu, S. (2012, October). Effect of kinematic hardening in stress calculations. *Advanced in Civil Engineering*, Ankara.

Fujino, Y., & Yoshida, Y. (2002). Wind-induced vibration and control of trans-tokyo bay crossing bridge. *Journal of Structural Engineering, 128*, 1012–1025. http://dx.doi.org/10.1061/(ASCE)0733-9445(2002)128:8(1012)

Honshu Shikoku Bridge Authority. (2005). Retrieved 17 January, 2005, from www.hsba.go.jp

Hoopah, W. (2004, August). Orthotropic decks for small and medium span bridges in France– evolution and recent trends. In *Proceedings of 2004 Orthotropic Bridge Conference, ASCE*. Sacramento, CA.

Huang, C., & Mangus, A. (2008, August). Redecking existing bridges with orthotropic steel deck panels. In *Proceedings of 2008 Orthotropic Bridge Conference, ASCE*. Sacramento, CA.

Huurman, et al. (2002, April). 3D-FEM for the estimation of the behaviour of asphaltic surfacings on orthotropic steel deck bridges. In *3rd International Symposium on 3D Finite Element for Pavement Analysis, Design & Research*, Amsterdam.

Jong, F. B. P. de (2007). Renovation techniques for fatigue cracked orthotropic steel bridge decks (Ph.D. Dissertation). Delft: Delft University of Technology.

Kennedy, D. J. L., Dorton R. A., & Alexander S. D. B. (2002, June). The sandwich plate system for bridge decks. In *International Bridge Conference*, Pittsburgh.

Korniyiv, M. (2004, August). Orthotropic deck bridges in Ukraine. In *Proceedings of 2004 Orthotropic Bridge Conference, ASCE*. Sacramento, CA.

Troitsky, M. S. (1987). *Orthotropic bridges theory and design* (2nd ed.). Cleveland, OH: The James F. Lincoln Arc Welding Foundation.

Virlogeux, M. (2004, June). The viaduct over the River Tarn. In *Conference Proceedings Steelbridge, OTUA*. Paris.

The influence of preliminary aerobic treatment on the efficacy of waste stabilisation under leachate recirculation conditions

Monika Suchowska-Kisielewicz[1]*, Andrzej Jedrczak[1] and Zofia Sadecka[1]

*Corresponding author: Monika Suchowska-Kisielewicz, Institute of Environmental Engineering, University of Zielona Gora, Licealna 9, 65-417 Zielona Gora, Poland

E-mail: m.suchowska-kisielewicz@iis. uz.zgora.pl

Reviewing editor: Roberto Revelli, Politecnico Di Torino, Italy

Abstract: This article presents the changes in the chemical composition of leachate and the concentrations and quantity of methane production in each individual decomposition phases, determined for untreated and after aerobic treatment of waste stabilised in anaerobic reactors with and without leachate recirculation.The research results demonstrate that leachate recirculation intensifies the decomposition of both aerobically treated and untreated waste. The methane production in the reactor with untreated, stabilised waste with recirculation was 28% higher; and in the reactor with aerobically treated waste, the methane production was 24% higher than in the reactors without recirculation. An important finding of the study is that aerobic treatment of waste prior to landfilling effectively reduces the quantity of pollutant emissions in leachate and biogas from waste and increases the availability for methane micro-organisms of organic substrates from difficult-to-decompose organic substances.

Subjects: Bioscience, Built Environment, Development Studies, Environment, Social Work, Urban Studies, Environmental Studies & Management, Science

Keywords: waste, aerobic decomposition, leachate recirculation, mechanical–biological waste treatment, methane

ABOUT THE AUTHORS

The major subject of our research is the area associated with the treatment and disposal of waste. This research can be used in industry. Currently, the waste is treated as a material which must be subjected to various technological actions tending to recovery (material and/or energy) and recycled, so that the amount of discharged waste in a landfill was minimised.

So far I carried out research related to the intensification of the biological degradation of waste under aerobic or anaerobic conditions. The main objective of my research was to determine the effects of different techniques of waste pretreatment on the efficiency of the biological treatment of waste and the production of methane. Currently, we are going to start research on energy production from waste using hydrogen–methane fermentation.

PUBLIC INTEREST STATEMENT

The article describes issues associated with the reduction of emissions from landfills. During the anaerobic degradation of municipal solid waste, most organic components are broken down by micro-organisms into simple ingredients, which are substrates for the production of landfill gas (mainly CO_2 and CH_4). In the EU-15, the contribution of landfill gas emissions to the whole anthropogenic greenhouse gas production is about 3%. Apart from gas emission, leachates—which can pollute ground water and soil—are also generated in landfills. These hazards can be reduced by using appropriate waste treatment techniques before landfilling and performing appropriate utilisation treatment to intensify the waste decomposition processes. An important finding of the study is that aerobic treatment of waste prior to landfilling re-effectively reduces the quantity of pollutant emissions in leachate and biogas from waste and increases the availability for methane micro-organisms of organic substrates from difficult-to-decompose organic substances.

1. Introduction

The most important task of waste management systems is the application of effective and economically justified technologies of waste disposal and treatment (low operating costs and low energy consumption) (Capela, Azeiteiro, Arroja, & Duarte, 1999). Landfilling continues to be the major method of municipal solid waste (MSW) disposal in the Poland and many other countries despite considerable efforts to limit its use (Siddiqui, Richards, & Powrie, 2012).The reason for the popularity of waste disposal by burial in landfills is the low cost of design and construction of landfills. This method is particularly attractive for developing countries. However, as shown Laner, Crest, Scharff, Morris, and Barlaz (2012), even some highly industrialised countries such as the US, Australia, the UK and Finland largely dispose off their waste in landfills. For example, the fraction of MSW landfilled in 2008 was 54% in the USA, 55% in the UK and 51% in Finland. In contrast, landfilling accounted for less than 5% of MSW management in 2008 in Germany, the Netherlands, Sweden, Denmark and Austria (Laner et al., 2012).

In traditional landfills, a large portion of the total costs of waste management, apart from the cost of investment, are those related to the monitoring of the landfills. This is because of the slow processes of decomposition in traditional landfills, which generate long-term pollution emissions that can even last for several decades.

During the anaerobic degradation of MSW, most organic components is broken down by microorganisms into simple ingredients, which are substrates for the production of landfill gas (mainly CO_2 and CH_4) (Zhang, Yue, Liu, He, & Nie, 2012). In the EU-15, the contribution of landfill gas emissions to the whole anthropogenic greenhouse gas production is about 3% (European Environment Agency [EEA], 2011). Apart from gas emission, leachates—which can pollute ground water and soil—are also generated in landfills.

These hazards can be reduced by using appropriate waste treatment techniques before landfilling and performing appropriate utilisation treatment to intensify the waste decomposition processes.

The content of organic substances in waste has a decisive influence on the amount and duration of pollutant emissions from landfills. The European Union Council Directive 1999/31/EC (1999) imposed on Poland, and other European countries, the obligation to reduce the amount of easily biodegradable organic waste deposited at landfills. Based on the amount of MSW generated in 1995, the Directive imposes a mandatory stepwise reduction of 25, 50 and 65%, respectively, by 2006, 2009 and 2016 (Di Maria & Sordi, 2013).

To achieve these goals many different strategies can be pursued (Di Maria & Sordi, 2013).

A method enabling a reduction in the content of organic substances in waste is mechanical-biological pretreatment of MSW which is being used increasingly in Europe.

The processes of mechanical-biological treatment (MBT) consist of a mechanical waste sorting system combined with an installation for the biological stabilisation of the biodegradable fraction of waste. The mechanical waste treatment stage can be placed either at the beginning of the MBT process (biostabilisation), or after the biological process (biodrying).

In biostabilisation technologies, in order to reduce the amount of biodegradable waste, aerobic or anaerobic decomposition is applied. The stabilised waste obtained in the MBT process is mainly disposed in the landfill, but it can also be used as fertiliser for non-agricultural land and for the reclamation of land for construction, or it can be burnt in municipal waste incineration plants (Białowiec, Bernat, Wojnowska-Baryła, & Agopsowicz, 2008; Erses, Onay, & Yenigun, 2008; Griffith & Trois, 2006).

The main purpose of this biostabilisation technology is to achieve the highest possible degree of stabilisation of organic waste, so that pollutant emissions from stabilised waste deposited at landfills are as low as possible. Additionally, MBT systems allow for increased reclamation of materials for recycling and a reduction in the quantity of deposited waste (Jędrczak, 2007; Lornage, Redon, Lagier, Hebe, & Carre, 2007; Robinson, Knox, & Bone, 2004).

In developing countries, the recommended method of biological waste stabilisation before landfilling is aerobic decomposition, which requires less investment and operating expenditures in comparison to anaerobic processes. An additional benefit resulting from using aerobic biostabilisation is the reduction in the time of pollutant emissions, due to the faster development of stable methane conditions in landfills with stabilised waste (Lornage et al. 2007; Rich, Gronow, & Voulvoulis, 2008).

Literature data indicates that under optimal aerobic stabilisation conditions, a reduction of up to 90% of the content of organic substances can be achieved, which corresponds to a reduction in landfill gas emission of up to as much as 15–20 m^3/Mg of waste and nitrogen in the range of 80–90% in relation to emissions from untreated waste (Robinson, Knox, Bone, & Picken, 2005).The typical production of biogas from untreated waste is 165 m^3/Mg of waste.

However, in stabilised waste obtained in this manner, there still remains the organic fraction (lignin, waxes and humic acids), difficult to decompose (Höring, Kruempelbeck, & Ehrig, 1999). Degradation of this waste in the landfill can be accelerated by using appropriate techniques which intensify decomposition. One of the most popular intensification methods used at landfills is leachate recirculation.

The transformations which take place in landfills with recirculation are similar to the anaerobic processes occurring in traditional landfills. In stabilised waste landfills with recirculation, the successive phases of stabilisation are conducted more intensively and in a more controlled manner than in landfills without recirculation (Öztürk et al. 1997). Recirculation increases the moisture content of waste, lowers the concentrations of potential methanogenesis inhibitors (Morris, Vasuki, Baker, & Pendleton, 2003) and supplies the nutrients and enzymes necessary for microbial growth (Sponza & Ağdağ, 2004). The result is acceleration in the decomposition of the organic content of waste, increase in methane production (Sanphoti, Towprayoon, Chaiprasert, & Nopharatana, 2006) and the removal of some of the pollutants present in the recirculated leachate.

The biological processes occur at the landfill in stages. Each of the phases has its environment and substrate requirements and ends with specific end products. In describing the progression of waste decomposition at landfills, the most often used division is one consisting of four phases:

- Phase I (hydrolysis) is characterised by high concentrations of organic substances in the leachate, aerobic conditions in the landfill and a high content of easily biodegradable organic substances in the waste. The end products of hydrolysis are monosaccharides, amino acids, long-chain organic acids and glycerol, which are the substrates for the acidic phase;
- Phase II (acidic) is characterised by still high concentrations of organic substances in the leachate, intensive production of short-chain organic acids, a pH drop (5.5–6.5) and methane production in the biogas at a low, practically undetectable level;
- Phase III (unstable methane) is characterised by a pH increase, a decrease in the redox potential to negative values, a significant drop in the concentrations of volatile fatty acids and organic substances in the leachate, a reduction of sulphates to sulphites and intensive methane production;
- Phase IV (stable methane) is characterised by relatively stable, low concentrations of organic substances in the leachate, an increase in the redox potential, decreased biogas production and the methane content of biogas remains at a relatively constant, high level of about 60–70%.

Temporal differentiation of the hydrolysis and acidic phases is difficult, due to the similar characteristics of the chemical composition of leachates. Hence, usually these phases are not differentiated with respect to time, treating them as one decomposition period.

This article presents the changes in the chemical composition of leachate and the concentrations and quantity of methane production in each decomposition phase, calculated for untreated waste and after aerobic treatment of stabilised waste in anaerobic reactors in leachate recirculation conditions.

2. Testing methodology

2.1. Test material

In the tests, MSW from Zielona Góra, from high-rise buildings with central heating was used: biologically untreated and treated (Table 1).

Biological waste treatment was conducted using the works of the Zielona Góra Municipal Waste Composting Plant. The process line of the Composting Plant's system consists of four aerobic, open reinforced concrete chambers, among which waste is transferred every 7–10 days. The total waste stabilisation time is about five weeks. Waste is aerated by sucking out the gasses from the bottoms of the ferroconcrete chambers.

Samples for the determination of the composition of the examined MSW were taken from randomly selected vehicles delivering waste to the composting plant's bunker. Samples of biodegradable solid waste (BSW) were taken at random from chosen batches of composted waste.

Characterisation of the morphological composition involved the screening of the waste through a 10 mm mesh sieve and weighing of the fractions obtained: 0–10 mm and >10 mm according to Polish Standards PN-93/Z-15006.

Table 1. Properties and morphological composition of MSW and BSW

Specification	Waste type	
	MSW	BSW
Waste properties		
Moisture, %	40.2 (22)*	34.5 (11)*
Loss on ignition, % of dry matter	58.5 (15)*	52.2 (12)*
Total organic carbon, kg/kg of dry matter	0.38 (19)*	0.27 (10)*
Biodegradable carbon, kg/kg of dry matter	0.11 (12)*	0.046 (14)*
Morphological composition, %		
Kitchen and garden waste + 10–20 mm fraction	42.5 (16)*	32.6 (8)*
Paper and cardboard	17.5 (11)*	14.9 (6)*
Glass	9.8 (24)*	12.5 (21)*
Plastics	13.5 (18)*	17.7 (16)*
Textiles	2.5 (16)*	2.9 (11)*
Multi-material packaging	2.5 (9)*	3.6 (12)*
Wood	0.1 (14)*	0.1 (10)*
Metals	1.6 (10)*	1.9 (6)*
Mineral waste, including <10 mm fraction	10.0 (19)*	13.8 (10)*
Total	100	100

*Coefficients of variation in percentage.

The group composition of the waste was determined in the >10 mm fraction by separating the following morphological constituents: kitchen and garden waste, paper and cardboard, glass, plastics, textiles, multi-material packaging, wood, metals and mineral waste. The group composition of the fraction was expressed as a percentage of a constituent in the general mass of the waste, in % (v/v).

Chemical analysis was conducted on samples of waste without large-size constituents and metals. The scope of the analysis included the following parameters: moisture, volatile substances and organic carbon (Table 1). Moisture content was determined by heating the ground sample at 105°C for 24 h according to PN-Z-15008-02:1993. Volatile solids (VS) content, assimilated to the ignition loss at 550°C by PN-EN 15169:2007.

Organic carbon in the samples was determined after the determination of moisture and crushed to grains <1 mm. The content of carbon determined using gas chromatograph GC17A Shimadzu, according to PN-Z-15011-3:2001. Biodegradable organic carbon was determined according to the methodology recommended in the Guidance on monitoring MBT (2005) and other pretreatment processes for the purposes of the landfill allowances schemes (Environment Agency, 2005).

2.2. Testing stations

The laboratory tests were conducted in four reactors made from PVC pipes 0.15 m in diameter and 1.30 m in height. In the bottom of each reactor, a stub (a pipe with a valve) was installed for draining the leachate, and in the lid a stub for discharging biogas, and a stub for dosing water, in order to simulate precipitation. The collected leachate was stored in a tank with a capacity of 20 L. The gas stub in the reactor's lid was connected by 10 mm flexible pipelines to gas burettes (cylinders with an internal diameter of 85 mm with graduation). The other stub in the reactor's lid was connected to a tank with water simulating precipitation.

Before filling the reactors with waste, a 0.15 m layer was placed on their bottoms and thermocouples were installed. Into each reactor was poured 10 kg of waste crushed to grain size <40 mm and thoroughly mixed. After filling the reactors with waste and closing them hermetically, dosage of water began for each of them, at the amount of 1 L/day, in order to saturate the waste with water.

After the first quantities of leachate have appeared (after 7 days), the daily dosage of tap water began for each of the reactors, in the amount simulating real precipitate, determined based on monthly reports of the Zielona Góra Institute of Meteorology.

To two reactors were connected pipelines for leachate recirculation: one filled with MSW (SR reactor) and one filled with waste after biological treatment (PR reactor). The remaining two were control reactors (S, P). Leachate was recirculated once a week in the amount of 1 L. The temperature in the room with the reactors ranged between 20 and 25°C. The quantities of leachate and biogas produced were registered daily. Analyses of the chemical composition of biogas and leachate were conducted once a week.

2.3. Process control

The process control involved the testing of the quantity and quality of leachate and biogas. Quantitative tests were conducted every day and qualitative tests once a week. In the leachate, the concentrations of chemical oxygen demand (COD), total organic carbon (TOC), biological oxygen demand (BOD_5), volatile fatty acid (VFA), nitrogen, chlorides and sulphates, and their pH were measured. Collection and storage of samples and chemical composition of leachate were performed in accordance with applicable Polish Standards PN-EN ISO 5667-3: 2002. BOD, COD, nitrogen, chlorides and sulphates were measured according to standard methods APHA (1995). pH and total alkalinity determined potentiometrically according to PN-90/C-04540/02; volatile fatty acids were determined by direct distillation according to PN-75/C-04616; TOC was determined using gas chromatograph GC17A Shimadzu, according to PN-EN 1484. The biogas was tested for its methane and carbon dioxide content using analyser GA 2000 PLUS, Geotechnical Instruments.

3. Test results

3.1. The properties and morphological composition of the waste

The properties and morphological composition of the MSW samples and stabilised waste are presented in Table 1.

MSW was characterised by a greater content of moisture and organic carbon than stabilised waste. The TOC content in aerobically treated waste was lower by 29% and biodegradable carbon by 59%. It resulted from the fact that TOC content does not define susceptibility to biological decomposition of waste since it takes into account not only the biodegradable organic matter but also the plastics, the lignin and the humic substances. The dominant constituents in waste which were not treated biologically were kitchen and garden remains (42.5%) and paper and cardboard (17.5%). The percentages of these portions in the MSW, compared to those in the BSW, were higher by 23.3 and 14.8%, respectively. The percentages of the other constituents were lower than in the BSW by: glass—21.6%, plastics—23.7%, textiles—13.8%, multi-material packaging—30.5%, metals—15.7% and fine fractions below 10 mm—27.5%. The coefficients of variation parameter determined for untreated and treated waste varied in the range from 20 to 40%, which indicates the average diversity of results.

3.2. The amount and chemical composition of biogas and leachate

Figure 1 shows summation curves of the volumes of leachate produced in the reactors and added water and water with recirculate.

The total volume of leachate from reactor S was similar to the volume of dosed water simulating precipitation (greater only by 3.6%), and from reactor P lower only by 11%. The highest leachate volume was registered in the reactors with recirculation. The total net leachate volume (after deducting the volume of recirculated leachate) from reactor SR was 14.0% greater than the volume from the S reactor, and the volume of leachate from reactor PR was 16.6% greater than the volume from the P reactor.

Figures 2–5 show the changes in the chemical composition of leachate and the concentrations and production of methane in the reactors S, SR, P, PR, in the following phases; I—hydrolysis and acidic, II—unstable methane and III—stable methane. The duration of each phase was determined based on the changes in the chemical composition of leachate and the methane concentration in the biogas.

The durations of each of the decomposition phases determined in the tests indicate a beneficial influence of recirculation which, in stabilised waste reactors with recirculation, accelerated the establishment of the phases: unstable and stable methanogenic.

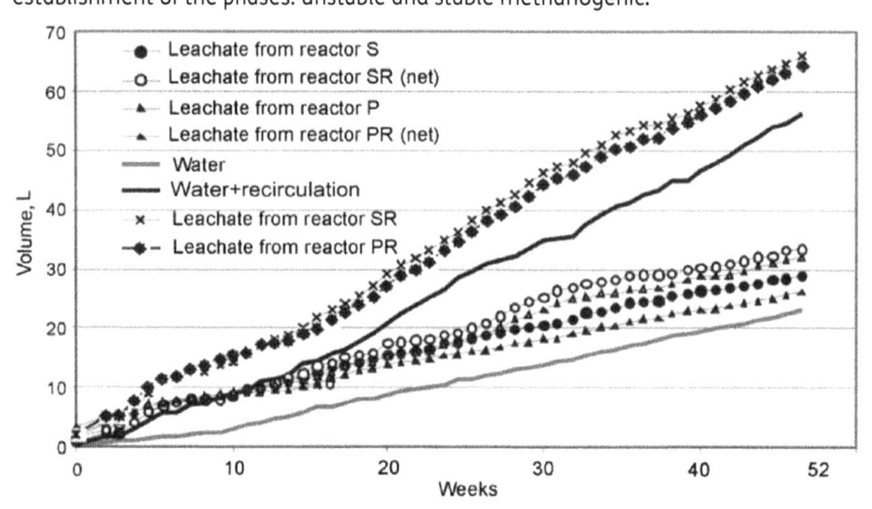

Figure 1. Summary of volumes of water simulating precipitation, recirculate and leachate produced in the reactors S, SR, P and PR.

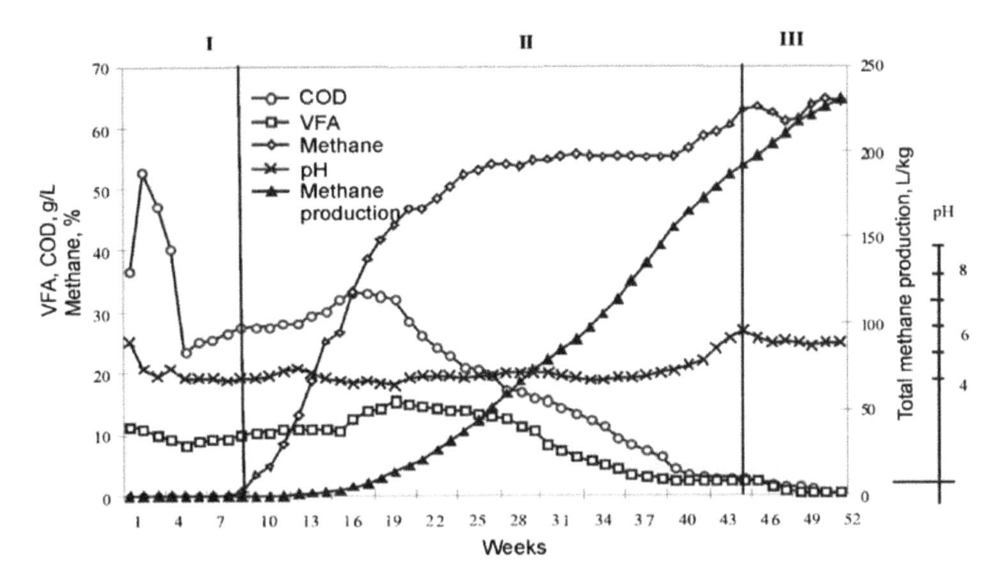

Figure 2. Changes in the chemical composition of leachate and the concentrations and production of methane in the S reactor.

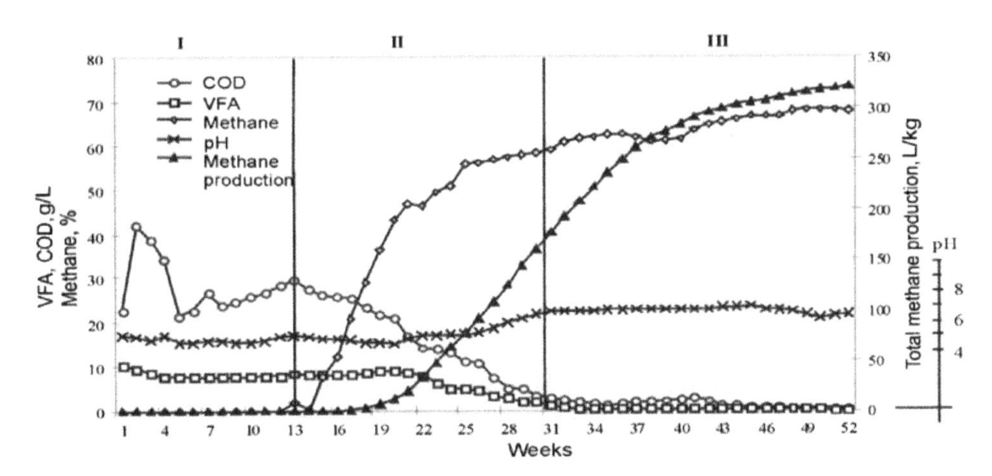

Figure 3. Changes in the chemical composition of leachate and the concentrations and production of methane in the SR reactor.

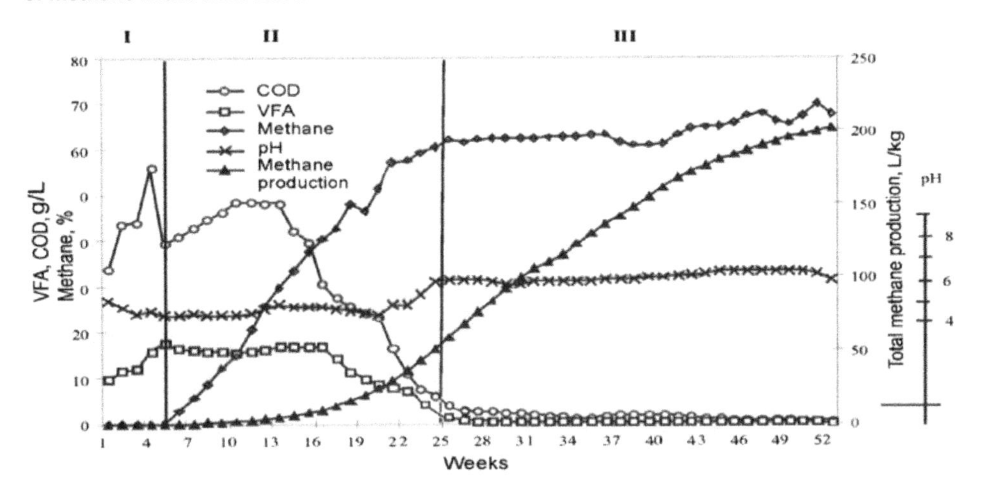

Figure 4. Changes in the chemical composition of leachate and the concentrations and production of methane in the P reactor.

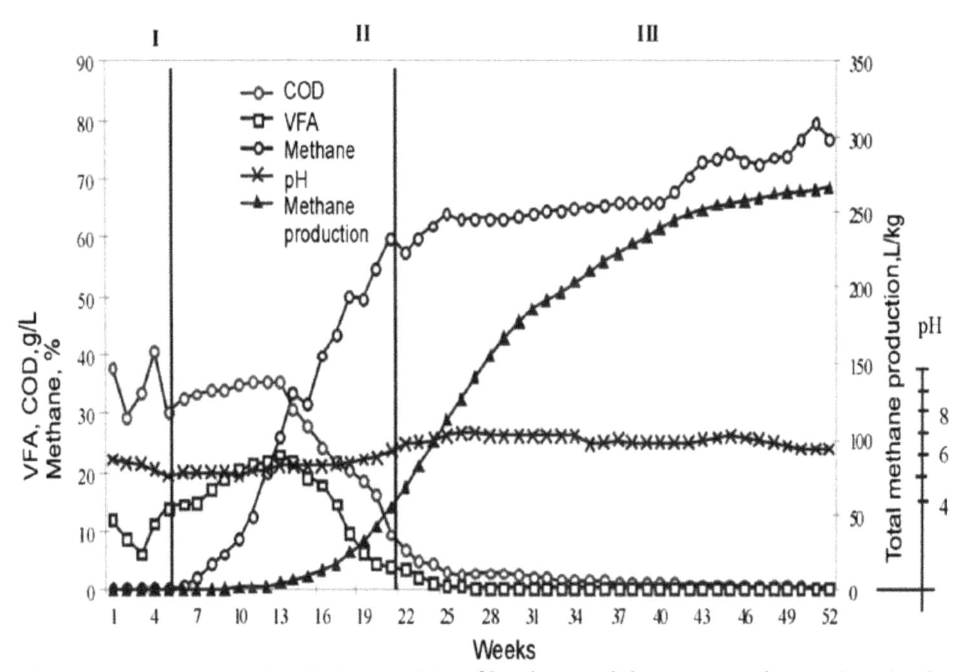

Figure 5. Changes in the chemical composition of leachate and the concentrations and production of methane in the PR reactor.

The waste decomposition progressed most efficiently in the PR reactor. A stable methane phase was recorded in this reactor in the 21st week of the process, whereas in reactor SR it was in the 31st week, in reactor S in the 44th week and in reactor P in the 24th week (Table 2).

The mean average COD concentration was determined for the entire study period, and in the SR and PR reactors, it was lower than the ones recorded in the S and P reactors by 34 and 27%, respectively, and the average methane production was greater by 2 and 12%. The mean COD concentrations determined in reactor P were lower than those obtained in reactor S by 16% and the methane production was lower by 8%.

Table 2. Mean COD, VFA concentrations and mean concentrations and methane production, and decomposition phase durations determined for the reactors S, SR, P, PR

Reactor	Decomposition phases	Phase durations	COD, g/L	VFA, g/L	CH_4, %	CH_4, L/kg
S	I	9	40.1	38.2	0.07	0.0
	II	35	18.8	9.1	40.3	65.9
	III	8	1.4	0.92	59.7	212.7
	I–III	**52**	**18.8**	**7.9**	**39.7**	**83.5**
SR	I	13	29.8	8.4	0.14	71.2
	II	19	14.5	5.6	17.7	78.1
	III	20	1.3	0.21	60.4	99.1
	I–III	**52**	**12.4**	**4.10**	**40.4**	**85.01**
P	I	6	43.3	13.4	0.4	78.1
	II	18	29.8	12.1	26.5	80.1
	III	28	1.5	0.3	62.3	99.5
	I–III	**52**	**15.8**	**5.6**	**46.9**	**90.7**
PR	I	6	34.2	10.4	0.11	81.5
	II	15	25.8	15.3	6.7	82.5
	III	31	1.8	0.42	61.1	99.4
	I–III	**52**	**11.8**	**5.3**	**48.8**	**92.9**

4. Discussion

The quantity of methane production and the emission of the pollutant content and its leaching rate are dependent on the waste biodegradation intensity.

In the tests, greater rates of pollutant leaching from aerobically treated waste than from untreated waste (Table 3) were observed. The beneficial effect of preliminary aerobic waste stabilisation before landfilling on the "fluidisation" (hydrolysis) under anaerobic conditions, of the remaining organic constituents present in the waste is a known phenomenon, described in the literature (Capela, 1999). The high dissolution level of organic substances in aerobically treated waste and the high vulnerability to methanation of the products of their hydrolysis influenced a faster establishment of a stable methanogenic phase in the landfill with this waste (Table 2).

The summary pollutant contents removed with leachate from a reactor containing aerobically treated waste stabilised with recirculation were lower than from a reactor without recirculation and the leaching rates were also lower. This was the result of the transformation of a substantial portion of dissolved organic substances into methane (the highest methane volumes were produced in the treated waste reactor with recirculation) (Table 2).

The amount of methane produced in reactor SR was 28% greater than in the S reactor, and, 24% greater in reactor PR than in the P reactor.

The methane production rate in reactors with recirculation was higher than in reactors without recirculation. The maximum daily methane production in the SR chamber was 0.433 dm^3/kg of dry matter and, in the PR chamber it was 0.339 L/kg of dry matter.

The test results confirmed the beneficial effect of the leachate recirculation on the improvement of the biogas and, as a result, on the profitability of its recovery and its utilisation in power

Table 3. The summary pollutant contents, their leaching rate as well as the total methane production and methane production rate in the reactors S, SR, P, PR

Reactor/ Decomposition phases		Summary decomposition contents, g/kg		Pollutant leaching rate, mg/(kg d)		Total methane production throughout 52 weeks, L	Maximum methane production rate, L/kg
		COD	TOC	COD	TOC		
S	I	46.9	14.1	146	43.6	230	0.176
	II	20.4	6.23	43.1	13.2		
	III	0.41	0.17	4.24	2.32		
	I + II + III	67.7	20.5	75.9	22.9		
SR	I	40.2	11.6	129	37.3	321	0.433
	II	10.3	3.48	43.6	14.7		
	III	1.30	0.57	3.79	1.66		
	I + II + III	51.8	15.7	58.1	17.6		
P	I	39.2	11.5	166	48.5	202	0.185
	II	13.0	4.78	63.0	23.2		
	III	1.51	0.50	3.37	1.12		
	I + II + III	53.7	16.8	60.3	18.8		
PR	I	34.9	10.2	147	43.2	267	0.339
	II	10.0	3.20	47.9	15.3		
	III	2.4	1.48	5.39	3.31		
	I + II + III	47.3	14.9	53.0	16.7		

generation (Warith, 2002). Both in the case of aerobically treated and untreated waste, biogas recovered from the reactors with recirculation contained 7% more methane in the stable methane digestion phase than from the reactors without recirculation.

Leachate recirculation reduced the time of intensive waste treatment. During the last week, the biogas production rate in reactor PR was nearly two times lower than in reactor SR and four times lower than in the S reactor.

A quicker determination of the methane conditions in the reactor with preliminarily aerobically treated waste was the result of the removal of biodegradable organic substances from the waste in the process of preliminary, aerobic treatment. On the other hand, the observed high methane production rate in the stable methane digestion phase in the P and PR reactors suggests partial decomposition of difficult-to-decompose organic substances remaining in the waste after preliminary aerobic treatment. The dominant organic constituents in the waste are: cellulose—50%, lignins—15%, hemicellulose—10%, proteins—5% and starch, pectins and other dissolved sugars (Barlaz, 1992). The main sources of carbon consumed by methane micro-organisms are cellulose and hemicellulose, which are classified as difficult-to-decompose materials in anaerobic conditions. Aerobic treatment of these compound organic substances before their anaerobic stabilisation leads to the decomposition of some of these organic compounds into forms which are more easily absorbed by methane organisms (Schön, 1994). Increased activity of methane bacteria can be recognised by a gradual drop in the concentration of leachate, with relatively high methane production (Figures 2–5) (Schroeder, Morgan, Wolski, & Gibson, 1984).

5. Conclusions

A large amount of biodegradable waste production will still be generated even in zero waste approach and strategies (Di Maria & Sordi, 2013). In order to reduce the environmental and social impact caused by a landfill, some management strategies can be adopted (Boni, Leoni, & Sbaffoni, 2007):

- waste mechanical-biological and/or thermal pretreatment, leading to a significant reduction in the COD, BOD_5 and ammonium nitrogen release in the leachate and to an acceleration of the landfill gas production;
- in situ aeration, providing a rapid and significant oxidation of the organic fraction as well as ammonium nitrogen consumption through nitrification;
- in situ water supply allowing a faster reduction of the leachate organic load; but also a higher leachate quantity to be treated is produced; and
- leachate recirculation.

MBT process is widely used in many European areas that lack incineration and co-combustion plants. This stems from the fact that MBT investment and treatment costs are significantly lower than incineration (Siddiqui et al., 2012).

The research presented in this paper confirm the literature data that when aerobically pretreated MSW is buried in a landfill, the leachate pollution load and the biogas production potential will be diminished and the stabilisation process will be enhanced (Lornage et al., 2007; Rich et al., 2008; Zhang et al., 2012). Furthermore shown a beneficial effect of leachate recirculation on the size of production of methane and pollutant load reduction in leachate. This has been confirmed by many authors Morris et al. (2003), Öztürk et al. (1997), Sanphoti et al. (2006), Sponza and Ağdağ (2004).

The obtained results allowed to formulate the following general conclusions:

(1) Aerobic treatment of waste prior to landfilling effectively reduces the quantity of pollutant emissions in leachate and biogas from waste and increases the availability for methane micro-organisms of organic substrates from difficult-to-decompose organic substances.

Aerobic stabilisation of municipal waste accelerates the appearance of methane fermentation phase. Methane in the reactor with aerobically treated waste appeared 72 days earlier than from the reactor with untreated waste and the methane content in the reactor with treated waste in stable phase methane fermentation was 10% higher than in the reactor with untreated waste.

(2) Leachate recirculation intensifies the decomposition of both aerobically treated and untreated waste. The methane production in the reactor with untreated, stabilised waste with recirculation was 28% higher; and in the reactor with aerobically treated waste, the methane production was 24% higher than in the reactors without recirculation.

(3) Leachate recirculation reduced the time of intensive waste treatment. During the last week of study, the biogas production rate in the reactor with aerobically treated waste, stabilised with recirculation was nearly two times higher than in the reactor with untreated waste, stabilised with recirculation and four times lower than in the reactor with untreated waste, stabilised without recirculation.

Currently, there is little information on the long-term leachate quality and gas generating potential on landfills of treated wastes. This information is important because it is helpful for determining the way and the extent of pretreatment from the perspective of contaminant control. Also, there is little data on the impact of leachate recirculation on the decomposition remaining in the treated waste hardly degradable organic compounds.

Taking into account that leachate recirculation in landfills is a growing practice around the world due to high economically beneficial for leachate treatment (Clément, Oxarango, & Descloitres, 2011) studies to optimise this process are still very important.

Funding
The authors received no direct funding for this research.

Author details
Monika Suchowska-Kisielewicz[1]
E-mail: m.suchowska-kisielewicz@iis.uz.zgora.pl
Andrzej Jedrczak[1]
E-mail: a.jedrczak@iis.uz.zgora.pl
Zofia Sadecka[1]
E-mail: z.sadecka@iis.uz.zgora.pl
[1] Institute of Environmental Engineering, University of Zielona Gora, Licealna 9, 65-417 Zielona Gora, Poland.

References

APHA. (1995). *Standard methods for the examination of water and wastewater* (19th ed.). Washington DC: American Water Works Association.

Barlaz, M. A. (1992). Microbial, chemical and methane production characteristics of anaerobically decomposed refuse with and without leachate recycling. *Waste Management and Research, 10*, 257–267. doi:10.1016/0734-242X(92)90103-R

Białowiec, A., Bernat, K., Wojnowska-Baryła, I., & Agopsowicz, M. (2008). The effect of mechanical pretreatment of municipal solid waste on its potential in gas production. *Archive of Environmental Protection, 34*, 115–124.

Boni, M. R., Leoni, S., & Sbaffoni, S. (2007). Co-landfilling of pretreated waste: Disposal and management strategies at lab-scale. *Journal of Hazardous Materials, 147*, 37–47. doi:10.1016/j.jhazmat.2006.12.049

Capela, I. F., Azeiteiro, C., Arroja, L., & Duarte, A. C. (1999). Effects of pre-treatment (composting) on the anaerobic digestion of primary sludges from a bleached kraft pulp mill. In *II international symposium on anaerobic digestion of solid waste*. Barcelona.

Clément, R., Oxarango, L., & Descloitres, M. (2011). Contribution of 3-D time-lapse ERT to the study of leachate recirculation in a landfill. *Waste Management, 31*, 457–467. doi:10.1016/j.wasman.2010.09.005

Council Directive, 1999/31/EC of 26 April 1999 on the landfill of waste. (1999).

Di Maria, F., Sordi, A. (2013). Caterina Micale experimental and life cycle assessment analysis of gas emission from mechanically-biologically pretreated waste in a landfill with energy recovery. *Waste Management, 33*, 2557–2567. doi:10.1016/j.wasman.2013.07.011

Environment Agency. (2005, August). *Guidance on monitoring MBT and other pre-treatment processes for the landfill allowances schemes (England and Wales)*. Retrieved from www.environment-agency.gov.uk/commondata/acrobat/mbt_1154981.pdf

Erses, A., Onay, T., & Yenigun, O. (2008). Comparison of aerobic and anaerobic degradation of municipal solid waste in bioreactor landfill. *Bioresource Technology, 99*, 5418–5426. doi:10.1016/j.biortech.2007.11.008

European Environment Agency. (2011). *Greenhouse gas emission trends and projections in Europe 2011. Tracking progress towards Kyoto and 2020 targets* (EEA Report No. 4/2011). Copenhagen: EEA. ISSN 1725-9177 EEA.

Griffith, M., & Trois, C. (2006). Long-term emissions from mechanically biologically treated waste: Influence on leachate quality. *African Journals Online, 32*, 307–314. Retrieved from http://dx.doi.org/10.4314/wsa.v32i3.5275

Höring, K., Kruempelbeck, I., & Ehrig, H. J. (1999). Long term emission behavior of mechanical-biological

pre-treated municipal solid waste. In 7 *international waste management and landfill symposium.* Cagliari.
Jędrczak, A. (2007). *Biologiczne przetwarzanie odpadów [Biological treatment of waste].* Warsaw: PWN.
Laner, D., Crest, M., Scharff, H., Morris, J. W. M., & Barlaz, M. A. (2012). A review of approaches for the long-term management of municipal solid waste landfills. *Waste Management, 32*, 498–512.
Lornage, R., Redon, E., Lagier, T., Hebe, I., & Carre, J. (2007). Performance of low cost MBT prior to landfilling: Study of the biological treatment of size reduced MSW without mechanical sorting. *Waste Management, 27*, 1755–1764. doi:10.1016/j.wasman.2006.10.018
Morris, J. W. F., Vasuki, N. C., Baker, J. A., & Pendleton, C. H. (2003). Findings from long-term monitoring studies at MSW facilities with leachate recirculation. *Waste Management, 23*, 653–666. doi:10.1016/S0956-053X(03)-00098-9
Öztürk, I., Arikan, O., Demir, I., Demir, A., Inane, B., Kanat, G., & Yilmaz, S. (1997). *Solid waste characterization in Istanbul* (Report). Istanbul: Department of Environmental Engineering, Istanbul Technical University.
Rich, C. H., Gronow, J., & Voulvoulis, N. (2008). The potential for aeration of MSW landfills to accelerate completion. *Waste Management, 28*, 1039–1048. doi:10.1016/j.wasman.2007.03.022
Robinson, H. D., Knox, K., & Bone, B. D. (2004). *Improved definition of leachate source term from landfills Phase 1: review of data from European landfills.* Science Report Environment Agency. Retrieved from http//www.environment-agency.gov.uk
Robinson, H. D., Knox, K., Bone, B. D., & Picken, A. (2005). Leachate quality from landfilled MBT waste. *Waste Management, 25*, 383–391. doi:10.1016/j.wasman.2005.02.003
Sanphoti, N., Towprayoon, S., Chaiprasert, P., & Nopharatana, A. (2006). The effects of leachate recirculation with supplemental water addition on methane production and waste decomposition in a simulated landfill. *Journal of Environmental Management, 81*, 27–35. doi:10.1016/j.jenvman.2005.10.015
Schön, M. (1994). *Verfahren zur Vergärung organischer Rückstände in der Abfallwirtschaft* [Process for fermentation of organic residues in waste management]. Berlin: Verlag E. Schmidt GmbH.
Schroeder, P. R., Morgan, J. M., Wolski, T. M., & Gibson, A. C. (1984). *The hydrologic evaluation of landfill performance (HELP) model: Vol. I. User's guide for version 1* (Technical Resource Document EPA/530-SW-84-009). Cincinnati, OH: US Environmental Protection Agency.
Siddiqui, A. A., Richards, D. J., & Powrie, W. (2012). Investigations into the landfill behaviour of pretreated wastes. *Waste Management, 32*, 1420–1426. doi:10.1016/j.wasman.2012.03.016
Sponza, D. T., & Ağdağ, O. N. (2004). Impact of leachate recirculation and recirculation volume on stabilization of municipal solid wastes in simulated anaerobic bioreactors. *Process Biochemistry, 39*, 2157–2165. doi:10.1016/j.procbio.2003.11.012
Warith, M. (2002). Bioreactor landfills: Experimental and field results. *Waste Management, 22*, 7–17. doi:10.1016/S0956-053X(01)00014-9
Zhang, Y., Yue, D., Liu, J., He, L., & Nie, Y. (2012). Effect of organic compositions of aerobically pretreated municipal solid waste on non-methane organic compound emissions during anaerobic degradation. *Waste Management, 32*, 1116–1121. doi:10.1016/j.wasman.2012.01.005

10

DTALite: A queue-based mesoscopic traffic simulator for fast model evaluation and calibration

Xuesong Zhou[1*] and Jeffrey Taylor[2]

*Corresponding author: Xuesong Zhou, School of Sustainable Engineering and the Built Environment, Arizona State University, Tempe, AZ 85287, USA.

E-mail: xzhou74@asu.edu; xzhou99@ gmail.com

Reviewing editor: Filippo Pratico, University Mediterranea of Reggio Calabria, Italy

Abstract: A number of emerging dynamic traffic analysis applications, such as regional or statewide traffic assignment, require a theoretically rigorous and computationally efficient model to describe the propagation and dissipation of system congestion with bottleneck capacity constraints. An open-source light-weight dynamic traffic assignment (DTA) package, namely DTALite, has been developed to allow a rapid utilization of advanced dynamic traffic analysis capabilities. This paper describes its three major modeling components: (1) a light-weight dynamic network loading simulator that embeds Newell's simplified kinematic wave model; (2) a mesoscopic agent-based DTA procedure to incorporate driver's heterogeneity; and (3) an integrated traffic assignment and origin–destination demand calibration system that can iteratively adjust path flow volume and distribution to match the observed traffic counts. A number of real-world test cases are described to demonstrate the effectiveness and performance of the proposed models under different network and data availability conditions.

Subjects: Intelligent & Automated Transport System Technology, Systems Integration, Transport Engineering

Keywords: transportation network modeling, traffic simulation, traffic demand estimation, dynamic traffic assignment

ABOUT THE AUTHORS

Xuesong Zhou is an associate professor in the School of Sustainable Engineering and the Built Environment at Arizona State University. He has been the leading developer for the DTALite/NeXTA packages since he worked as an assistant professor at the University of Utah. His current research interests include analytical and computational modeling of transportation systems. He serves as a subcommittee chair for TRB Committee on Transportation Network Modeling (ADB30), Network Equilibrium Subcommittee. He is a member of IEEE and INFORMS.

Jeffrey Taylor received his BS in Civil and Environmental Engineering from the University of Utah in 2010 and he obtained his master's degree at the University of Utah in 2014. While most of his research work has been devoted to supporting and enhancing the DTALite/NeXTA packages, his research interests also extend to topics in transportation planning/modeling, transportation safety and traffic flow modeling. He holds memberships with IEEE, INFORMS, and ITE.

PUBLIC INTEREST STATEMENT

This paper describes the internal functions of DTALite, an open-source, light-weight dynamic traffic assignment (DTA) software package. DTALite can be used for large-scale transportation modeling applications to help planning/engineering organizations and public officials make transportation infrastructure investment decisions. Particularly important applications may include modeling the traffic impacts of work zones, proposed freeways/highways, and tolling facilities. DTALite's route choice model also allows organizations to test the effects of multiple strategies to improve traffic operations and manage travel demand. One of the highlights of this paper is DTALite's support for multi-threaded processing, which significantly reduces model run-times. This could allow users to evaluate more alternative strategies to solve a problem within a limited amount of time, providing more information to help decision-makers find better solutions to difficult problems.

1. Introduction

Motivated by a wide range of application needs, such as region-wide traffic analysis and route guidance provision, dynamic traffic assignment (DTA) models have been increasingly recognized as an important tool for assessing operational performance of those applications at multiple spatial resolutions (e.g. network, corridor, and individual segment levels). The advances of DTA in this aspect are built upon the capabilities of DTA models in describing the formation, propagation, and dissipation of traffic congestion in a transportation network.

DTALite, an open-source mesoscopic DTA simulation package, in conjunction with the Network eXplorer for Traffic Analysis (NeXTA) graphic user interface, has been developed to provide transportation planners, engineers, and researchers with a theoretically rigorous and computationally efficient traffic network modeling tool. This fully functional, open-source DTA model can be downloaded from http://code.google.com/p/nexta/. In general, the software suite of DTALite + NeXTA aims to:

(1) Provide an open-source code base to enable transportation researchers and software developers to expand its range of capabilities to various traffic management application

(2) Present results to other users by visualizing time-varying traffic flow dynamics and traveler route choice behavior in an integrated 2D/3D environment

(3) Provide a free, educational tool for students to understand the complex decision-making process in transportation planning and optimization processes.

Additionally, DTALite also adopts a new software architecture and algorithm design to facilitate the most efficient use of emergent parallel (multi-core) processing techniques and exploit the unprecedented parallel computing power newly available on both laptops and desktops.

This paper is organized as follows. Section 2 first introduces the overall system design and model structure of DTALite. This is followed by the related literature review and traffic flow modeling implementation in Section 3, and a queue-based traffic state estimation framework in Section 4. The paper closes with a case study using real-world networks (Section 5) and overall conclusions (Section 6).

2. System design and model structure

The software architecture designed in DTALite aims to integrate many rich modeling and visualization capabilities into an open-source DTA model. Using a modularized design, the software suite of simulation engine + visualization interface can also serve future needs by enabling transportation researchers and software developers to continue to build upon and expand its range of capabilities. The streamlined data flow from static traffic assignment models and common signal data interfaces aims to allow planners and engineers to rapidly apply the advanced DTA methodology, and further examines the effectiveness of traffic mobility, reliability, and safety improvement strategies. The overall structure, illustrated in Figure 1, integrates the four major modeling components highlighted in yellow.

DTALite's four major modeling components include:

(1) Time-dependent shortest path finding, based on a node-link network structure

(2) Vehicle/agent attribute generation, which combines an origin–destination (OD) demand matrix with additional time-of-day departure time profile to generate trips

(3) Dynamic path assignment module, which considers major factors affecting agents' route choice or departure time choice behavior, such as (i) different types of traveler information supply strategies (e.g. historical, pre-trip, and/or en route information, and variable message signs) and (ii) road pricing strategies where economic values are converted to generalized travel time

(4) A class of queue-based traffic flow models that can accept essential road capacity reduction or enhancement measures, such as work zones, incidents, and ramp meters.

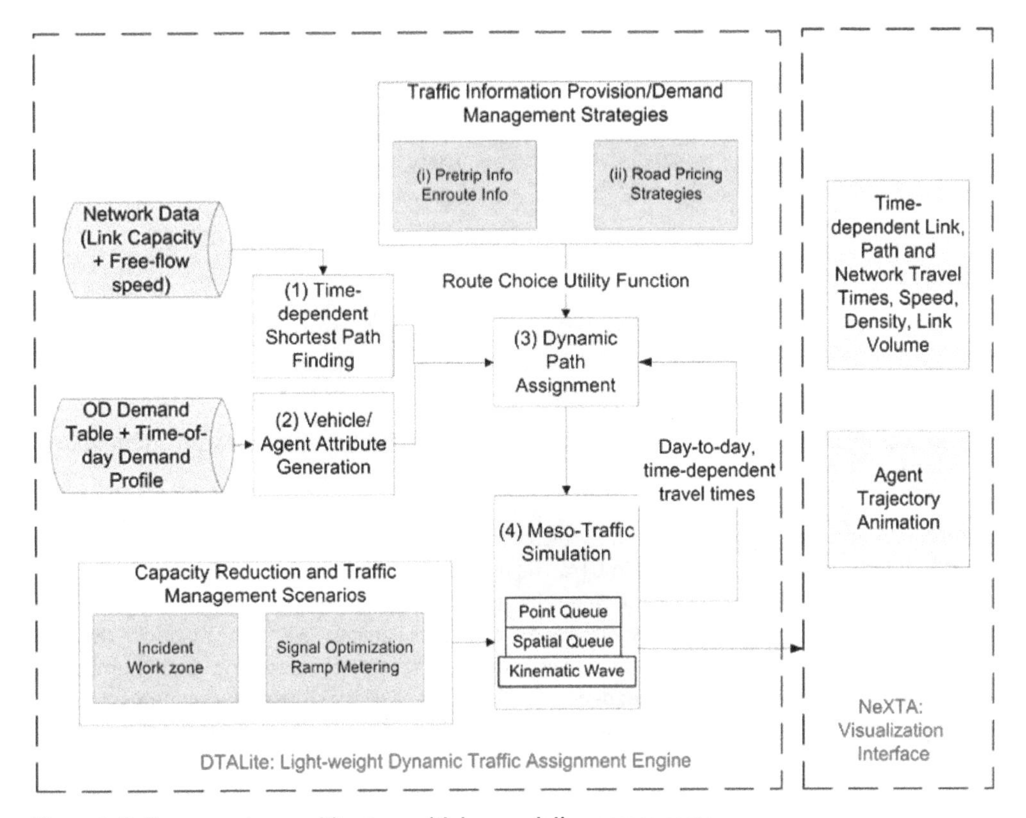

Figure 1. Software system architecture with key modeling components.

The major focus of this paper will be in describing the queue-based traffic simulation model and its application in an integrated origin-destination demand estimation procedure. The queue-based traffic simulation model in DTALite only requires basic link capacity and free-flow speed for operation, which are readily available from static traffic assignment models. Using simple input parameters, in addition to possible connections with common signal data interfaces, the proposed simulation package may enable state DOTs and regional MPOs to rapidly apply advanced DTA methodologies for large-scale regional networks, subareas, or corridors. Additionally, the modularized system design may help serve future needs by simplifying the process for transportation researchers and software developers to continue to build upon and expand its range of capabilities.

3. Queue-based traffic flow simulation model

3.1. Literature review on traffic simulation in a DTA context

In general, a simulation-assignment DTA model needs to read time-dependent OD demand matrices and then assign vehicles to different paths based on link travel time. This procedure is illustrated by the diagram in Figure 2. Given trip demand data and path information, the dynamic network loading (DNL) module in the DTA model then simulates the movement of vehicles through the network. The physical process of moving vehicles through a transportation network works in a two-step process. First, a vehicle is moved across a link by a link traversal model, and then moved between links at the node by a node transfer function. In particular, the link traversal model typically enforces a speed-density relationship and hard outflow capacity constraints, which are determined by link properties (e.g. number of lanes, link type). The node transfer model involves specific left-turn or through movement capacity determined by green time allocation at signalized nodes or other attributes. Link travel times from the traffic simulator are fed to the time-dependent shortest path model for path selection through a specified route choice utility

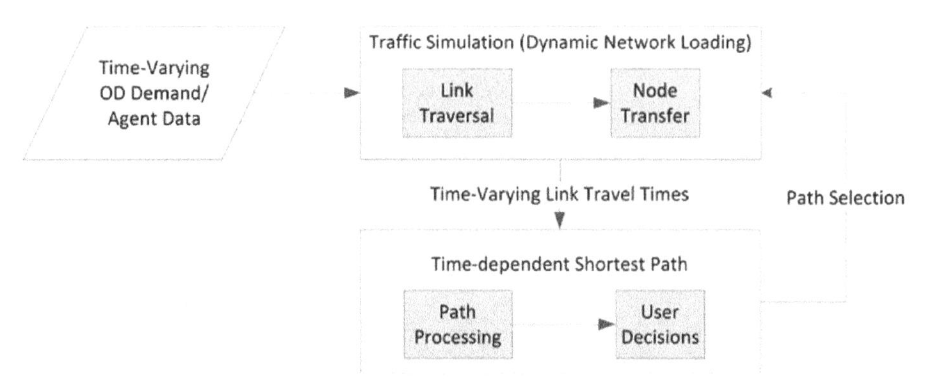

Figure 2. General simulation-based DTA modeling framework.

function or traffic assignment rules. The new paths are then fed back into the traffic simulator in the next iteration.

There are macroscopic, mesoscopic, and microscopic simulation-based methods for generating time-dependent travel time measures in DTA. The macroscopic models use continuum fluid representations, and the microscopic simulation models typically rely on detailed car following and lane changing models. In contrast, a mesoscopic model considers vehicles (with specific origin, destination, departure time, and routes) individually and moves vehicles according to a number of macroscopic traffic flow relations. Speed–density functions, like those used by Mahmassani, Hu, Peeta, and Ziliaskopoulos (1994), Mahmassani (2001), and Ben-Akiva, Bierlaire, Burton, Koutsopoulos, and Mishalani (2002) are commonly found in application.

To offer a computationally efficient traffic state simulator and estimator, this study focuses on how to use a queue-based mesoscopic traffic simulation methodology to capture complex traffic processes in realistic networks. Interested readers are referred to the survey paper by Peeta and Ziliaskopoulos (2001) for different models such as analytical optimization-based and optimal control-based formulations, and Nie, Ma, and Zhang (2008) for a comprehensive discussion on different types of DNL models and their trade-offs in capturing travel time.

3.2. Capacity and traffic flow models

To capture queue formation, spillback, and dissipation through simplified traffic flow models, DTALite uses a number of computationally simple but theoretically sound traffic queuing models (e.g. point queue model, spatial queue, and Newell's kinematic wave model) to track forward and backward wave propagation in its light-weight mesoscopic simulation engine. By doing so, traffic simulation in DTALite only requires a minimal set of traffic flow model parameters, such as outflow, inflow capacity, and storage capacity constraints, which are illustrated in Figure 3.

To capture the queue dynamics at typical bottlenecks (e.g. lane drop, merge, and weaving segments), the classical kinematic wave theory needs to be integrated with (1) flow conservation constraints, (2) traffic flow models that represent speed (or flow) of traffic as a function of density, and (3) partial differential equations. The flow conservation constraints typically follow a hyperbolic

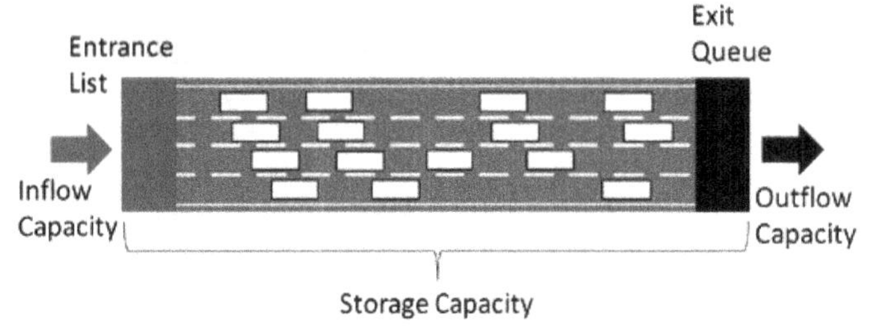

Figure 3. Modeling traffic dynamics through essential outflow, inflow capacity, and storage capacity constraints.

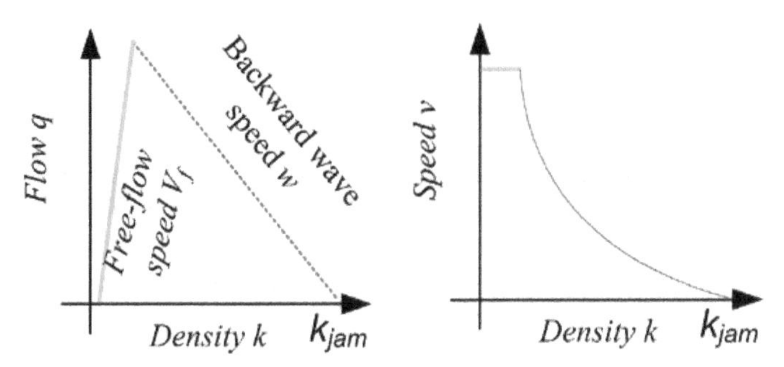

Figure 4. Triangular fundamental diagram.

system of conservation laws $\frac{\partial q}{\partial x}+\frac{\partial k}{\partial t}=g(x,t)$ where q and k are flow and density, respectively, and $g(x, t)$ is the net vehicle generation rate.

Various finite-difference approximation methods have been presented to solve these equations numerically. Based on a triangular flow–density relationship in Figure 4, there are two closely related finite deference-based numerical solution schemes to solve the first-order kinematic wave problem: (1) Newell's simplified model (1993) that keeps track of shock wave and queue propagation using cumulative flow counts on links and (2) Daganzo's cell transmission model (1995) that adopts a "supply–demand" or "sending–receiving" framework to model flow dynamics between discretized cells. DTALite implements Newell's solution, using cumulative flow counts to keep track of traffic entering and leaving each link. These cumulative flow counts are also useful when applied with queueing models.

3.3. Simple point queue and spatial queue models

The most simple queue model implemented in DTALite is the point queue model. By imposing a single outflow capacity constraint on each link, a point queue model aims to capture the effect of traffic congestion at major bottlenecks, although it does not take into account the queue spillback and the resulting delay due to storage capacity. Using a point queue model in the first few iterations and then applying a simplified kinematic wave model in the late assignment process, one can avoid unrealistic and unnecessary gridlock in the initial assignment process, and further allow agents to learn travel times from previous iterations and switch routes to achieve a smooth and close-to-reality traffic pattern.

Figure 5 represents a point queue as a vertical queue with a stack of vehicles, where some of the vehicles are mapped or "rotated" from the physical link (shaded) to the vertical stack queue. The other vehicles on the physical link (not shaded) correspond to the vehicles that move at free-flow speed. In this case, the length of the queue segment in this point-queue model is zero and

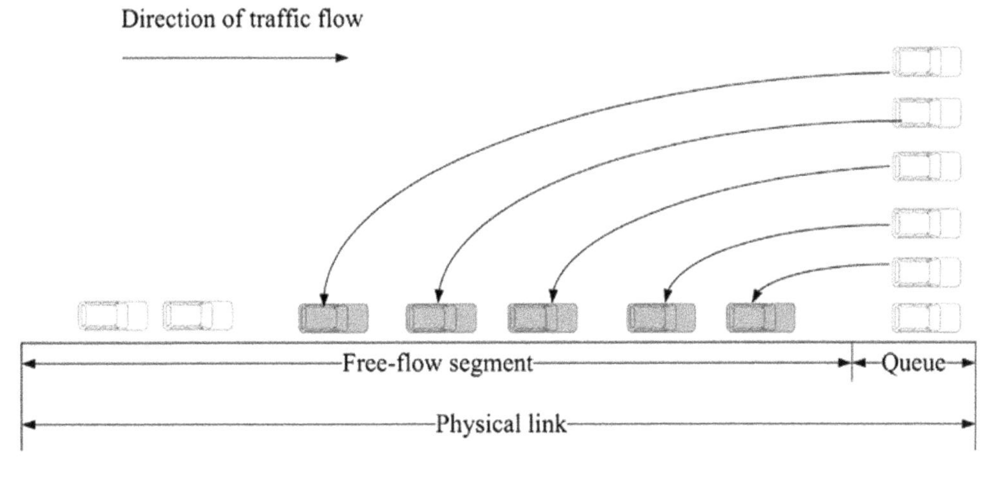

Figure 5. Point queue model represented as a vertical stack queue.

the link has unlimited storage capacity. Interested readers are referred to the paper by Hurdle and Son (2001) to examine the connection between physical spatial queues and vertical stack queues.

The realism of a point queue model can be enhanced by adding spatial storage capacity constraints, so that the resulting spatial queue model can capture queue spillbacks. This is accomplished in DTALite using link-specific jam densities, identifying how many vehicles can be stored on a link when no traffic is moving. Furthermore, by explicitly using the cumulative arrival and departure curves to track kinematic waves, Newell's flow model provides an effective means to realistically represent traffic congestion propagation and capture shockwaves as the result of bottleneck capacities.

Compared to other cell-based models that need to subdivide a long link into segments with short length, Newell's model can handle reasonably long links with homogeneous road capacity. Its simple form of traffic flow models and computational efficiency make it particularly appealing in establishing theoretically sound and practically operational DTA models for large-scale networks. There are a number of related studies on Newell's kinematic model, including model calibration research by Hurdle and Son (2000) and extensions to node merge and diverge cases by Yperman, Logghe, and Immers (2005) and Ni, Leonard, and Williams (2006), to name a few.

3.4. Simple queue-based DNL model

We first present a point-queue-based DNL model, implemented in DTALite, for a sequence of consecutive links of a freeway corridor that cover links $l = 1, 2, \ldots, L$. In the corridor without entrances or exits, traffic of a single OD pair is loaded to the first link $l = 1$ and leaves the network from the last link L.

Notation

N number of nodes in a corridor

n index of nodes, $n = 1, 2, \ldots, N$

L number of links in a corridor

l index of links, $l = 1, 2, \ldots, L$

Δt length of simulation interval (e.g. 6 s)

Δx link length (e.g. 1 miles)

$k_{l,t}$ prevailing density during the tth time step on link l

$q_{l,t}$ transfer flow rate from link l to link $l + 1$ during the tth time interval $[t, t + \Delta t]$

$cap_{l,t}^{out}$ outflow capacity on link l during the tth time interval $[t, t + \Delta t]$

v_{free} free-flow speed

k_{jam} jam density

3.4.1. Point queue model computational procedure

For time $t = 0$ to T

For link $l = 1$ to L

Step 1: Calculate flow ready to move from link l: $v_{free} \times k_l(t)$

Step 2: Calculate transfer flow from link l to link $l + 1$

$$q_{l,t} = \text{Min}\{v_{free} \times k_{l,t}, cap_{l,t}^{out}\}$$

Step 3: Update prevailing density at link l

$$k_{l,t+1} = k_{l,t} + q_{l-1,t} - q_{l,t} \times \Delta t / \Delta x$$

End for //link

End for //time

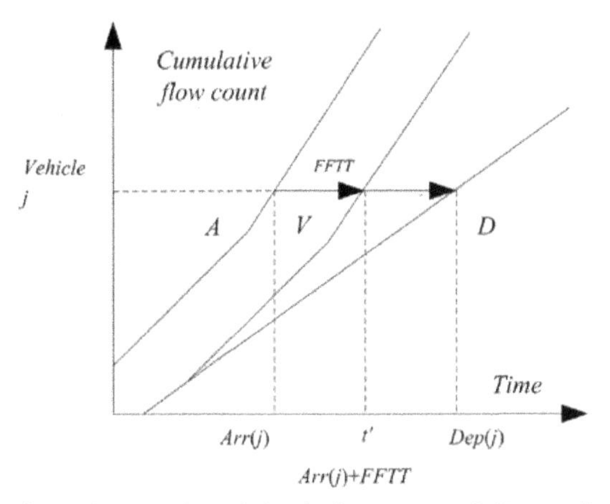

Figure 6. Event-based simulation process; A, D: cumulative arrival and departure flow counts; V: cumulative flow count of the vertical queue; FFTT is free-flow travel time of the link.

Within a mesoscopic simulation framework, the above fluid-based numerical computation scheme can be viewed as a pseudo event-based simulation process, as we do not simulate how a vehicle moves inside the link (at free-flow speed). As shown in Figure 6, at each simulation step, vehicle j is moved into a link and sets its arrival time at the upstream node of the link as Arr(j). Accordingly, the time entering the vertical queue at the stop bar is determined by Arr(j) + FFTT. As the simulation clock advances to t' = Arr(j) + FFTT, we need to check if the outflow link capacity $\text{cap}_{l,t'}^{\text{out}}$ is still available and vehicle j is in the beginning of the vertical queue. If the vehicle is at the beginning of the queue, this vehicle moves to the next link and its departure time stamp Dep(j) = t'. Otherwise, the vehicle must stay in the queue waiting for the available outflow capacity, which results in a departure time Dep(j) > t'. Thus, the entire link travel time of the vehicle is Dep(j) − Arr(j), which can be further used to infer space-mean speed of the link.

Similarly, the numerical calculation scheme of the spatial queue can be enhanced by changing Step 2 by including an additional storage capacity constraint as follows:

$$q_{l,t} = \text{Min}\left\{ v_{\text{free}} \times k_{l,t}, \text{cap}_{l,t}^{\text{out}}, \left(k_{\text{jam}} - k_{l+1,t}\right) \times \Delta x \right\}$$

where $(k_{\text{jam}} - k_{l+1,t}) \times \Delta x$ is the physical space availability at downstream link $l + 1$.

As the proposed model strictly satisfies the first-in-first-out (FIFO) constraint, if a vehicle in the beginning of the vertical queue is blocked due to unavailable outflow capacity (in the case of spatial queue or kinematic wave model), then the vehicles arriving later in the queue will be blocked as well. To model complex geometric features such as short left-turn bays on a multi-lane facility, one needs to decompose a link into multiple connected cells with each cell satisfying FIFO constraints (Reynolds, Zhou, Rouphail, & Li, 2010).

3.5. Modeling forward and backward wave propagation using Newell's simplified kinematic wave model

To describe traffic congestion propagation realistically by including the phenomena of queue build-up, spillback, and dissipation along freeway corridors, DTALite incorporates Newell's (1993) simplified kinematic wave model, which is built upon the assumption of a triangular flow-density relationship. Newell's model is implemented using cumulative flow counts on each link. As illustrated in Figure 7, $A(t)$ represents the cumulative number of vehicles moving into a link, $D(t)$ represents the cumulative number of vehicles leaving a link, and $V(t)$ is the cumulative number of vehicles in the vertical queue, waiting to exit the link.

Hurdle and Son's framework (2000) is adopted in this research to explain how Newell's model is able to model forward and backward waves using the cumulative flow counts. Let x be the location

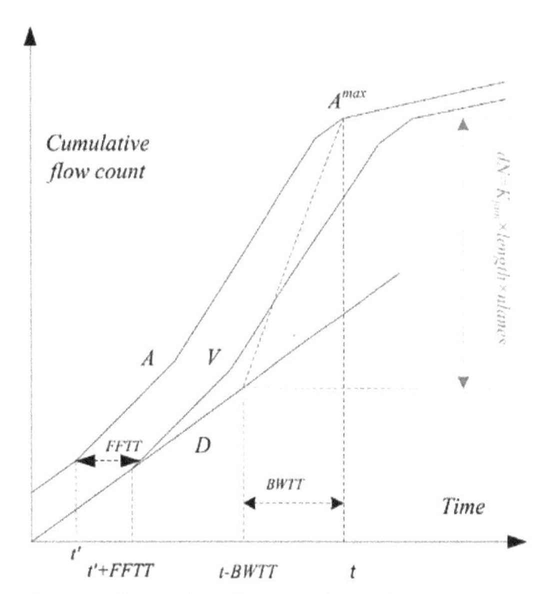

Figure 7. Illustration of cumulative arrival and departure curves *A*(*t*) and *D*(*t*), and the shifted arrival curve *V*(*t*).

along the corridor, and $N(x, t)$ is the cumulative flow count at location x and time t of a link. The change of $N(x, t)$ along a characteristic line (wave) is represented as follows:

$$dN(x,t)=\frac{\partial N}{\partial x}dx+\frac{\partial N}{\partial t}dt=qdt-kdx \tag{1}$$

A wave represents the propagation of a change in flow and density along the roadway, and the wave speed is the slope of the characteristics line $w=\frac{\partial q}{\partial k}=\frac{dx}{dt}$. Along the movement of a wave, we substitute $dt=\frac{dx}{w}$ into the above equation, so that we can link the difference of cumulative flow counts together through

$$dN(x,t)=qdt-kdx=\left(-k+\frac{q}{w}\right)dx \tag{2}$$

For the triangular-shaped flow-density relationship with constant forward and backward wave speeds, it is easy to verify that, when the speed of the forward wave is v_f, the general cumulative flow count updating formula reduces to $-k+\frac{q}{v_f}=-k+k=0$. Under congested traffic conditions with a constant backward wave speed w_b, we have $-k+\frac{q}{w_b}=-k_{jam}$, and this equation can be rewritten as:

$$dN=\left(-k+\frac{q}{w_b}\right)dx=-k_{jam}(a)\times \text{length}(a)\times nlanes(a) \tag{3}$$

The above equation is used to describe how a backward wave travels through the link. Under the queue spillback condition, the difference of $k_{jam}(a)\times \text{length}(a)\times nlanes(a)$ between the cumulative arrival and departure counts is illustrated by the vertical dashed line in Figure 7. Now consider a situation with two consecutive links, upstream link a and downstream link b, as shown in Figure 8. Newell's model is concerned about three-state variables on upstream link a: (1) cumulative flow count $A(a, t)$ for vehicles moving into link a through the upstream node; (2) cumulative flow count $V(a, t)$ for vehicles waiting at the vertical queue of the downstream node of link a at time t; and (3) cumulative flow count $D(a, t)$ for vehicles moving out of link a through the downstream node.

As shown in Figure 8, when a queue spills back from the downstream to the upstream, the arrival and departure cumulative flow counts at two ends of a link (at timestamps t and time t-BWTT(a)) need to ensure a constant difference of $dN = k_{jam}(a)\times \text{length}(a)\times nlanes(a)$, and the capacity restriction is propagated throughout the link using a time duration of BWTT(a) = length(a)/$w_b(a)$.

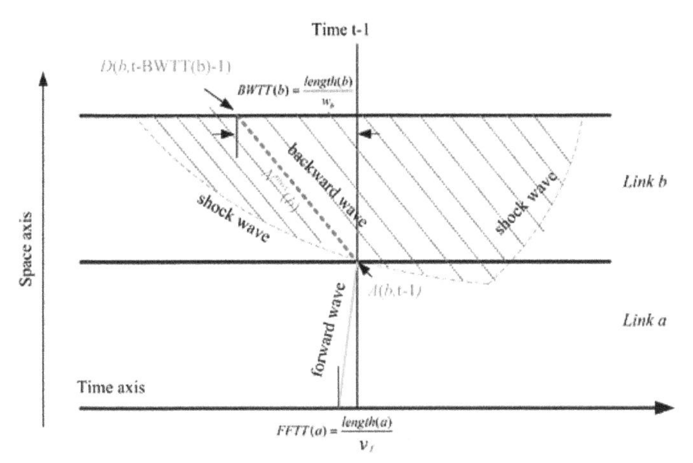

Figure 8. Illustration of forward and backward wave representation in Newell's simplified KW model.

The space–time plot in Figure 8 clearly demonstrates how Newell's model uses cumulative counts to model the forward and backward waves in describing queue phenomenon on links *a* and *b*. Exactly at time $t-1$, the tail of the queue at the downstream link *b* propagates to its upstream node. That is, if the cumulative outflow count at a lagged time stamp $t-\text{BWTT}(b)-1$ on link *b* is equivalent to the cumulative inflow count at time $t-1$, then a queue spillback occurs as a backward wave is able to propagate through the congested time-space "mass" of the link. Compared to the conventional vertical queue model, where the inflow rate of a link is not constrained by the available physical length, the inflow rate in Newell's model is governed by the discharge flow of a link through the backward wave propagation process. Furthermore, the number of vehicles on a lane should not be greater than $k_{\text{jam}} \times \text{length}(b)$. As a result, the proposed model is able to detect and capture queue spillbacks to upstream links.

3.6. Implementing Newell's simplified kinematic wave model as an advanced queue model

Notation

Link attribute variables

$\text{length}(l)$	length of link l
$\text{nlanes}(l)$	number of lanes on link l
$q^{\max}(l, t)$	maximum flow rate on link l between time $t-\Delta T$ and time t
$v_f(l)$	free-flow speed (or forward wave speed) on link l
$w(l)$	constant backward wave speed for link l
$k_{\text{jam}}(l)$	jam density on link l
$\text{FFTT}(l)$	free-flow travel time on link l, i.e. $\text{length}(l)/v_f(l)$
$\text{BWTT}(l)$	backward wave travel time on link l, i.e. $\text{length}(l)/w(l)$

Capacity constraint variables

$\text{cap}^{\text{in}}(l, t)$	inflow capacity of link l between time $t-\Delta T$ and time t
$\text{cap}^{\text{out}}(l, t)$	outflow capacity of link l between time $t-\Delta T$ and time t

Cumulative flow count variables

$A(l, t)$	cumulative number of vehicles entering link l at time t
$V(l, t)$	cumulative number of vehicles waiting at the vertical queue of link l at time t
$D(l, t)$	cumulative number of vehicles departing/leaving link l at time t

For simplicity and clarity, we present the proposed DNL model for a sequence of consecutive, directional links of a freeway corridor that covers links $l = 1, 2, \ldots, L$. Each link l is attached to an upstream node and a downstream node. In the corridor, traffic of a single OD pair is loaded to the first link $l = 1$ and leaves the network from the last link L.

Step 1 (Initialize variables and boundary conditions)

For each link, initialize cumulative flow counts for the entering queue, vertical queue, and exit queue.

$$A(l,t=0)=0, V(l,t=0)=0, D(l,t=0)=0$$

Link attributes, including $q^{max}(l, t)$, $k_{jam}(l)$, nlanes(l), FFTT(l), and BWTT(l), are assumed to be known for all links. $q^{max}(l, t)$ is dependent on node flow management rules described in the following section.

Two boundary conditions are assumed to be known for all time intervals: the demand flow, $A(1,t)$, on the first link, and the destination capacity $q^{max}(L, t)$ on the last link.

Step 2 (Update time-dependent traffic states at time t)

For current time stamp t, given $A(l, t')$, $D(l, t')$, and $q^{max}(l, t)$, such that $t' = 0, \Delta T, 2\Delta T, \ldots, t-\Delta T$, perform Steps 2.1–2.3 sequentially for each link $l = 1, 2, \ldots, L$.

Step 2.1 (Forward wave propagation): For each link l, move cumulative arrival flow counts of link l to its vertical queue.

$$V(l,t)=A(l,t-FFTT(l))$$

Step 2.2 (Backward wave propagation): For each link l, calculate the maximum possible cumulative arrival count for the flow to move into link l at time t.

$$A^{max}(l,t)=D(l,t-\text{BWTT}(l))+k_{jam}(l)\times \text{length}(l)\times \text{nlanes}(l)$$

Step 3 (Calculate capacity constraints)

Determine the inflow and outflow capacity constraints for each link at time t.

Step 3.1 (Determine inflow capacity): For each link $l = 1, 2, \ldots, L$, determine the inflow capacity of link l at time t.

$$\text{cap}^{in}(l,t)=A^{max}(l,t)-A(l,t-\Delta T)$$

Step 3.2 (Determine outflow capacity): For links $l = 1, 2, \ldots, L-1$, determine the outflow capacity of link l at time t.

$$\text{cap}^{out}(l,t)=\min\{q^{max}(l,t),\text{cap}^{in}(l+1,t)\}$$

$$\text{cap}^{out}(L,t)=q^{max}(L,t) \text{ for the last link L}$$

Step 4 (Transfer flow between links)

For each node n = 1, 2, ..., N, identify incoming and outgoing links, and identify agents/vehicles ready to move between those incoming and outgoing links at time t.

Step 4.1 (Update cumulative departure count): For each link upstream from node n, update the cumulative departure counts.

$$D(l,t)=\min\{V(l,t),D(l,t-\Delta T)+\text{cap}^{out}(l,t)\}$$

Step 4.2 (Update cumulative arrival count): For each link downstream from node n, update the cumulative arrival counts.

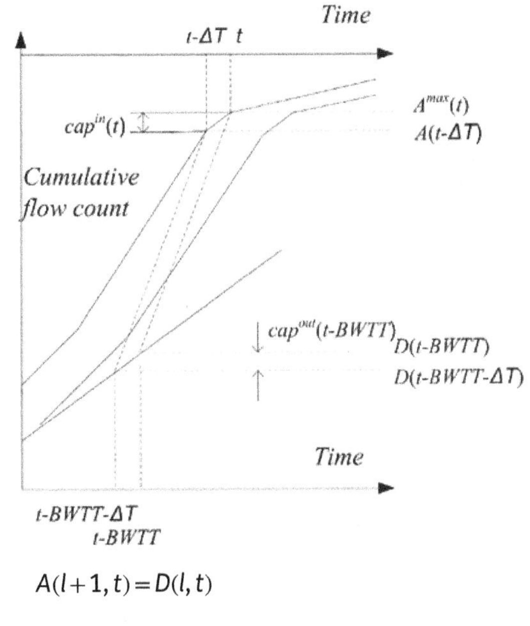

$$A(l+1, t) = D(l, t)$$

Step 5 (Advance simulation clock):

Advance $t = t + \Delta T$, go back to Step 2.

Figure 9. Illustration of queue spillback and propagation of outgoing flow constraint from the downstream end at time *t*-BWTT to the upstream *end* at time *t*.

Given existing traffic states in the previous time interval $t-\Delta T$, the DNL model follows a multi-step process for moving agents/vehicles along a given path between specific origin and destination nodes, subject to capacity constraints and time-dependent traffic states defined by shockwave propagation in Newell's simplified kinematic wave model. The DNL first updates the values of traffic state variables in time interval *t* in Step 2, and determines the maximum number of agents/vehicles which can be transferred between links (at nodes) in Step 3. Those agents/vehicles are then moved between links in Step 4.

As a detailed illustration, Figure 9 demonstrates how to calculate the maximum inflow capacity and outflow capacity in a queue spill-back case. The maximum inflow capacity $cap^{in}(t)$ at the upstream end of a link is calculated using the difference of cumulative arrival flow counts at two consecutive time stamps $t-\Delta T$ and t, which is determined by the outflow capacity $cap^{out}(t-\text{BWTT})$ at the downstream end between time stamps $t-\text{BWTT}-\Delta T$ and $t-\text{BWTT}$.

3.7. Node management/control models

Within a typical network, different node types (i.e. origin, merge, and signalized intersections) may require unique methods for moving agents between different links. In the proposed DNL model, agents are loaded into the network at the origin node using a loading buffer, rather than being loaded directly into link $l = 1$. Three separate models, based on previous studies in the literature, can be used to allocate capacity at merge junctions. Lastly, outflow capacity for incoming links at signalized intersections is allocated based on a simplified model using cycle times and link effective green times. Diverge junctions require no special handling because the paths are known for agents/vehicles traveling through the network (Figure 10).

3.7.1. Origin nodes and the loading buffer

As stated in kinematic wave-based DNL algorithm, the demand flow, $A(1, t)$, on the first link is assumed to be known for all time intervals. As a result, each agent/vehicle is assigned to enter the network at a specific departure time based on some semi-random time interval between arrivals. At the origin node, located upstream from the first link, agents/vehicles first enter the loading buffer, which acts as a temporary storage queue. In each time interval *t*, if an agent's departure time ≤ *t* and

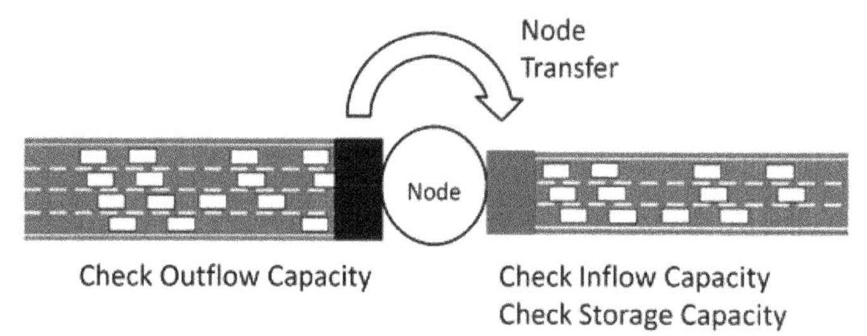

Figure 10. Illustration of node transfer process.

there is inflow capacity available, those agents/vehicles are moved from the loading buffer to the entrance queue on link $l = 1$.

3.7.2. Merge junctions

Consider a merge junction with incoming link 1 and 2 merging into downstream link 3, as shown in Figure 11. If the total demand from all incoming links (denoted as $d1$ and $d2$ on the two links) is greater than the available inflow capacity on link 3, denoted as cap^{in}, then the available inflow capacity must be distributed to each incoming link.

To determine the outflow capacity for mainline link 1 and onramp link 2, namely $cap^{out}(1)$ and $cap^{out}(2)$, we use Daganzo's priority-based merge model (1994) in this study to maximize the utilization of downstream inflow capacity, while upstream outflows satisfy feasibility constraints.

If $d1 + d2 < cap^{in}$, then $\begin{cases} cap^{out}(1) = mid\{d1, cap^{in} - d1, p1 \times cap^{in}\} \\ cap^{out}(2) = mid\{d2, cap^{in} - d2, p2 \times cap^{in}\} \end{cases}$ where

$p1 = \dfrac{nlanes(1)}{nlanes(1) + nlanes(2)}, p2 = \dfrac{nlanes(2)}{nlanes(1) + nlanes(2)}$ according to a lane-based proportional distribution rule, and the mid function takes the middle value of all three input parameters.

Illustrated in Figure 11, in case (i), the solution corresponds to point e. The solution to case (ii) should be point g rather than point f. While point g optimizes the overall inflow capacity allocation, point f underutilizes the inflow capacity by strictly following the lane-based distribution rule.

3.7.3. Flow conservation at diverge junctions

In continuous flow-based traffic simulation, many macroscopic fluid-based models use continuous ratios of destination distribution to decide the amount of traffic flow to different destinations. As the mesoscopic DNL model moves agents/vehicles with OD and path information, the proportions of vehicles moving out of a link to individual outgoing links are in fact determined directly by the paths (downstream node sequences) associated with vehicles at this link, so it is very easy to decide which

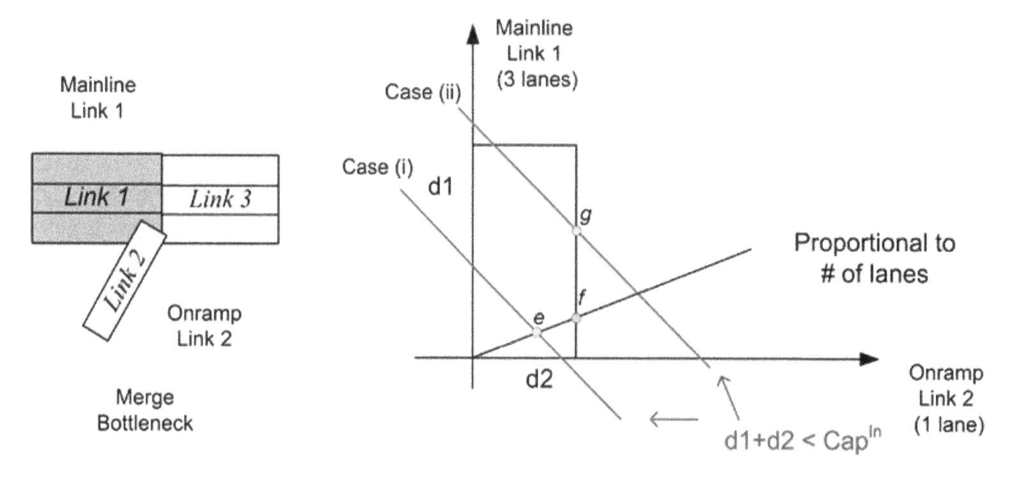

Figure 11. Capacity allocation at merge bottleneck.

sets of mesoscopic agents to be moved to a specific destination at any intersection or diverge point associated with an off ramp.

3.7.4. Signalized intersections

The proposed DNL model considers the influence of traffic signals using a simplified method for estimating the link outflow capacity. Assuming that the effective green time and saturation flow rate are known for link l, upstream from node n, and the cycle time is known for the signal at node n, the link outflow capacity on link l is calculated using the equation below.

$$\mathrm{cap}^{\mathrm{out}}(l,t)=q^{\mathrm{saturation}}(l,t)\frac{\text{effective green time}(l)}{\text{cycle length}(n)} \tag{4}$$

To achieve an effective coupling of a signal timing estimation model and DTA in feedback loops, Zlatkovic and Zhou (2014) integrated a quick estimation method to link critical movement analysis methodologies with the DTALite simulation package.

4. Queue-based dynamic OD demand estimation

To allow consistent OD demand estimation for fast model calibration, DTALite embeds a path flow-based optimization model that utilizes sensor data (i.e. observed link flows and densities) and target OD demands $\bar{d}(i,j,\tau)$to estimate a set of path flow volumes. This approach combines OD estimation/adjustment with traffic assignment seamlessly. That is, under this ODME model, DTALite first runs K assignment iterations (e.g. 40 iterations) to generate likely paths, and then performs anther K' iterations (e.g. 100 iterations) with ODME enabled to provide the final solution as a set of path flow patterns satisfying "tolled user equilibrium." In our discussion below, we want to highlight the use of the approximate gradient method in DTALite, which utilizes a queue model to calculate link flow–density change due to incoming path flow change. The detailed mathematical formulation and solution algorithm can be referred to a recent paper by Lu, Zhou, and Zhang (2013).

4.1. Mathematical model formulation

The problem statement and notations of ODME can be summarized as follows. Using path flow vector $\mathbf{r}(i,j,\tau,k)$ for OD pair (i,j) at departure time at time k, the model minimizes a set of deviation functions with respect to target OD demands $\bar{d}(i,j,\tau)$ and observed link flows and densities, subject to (i) DNL constraints that describe traffic flow propagation with multiple OD pairs and (ii) a gap function-based constraint that measures the deviation from the dynamic user equilibrium conditions.

$$\mathrm{Min}\,Z=\beta_d\sum_{i,j,\tau}\left[d(i,j,\tau)-\bar{d}(i,j,\tau)\right]^2+\sum_{a\in S}\sum_{t\in H_o}\left\{\beta_q[q(a,t)-\bar{q}(a,t)]^2+\beta_k[k(a,t)-\bar{k}(a,t)]^2\right\} \tag{5}$$

where $q(a,t)$ and $\bar{q}(a,t)$ are the simulated and observed link flows, respectively, on link a at time interval $[t, t+1]$; $k(a,t)$ and $\bar{k}(a,t)$ are the simulated and observed link densities, respectively, on link a at time t; S is the set of links with observations; H_o is the set of link intervals with observations; and β_d, β_q, β_k is the weights reflecting different degrees of confidence on target OD demands and observed link flows and densities, respectively.

In the estimation step, within a column generation-based framework, a gradient-projection-based descent direction method (Lu, Mahmassani, & Zhou, 2009) is used to update path flows $r^{(m+1)}$ at step $m+1$. Specifically,

$$\begin{aligned} r(i,j,\tau,k)^{m+1} \;=\; & \mathrm{Max}\Big\{0, r(i,j,\tau,k)^m-\gamma^{(m)}\Big[\beta_d\nabla h^d(r)|_{r=r^{(m)}} \\ & +\beta_q\nabla h^q(r)|_{r=r^{(m)}}+\beta_k\nabla h^k(r)|_{r=r^{(m)}}+\lambda^{(n)}\nabla g(r,\pi)|_{r=r^{(m)}}\Big]\Big\} \end{aligned} \tag{6}$$

where $\gamma^{(m)}$ is the step size, and the gradients that consist of the first-order partial derivatives with respect to a path flow variable $r(i, j, \tau, k)$ are discussed as follows, based on an queue-representation in Figure 13.

The solution framework integrates a gradient-projection-based path flow adjustment method. The simple queuing model allows this tool to also derive analytical gradient formulas for the changes in link flow and density due to the unit change of time-dependent path inflow in a general network under congestion conditions, as shown in Figure 12.

In this study, the link partial derivative is referred to as the change in link flow and density due to an additional unit of link/path inflow. Ghali and Smith (1995) presented an analytical approach to evaluate the (local) link marginal travel time (or delay) on a congested link, based on link cumulative flow curves of a simple point queue model. We illustrate the use of the queue-based link partial gradient calculation in Figure 13, which shows the cumulative arrival and departure curves $A(t)$ and $D(t)$, and the dashed line represents the cumulative vertical queue. Given (outflow) capacity of the link as c, the additional change in the A/D curve is equivalent to the gray area. In this example, FFTT(l) = 15 min. The queue starts at t_l^{qs} = 7:00, and an additional unit of flow, vehicle n' enters link l

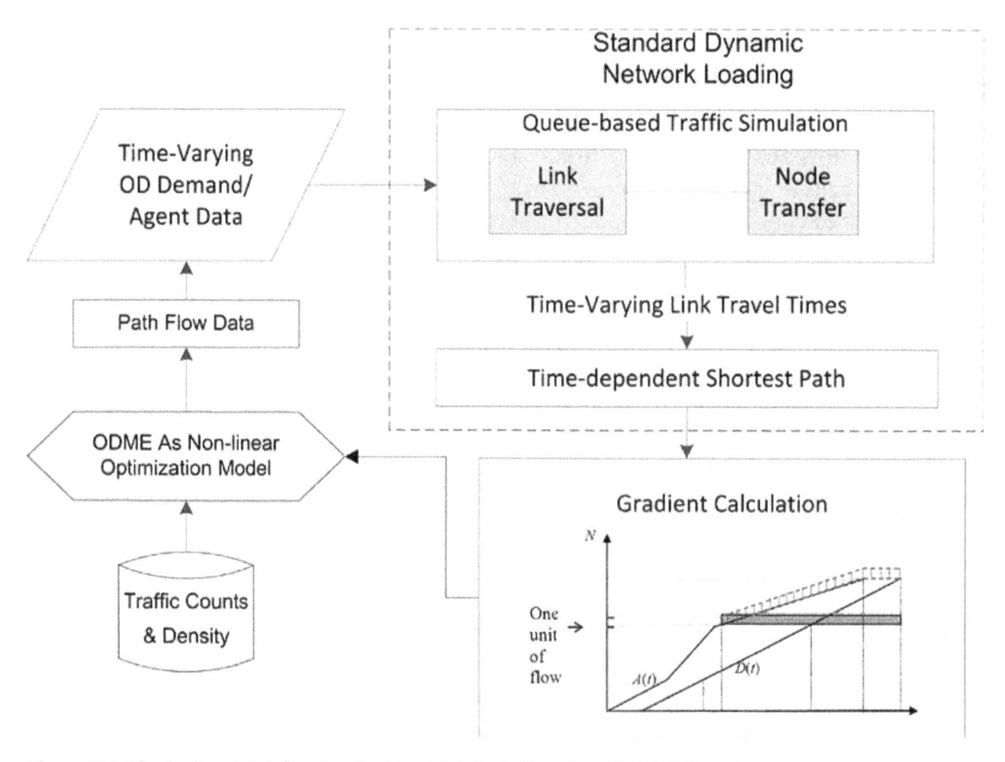

Figure 12. Single-level Origin -Destination Matrix Estimation (ODME) flowchart.

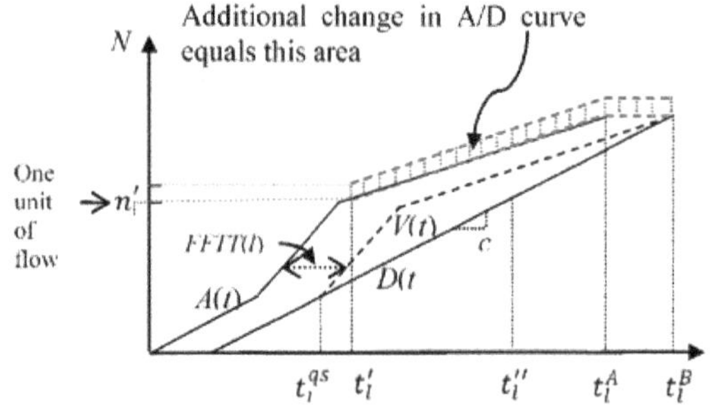

Figure 13. Illustration of link partial gradient on a congested link.

at time $t_l' = 7{:}10$. This vehicle leaves the link at time $t_l'' = 7{:}50$, the queue dissipates on link l at $t_l^B = 8{:}30$, and $t_l^A = 8{:}15$.

We can easily analyze the marginal effects of an additional unit of flow on link flow (inflow and outflow) and the number of vehicles (and corresponding density). In particular, under partially congested conditions shown in Figure 13, the link inflow and outflow increase by one at times t_l' and t_l^B, respectively, and the flow rates at the other time intervals do not change due to the assumed constant capacity from t_l' to t_l^B, while the number of vehicles in the link increases by one for the same time period.

5. Numerical experiments

5.1. Verifying shock wave propagation speed on a hypothetic corridor

The proposed queue-based mesoscopic DNL model and the ODME method have been implemented in DTALite in C++, using an OpenMP as the application programming interface for parallel computing. We first consider a hypothetic corridor with a lane drop bottleneck. Based on the flow-density relationship in Figure 14, the theoretical shock wave propagation speed is (1560−1200)/(80−26) = 6.67 mph. Figure 15 shows the dynamic density contour generated from DTALite simulation results. The congestion starts at 7:12 at node 9 and propagates to the upstream node 1 at around 8:15 along this 7 mile stretch. Accordingly, we can calculate the simulated shockwave propagation speed from DTALite as 7 mile/63 min = 6.67 mph, which is nicely consistent with the theoretical value.

5.2. Large real-world network calibration using OD demand calibration

As shown in Figure 16, this real-world network is the Triangle Regional Model network containing most of Raleigh, North Carolina, USA. This large-scale regional network has 2,389 zones, 20,259 links, and about 2,000 signalized intersections. Provided by the local metropolitan planning agency, the

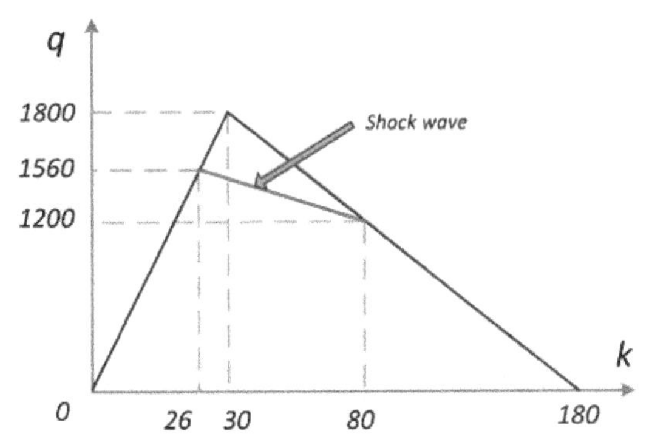

Figure 14. Flow–density relationship and shock wave on a lane drop bottleneck.

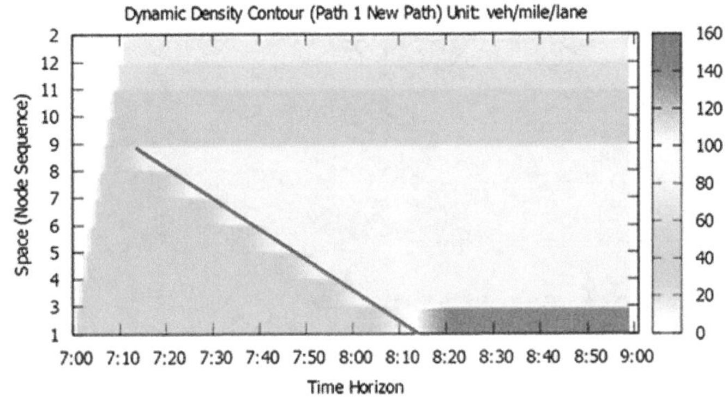

Figure 15. Dynamic density contour on a hypothetic corridor with a lane drop bottleneck on node 9.

Figure 16. Triangle Regional Model, NC network with 2,389 zones, 20,259 links, about 2,000 signalized intersections, 1.06 million vehicles in morning peak hours.

morning peak-hour demand matrix has about 1.06 million vehicles, covering a time period from 6 am to 10 am. There are about 16 and 14 sensors, respectively, on freeway and arterial links, producing a total of 120 hourly link count observations. The experiments were performed on a PC with 16 GB memory and 4-core processors with hyperthreading, running at 2.70 GHz. Using the open-source DTA package DTALite, the running time is about 2 min and 45 s per iteration. When incorporating the additional path flow adjustment process, the average running time increases to 5 min and 3 s per iteration. The iterative sequential adjustment converged after 140 iterations, which required a total of 12 h of CPU time. The scatter plots in Figures 17 and 18 show additional MOEs—R^2, total estimated vs. observed flow ratio. The average absolute link flow deviations are 435.15 and 212.21 vehicles per hour per link, respectively, on freeway and arterial links, which further

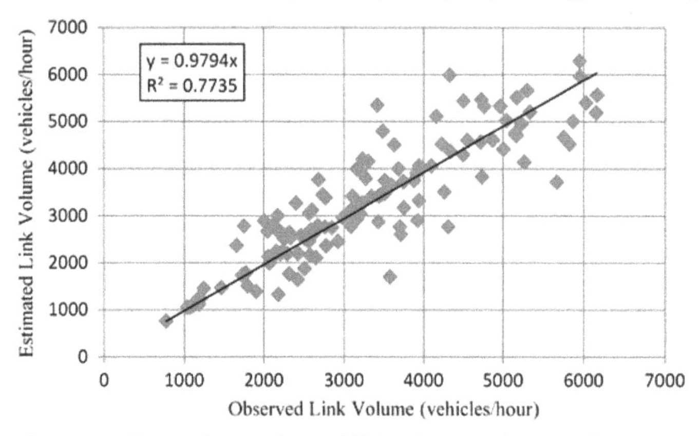

Figure 17. Observed vs. estimated link volume on freeway links on the Triangle Regional Model, NC network.

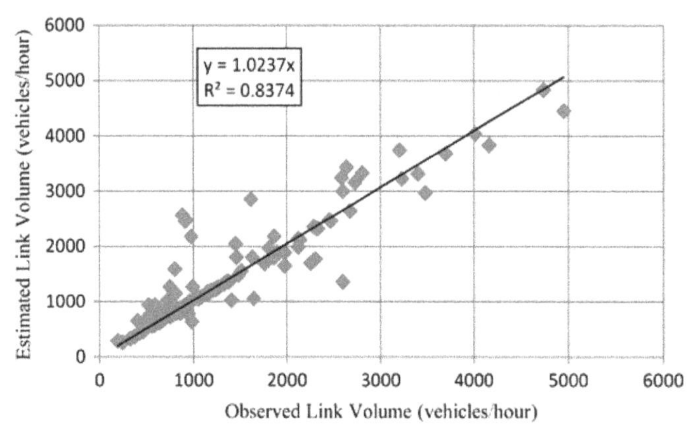

Figure 18. Observed vs. estimated link volume on highway and arterial links on the Triangle Regional Model, NC network.

demonstrate the effectiveness of the proposed method on both freeway and arterial links of the large-scale network.

6. Conclusions

In this paper, a queue-based mesoscopic DNL model is proposed to allow rapid simulation of traffic congestion propagation in a network with various bottlenecks. This mesoscopic DNL model, based on point queues, spatial queues, and Newell's simplified kinematic wave theory, is proposed to realistically estimate dynamic flow and travel time performances for an estimated OD demand table.

Based on the cumulative flow counts from the queueing simulation results, we can analytically derive partial derivatives of link flow and density with respect to path inflow perturbations, and these results can provide essential gradient information to determine the feasible descent direction in the OD demand estimation algorithm. Based on a class of queueing models, the seamless integration between the OD flow estimation and DNL model offers an efficient way to utilize available sensor data sources.

The presented queue-based mesoscopic traffic simulation model has considerable potential for generalizing the modeling framework into the field of real-time traffic state estimation and prediction. In our future research, we will further examine different ways for calibrating the maximum queue discharge rates, utilizing end-to-end travel times, and considering an agent-based learning framework to fully consider behavioral heterogeneity.

Acknowledgments

The first author appreciates Prof. Hani S. Mahmassani's instructions and guidance for the fundamental knowledge of DTA. The first author would also like to thank Jason Chung-Cheng Lu and Kuilin Zhang for their help in developing the dynamic origin destination demand estimation model. Both authors, when working at the University of Utah, were partially supported through a FHWA project titled "An Open-Source DTA Tool for Assessing the Effects of Roadway Pricing and Crash Reduction Strategies on Recurring and Non-Recurring Congestion." Special thanks to our colleagues Nagui Rouphail and Anxi Jia at the North Carolina State University, Wayne Kittelson and Brandon Nevers at Kittelson & Associates, Inc, Brian Gardner from FHWA, for their constructive comments. The work presented in this paper remains the sole responsibility of the authors.

Funding

Xuesong Zhou and Jeffrey Taylor, when working at the University of Utah, were partially supported through a FHWA project titled "An Open-Source DTA Tool for Assessing the Effects of Roadway Pricing and Crash Reduction Strategies on Recurring and Non-Recurring Congestion."

Author details

Xuesong Zhou[1]

E-mail: xzhou74@asu.edu; xzhou99@gmail.com

Jeffrey Taylor[2]

E-mail: jeff.d.taylor@utah.edu

[1] School of Sustainable Engineering and the Built Environment, Arizona State University, Tempe, AZ 85287, USA.

[2] Department of Civil and Environmental Engineering, University of Utah, Salt Lake City, UT 84112-0561, USA.

References

Ben-Akiva, M. E., Bierlaire, M., Burton, D., Koutsopoulos, H. N., & Mishalani, R. (2002). Real-time simulation of traffic demand–supply interactions within DynaMIT. In M. Gendreau, & P. Marcotte (Eds.), *Transportation and network analysis: Current trends. Miscellenea in honor of Michael Florian* (pp. 19–36). Boston, MA: Kluwer.

Daganzo, C. F. (1994). The cell transmission model: A dynamic representation of highway traffic consistent with the hydrodynamic theory. Transportation Research Part B: Methodological, 28, 269–287. http://dx.doi.org/10.1016/0191-2615(94)90002-7

Daganzo, C. F. (1995). The cell transmission model, part II: Network traffic. Transportation Research Part B: Methodological, 29, 79–93. http://dx.doi.org/10.1016/0191-2615(94)00022-R

Ghali, M. O., & Smith, M. J. (1995). A model for the dynamic system optimum traffic assignment problem. *Transportation Research Part B, 29,* 155–170. doi:10.1016/0191-2615(94)00024-T

Hurdle, V., & Son, B. (2000). Road test of a freeway model. *Transportation Research Part A: Policy and Practice, 34,* 537–564.

Hurdle, V., & Son, B. (2001). Shock wave and cumulative arrival and departure models: Partners without conflict. *Transportation Research Record: Journal of the Transportation Research Board, 1776,* 159–166. http://dx.doi.org/10.3141/1776-21

Lu, C.-C., Mahmassani, H. S., & Zhou, X. (2009). Equivalent gap function-based reformulation and solution algorithm for the dynamic user equilibrium problem. *Transportation Research Part B: Methodological, 43,* 345–364.

Lu, C.-C., Zhou, X., & Zhang, K. (2013). Dynamic origin–destination demand flow estimation under congested traffic conditions. *Transportation Research Part C: Emerging Technologies, 34,* 16–37. http://dx.doi.org/10.1016/j.trc.2013.05.006

Mahmassani, H. S. (2001). Dynamic network traffic assignment and simulation methodology for advanced system management applications. *Networks and Spatial Economics, 1,* 267–292. http://dx.doi.org/10.1023/A:1012831808926

Mahmassani, H. S., Hu, T.-Y., Peeta, S., & Ziliaskopoulos, A. (1994). *Development and testing of dynamic traffic assignment and simulation procedures for ATIS/ATMS applications*. McLean, VA: US DOT, Federal Highway Administration.

Newell, G. F. (1993). A simplified theory of kinematic waves in highway traffic, part I: General theory. *Transportation Research Part B: Methodological, 27*, 281–287. http://dx.doi.org/10.1016/0191-2615(93)90038-C

Ni, D., Leonard, J. D., & Williams, B. M. (2006). The network kinematic waves model: A simplified approach to network traffic. *Journal of Intelligent Transportation Systems, 10*(1), 1–14. http://dx.doi.org/10.1080/15472450500455070

Nie, Y., Ma, J., & Zhang, H. M. (2008). A polymorphic dynamic network loading model. *Computer-Aided Civil and Infrastructure Engineering, 23*, 86–103.

Peeta, S., & Ziliaskopoulos, A. K. (2001). Foundations of dynamic traffic assignment: The past, the present and the future. *Networks and Spatial Economics, 1*, 233–265. http://dx.doi.org/10.1023/A:1012827724856

Reynolds, W. L., Zhou, X., Rouphail, N. M., & Li, M. (2010). Estimating sustained service rates at signalized intersections with short left-turn pockets. *Transportation Research Record: Journal of the Transportation Research Board, 2173*, 64–71. http://dx.doi.org/10.3141/2173-08

Yperman, I., Logghe, S., & Immers, B. (2005). *The link transmission model: An efficient implementation of the kinematic wave theory in traffic networks* (pp. 122–127). Paper presented at the Proceedings of the 10th EWGT Meeting, Poznan, Poland.

Zlatkovic, M., & Zhou, X. (2014). *Effective coupling of signal timing estimation model and dynamic traffic assignment in feedback loops: System design and case study*. Paper presented at the Transportation Research Board 93rd Annual Meeting, Washington, DC, USA.

11

Assessment of pile response due to deep excavation in close proximity—A case study based on DTL3 Tampines West Station

C.G. Chinnaswamy[1]* and David N.G. Chew Chiat[1]

*Corresponding author: C.G. Chinnaswamy, Meinhardt Infrastructure Pte Ltd, Singapore, Singapore.
E-mail: cgc@meinhardt-infra.com.sg

Reviewing editor: Gang Zheng, Tianjin University, China

Abstract: Ground movements during deep excavations and tunnelling, especially in urban areas, may potentially have major impact on adjacent buildings, structures and utilities. This impact on buildings and structures needs to be assessed by considering the horizontal and vertical displacements induced by deep excavations to determine the necessary mitigation measures. One major factor affecting the degree of severity the impact due to deep excavation may have on the buildings and structures is the type of foundation systems. While methodology in determining the damage category for the buildings on shallow foundation has been quite well established, the methodology for assessing the impact on the pile foundation is not straightforward due to the geometry and complexity of soil structure interaction. Often simplified two-dimensional (2D) or comprehensive three-dimensional (3D) finite element analyses would be carried out for the stage excavation to predict the displacement and stresses in the piles. Suitable protective and preventive measures would need to be designed and implemented for the existing buildings/structures if the damage category falls within the unacceptable range. This paper discusses the analysis and methodology to assess the effect on the pile foundation of a high-rise building due to the deep excavation of the Down Town Line Stage 3 (DTL3) Tampines West (TPW) Station. The approach to assess the geotechnical capacity of the pile as a result of the deep excavation is presented in this paper. Based on the assessment

ABOUT THE AUTHOR

C.G. Chinnaswamy completed his PhD from IIT Delhi in 1988 and continued his postdoctoral studies in Colorado University, Boulder, CO, USA in 1991. He has 25 years of industrial experience in infrastructure projects across many countries. He has published more than 25 papers in referred journals, conferences and symposiums. His key research areas are non-linear finite element analysis, iterative and multigrid solvers mesh generation, slope stability, 3D CAD-CAE, geostatistical interpolation, etc. Currently, he is developing of a novel algorithm for accurate determination of non-circular slope failure profile in CAD.

PUBLIC INTEREST STATEMENT

The assessment of construction impact due to tunnelling and underground excavations on the existing structures/buildings, especially in urban areas, is of prime importance and the stagewise assessment procedure is briefly covered in this paper. In that process, behaviour of piles supporting these adjacent buildings has also been studied. Any pile, being close to the retaining system for a deep excavation, will be subjected to reduction in mobilized effective normal pressure on the pile and thus the shaft friction, which of course depends upon on its distance away from the retaining system, pile toe level with respect to the final excavation level, etc. Considering the overall force equilibrium, this reduction in shaft friction will be distributed to soil around the pile toe or adjacent row of piles.

of pile response, predicted movement, structural and geotechnical capacities of the pile, it was found to be within the acceptable limit and the pile foundation has adequate factor of safety with the deep excavation in close proximity.

Subjects: Civil, Environmental and Geotechnical Engineering; Geomechanics; Soil Mechanics; Tunnelling & Underground Engineering

Keywords: underground excavation; damage assessment; prediction of pile behaviour; numerical analysis

1. Introduction

Ground movements during deep excavations and tunnelling, especially in urban areas, may potentially have major impact on adjacent buildings, structures and utilities. Figure 1 shows the crack on the external walls and columns of the buildings adjacent to deep excavation projects. Hence, it is critical to assess this impact on buildings and structures by considering the horizontal and vertical displacements induced by deep excavations to determine the necessary mitigation measures.

One major factor affecting the degree of severity the impact due to deep excavation may have on the buildings and structures is the type of foundation systems. While methodology in determining the damage category for the buildings on shallow foundation has been quite well established, the methodology for assessing the impact on the pile foundation is not straightforward due to the geometry and complexity of soil structure interaction. Often, simplified two-dimensional (2D) or comprehensive three-dimensional (3D) finite element analyses would be carried out for the stage excavation to predict the displacement and stresses in the piles. Suitable protective and preventive measures would need to be designed and implemented for the existing buildings/structures if the damage category falls within the unacceptable range.

This paper discusses the case study of analysis and methodology to assess the effect on the pile foundation of a high-rise building due to the deep excavation of the Down Town Line Stage 3 (DTL3) Tampines West (TPW) Station. Figure 2 shows the location map of the TPW station in relation to the DTL3 alignment. It is located in the eastern part of Singapore. TPW Station is located in close proximity to the existing HDB Blocks as shown in Figure 3. TPW Station is a three-level Civil Defence underground station of about 160 m length, 40 m maximum width and 22.6 m deep. The geological formation at the site is mainly old alluvium (OA) soil with overlying fill. The engineering properties of OA material were comprehensively described by Wong et al. (2001), Chiam et al. (2003) and Chu, Goh, Pek, and Wong (2003). Table 1 shows the summary table of the design parameters for the soils. Figure 4 shows the geological profile along the TPW Station.

Based on geological survey by PWD (1976), the OA is an alluvial deposit that has been variably cemented, often to the extent that it has the strength of a very weak or weak rock. The upper zone of the OA has typically been affected by weathering and has typically penetrated as a discernible front from the surface. All five classes of weathering classification of the OA are encountered at this site.

Figure 1. Cracks on external walls and columns of buildings caused by deep excavation.

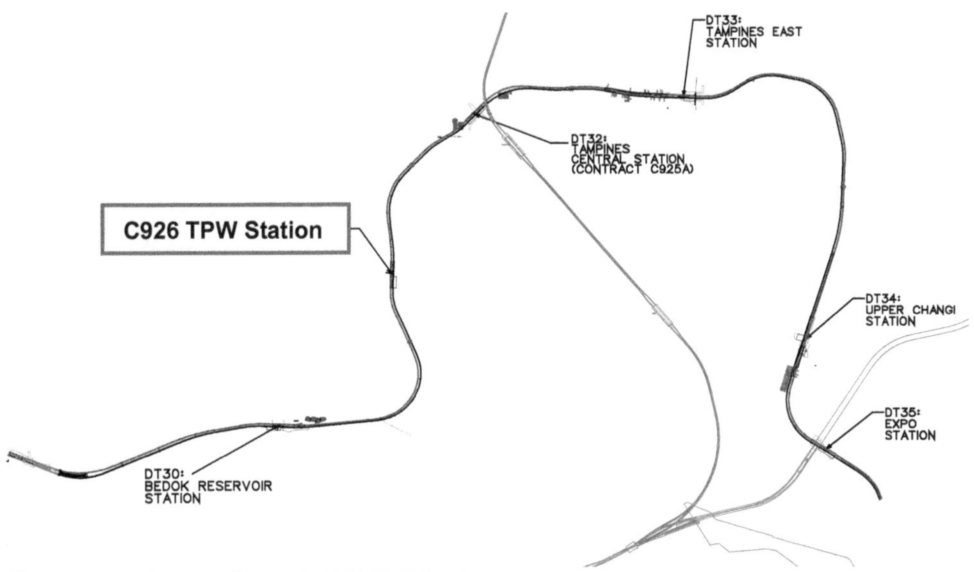

Figure 2. Location map for DTL3 C926 TPW Station.

2. Stagewise damage assessment

Damage assessment of buildings or structures adjacent to deep excavations is a major design consideration in densely built-up areas. These excavations are designed with earth retaining and stabilizing structures (ERSS), which must be robust enough to prevent and minimize any damage to the adjacent structures. It is necessary to predict the extent of ground movements that may cause damage to the structures. For buildings and structures on pile foundations, the following steps are part of the damage assessment procedure:

(1) Predicting the vertical and horizontal movements of building and foundations which are determined from the numerical analysis by considering the foundation contribution in the continuum model or by empirical methods using Gaussian Settlement curve for the case of bore tunnelling works.

(2) Damage assessment of the structure based on the predicted vertical and horizontal movements and assuming greenfield conditions with buildings as masonry structures.

(3) Study on pile behaviour and response based on the reduction in pile skin friction due to change in stress-field in soil and thus the impact on the geotechnical capacity of the pile. The additional pile displacements, bending moments and shear forces induced on the pile due to the excavation would also be studied.

A case study based on DTL3 C926 TPW Station deep excavation effect on the adjacent pile foundation for a high-rise building is presented in the following sections. In general, the damage assessment procedure as described in Step 2 above for buildings and structures are carried out in three stages, Stages I–III which are discussed in the following paragraphs.

Stage I assessment is a preliminary assessment based on the allowable settlement or rotation according to CIRIA PR 30 (1996). If the predicted settlement contours shows more than 10 mm at the building location or if the settlement gradient is more than 1/500, the building or structure should be subjected to Stage II assessment. The predicted settlement is not only due to ERSS's direct deformation effect, but also the settlement due to ground water draw-down and consolidation settlement in case of clayey soils overlying highly pervious soils. Figure 5 shows the settlement contour around the TPW Station due to deep excavation. HDB Blocks No. 802, 803 and 933 fall within the settlement zone of more than 10 mm. Hence, they are subjected to Stages II and III of the damage assessment procedure.

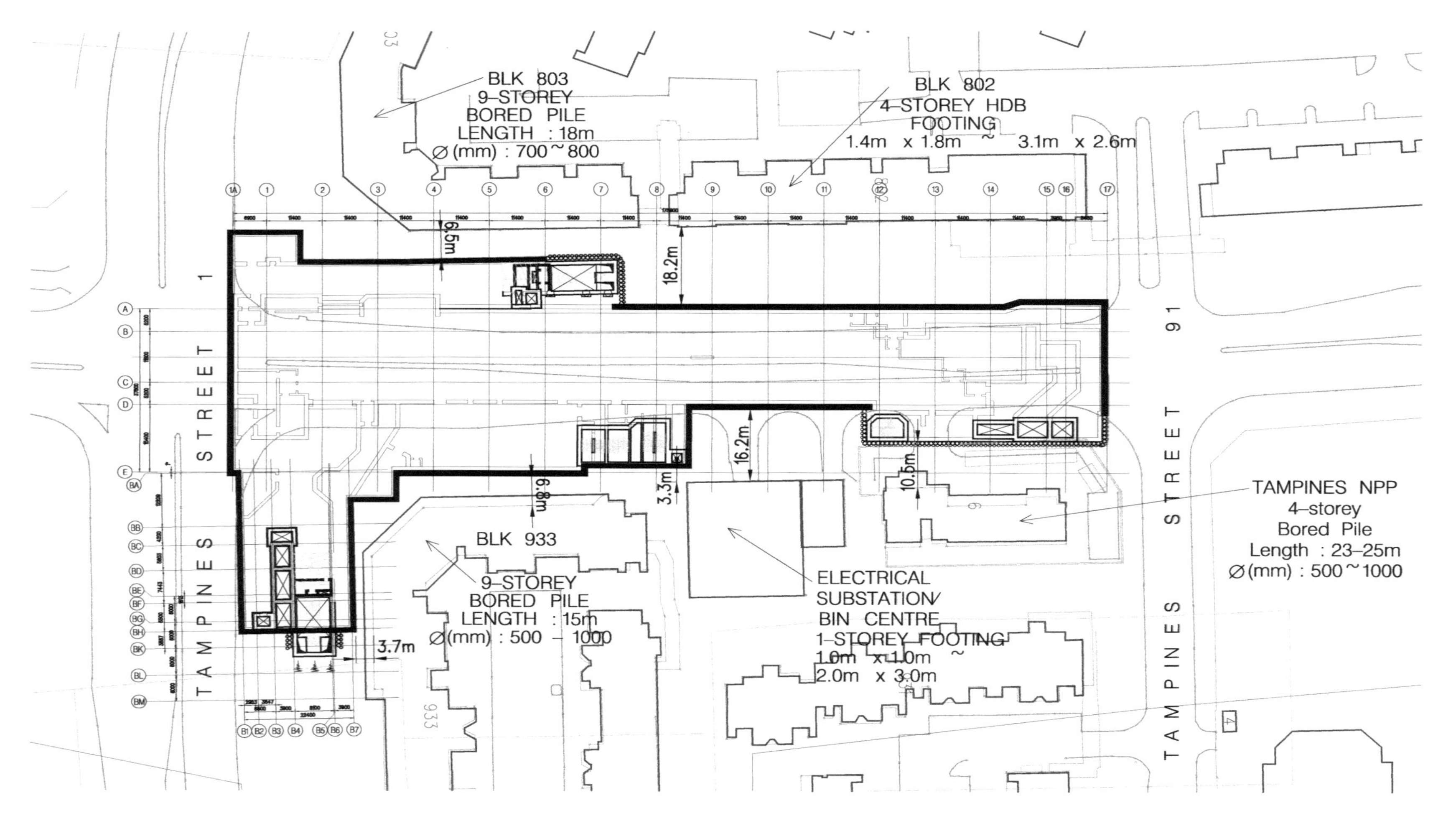

Figure 3. Plan showing proximity of the TPW Station to the HDB Blocks.

Table 1. Summary table of soil parameters								
Material	**Unit weight (kN/m³)**	**Strength parameters**			**Undrained modulus, E_u (MN/m²)**	**Drained Modulus, E' (MN/m²)**	**Coefficient of earth pressure at-rest, K_o**	**Permeability (m/s)**
		Total stress	**Effective stress**					
		S_u (kN/m²)	**c′ (kN/m²)**	**φ' (°)**				
Fill	20	30	0	-	-	8.7	0.5	10^{-7}
E	15	0.75z + 16.25 (20 ≤ S_u ≤ 35)	0	15	0.2S_u	E_u/1.2	1.0	10^{-9}
F1	20.5	-	0	30	-	8.7	0.7	10^{-5}
F2	19	1.5z + 12.5 (20 ≤ S_u ≤ 50)	5	25	0.2S_u	E_u/1.2	1.0	10^{-9}
M	16	1.285z + 3.575 for 10≤ S_u ≤55	0	22	0.3S_u	E_u/1.2	1.0	10^{-9}
Old Alluvium								
OA (E) (N < 10)	20	5 N	0	30	1.0	E_u/1.2		10^{-7}
OA (D) (10 ≤ N < 30)	20	5 N	5	32	2 N	E_u/1.2		10^{-7}
OA (C) (30 ≤ N < 50)	21	5 N	10	32	2 N	E_u/1.2	0.7[5]	10^{-7}
							1.0	
OA (B) (50 ≤ N < 100)	21	3 N + 100	10	35	1.2 N + 40	E_u/1.2		10^{-7}
OA (A) (N ≥ 100)	21	400	20	35	160	E_u/1.2		10^{-7}

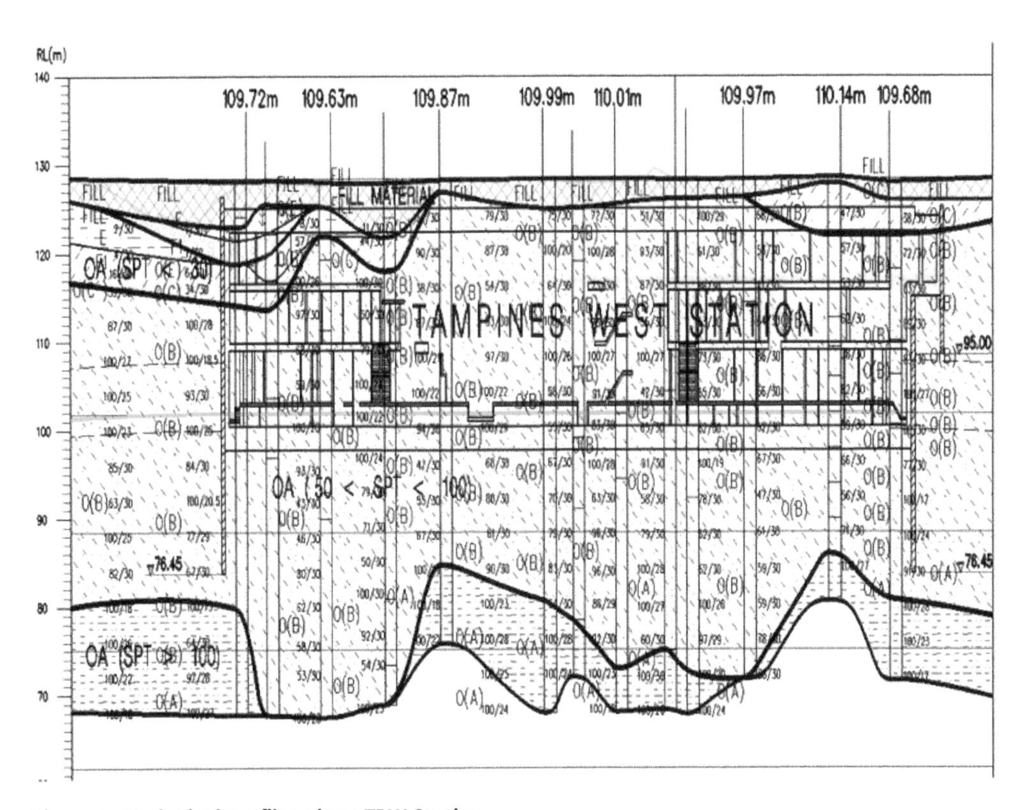

Figure 4. Geological profiles along TPW Station.

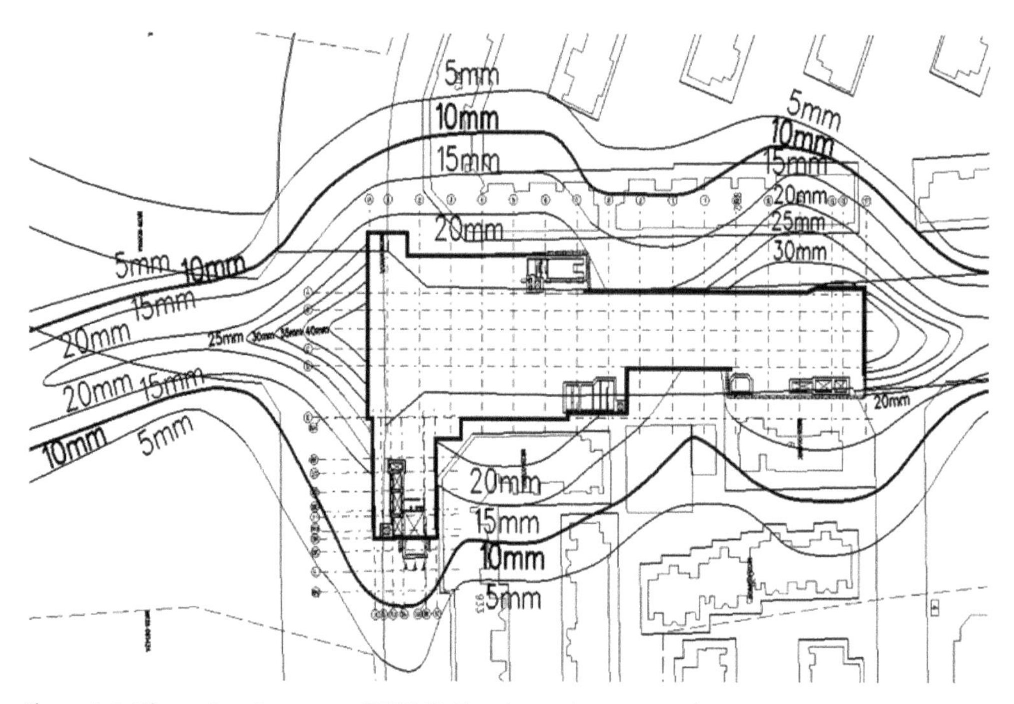

Figure 5. Settlement contour around TPW Station due to deep excavation.

Stage II assessment is based on limiting tensile strain approach adopted by Burland and Wroth (1974), Boscardin and Cording (1989) and Burland (1997, 2008), where the building is idealized as an equivalent deep beam of length, L and height H as shown in Figure 6, with an assumption that the building follows the settlement trough and also the lateral movements induced by deep excavation works as in Figure 7.

The deflection ratio, h/l_h and s/l_s, where suffix "s" is for sagging and "h" is for hogging, is a measure of curvature and various induced strains viz., maximum extreme fibre strain, $b_{(max)}$ (bending), maximum diagonal strain, $d_{(max)}$ (shear) and their respective resultants, ε_{br} and ε_{dr} when combined with horizontal strain, h. The procedure to estimate various strains is well described in publications of Burland (2008) and Loganathan (2011).

The maximum induced strain among these resultant strains is set as the limiting tensile strain, ε_{lim} and a range in the limiting tensile strain is used to categorize the damage to the building from being at "Negligible" through to "Very Severe" risk as shown in Table 2. Description of typical damage according to degree of severity with particular reference to ease of repair of plaster and brickwork or masonry can be seen from references by Burland (2008), Civil Design Criteria of LTA (2008), and Loganathan (2011).

Stage III assessment is a detail assessment of the structures with numerical analyses. In general, all structures that have been classified in the "Moderate" or higher damage risk categories during Stage II assessment are classified as "Sensitive Structures". However, historical and sensitive structures with a "Slight" damage category and any structures on pile and mixed foundations will also be subjected to Stage III assessment according to LTA Civil Design Criteria (2008). The Stage III assessment can be performed by using either the method proposed by Potts and Addenbrooke (1996) or using numerical modelling by incorporating the building stiffness as well. In addition, all reinforced concrete structures will be assessed based on their service-ability limits. Two-dimensional analyses are usually carried out instead of 3D analyses due to the complexity of the 3D modelling procedures. While selecting the 2D numerical approach for Stage III assessment, the following factors should be kept in view:

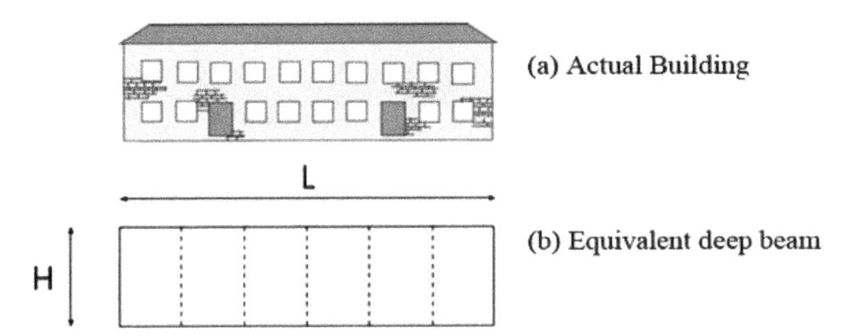

Figure 6. Building idealization for Stage II damage assessment.

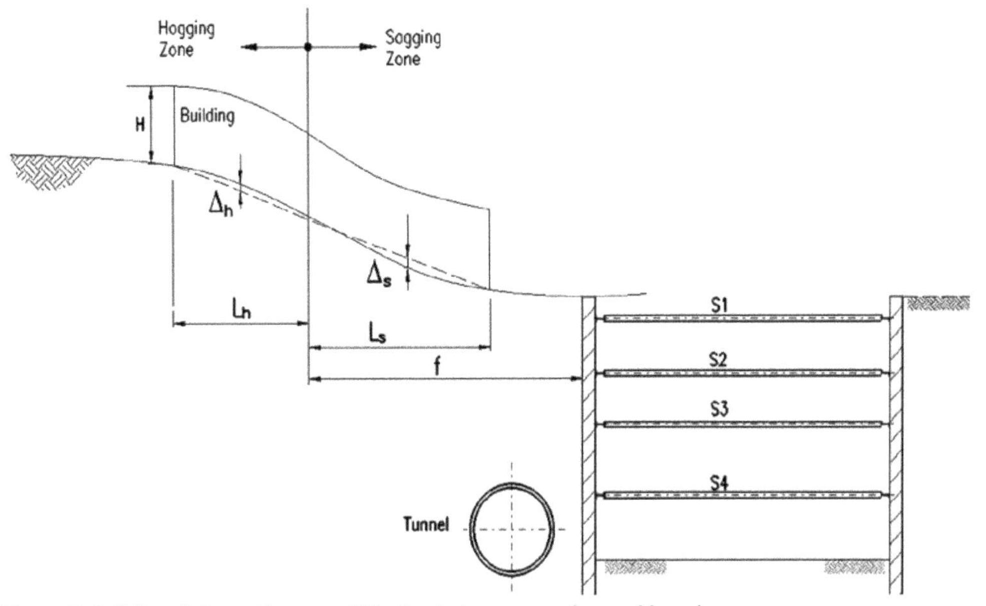

Figure 7. Building deformation—partitioning between sagging and hogging.

Table 2. Relationship between category of damage and limiting tensile strain

Category of damage	Normal degree of severity	Limiting tensile strain, ε_{lim} (%)
0	Negligible	0–0.05
1	Very slight	0.05–0.075
2	Slight	0.075–0.15
3	Moderate	0.15–0.3
4 and 5	Severe to very severe	>0.3

- Since piles are discrete elements, smearing of pile stiffness should be suitably considered.
- If as-built information is not suitable like pile length, geophysical survey to ascertain the pile length and a set of parametric study needs to be carried out for a possible range of pile lengths to make it reasonably conservative.

3. Assessment of impact on pile foundation by numerical analyses

This section will discuss the numerical analyses aimed to check for the changes in geotechnical pile capacities which are quite likely to happen due to the settlements and changes in stress field in the soil surrounding the piles due to wall deflection and base heave during excavation. This stress field changes will lead to a reduction in the effective normal stress and thus the skin friction on the piles, which in turn will increase in end bearing pressure. These changes need to be checked for the ultimate end bearing capacity of the pile. The typical section of ERSS for TPW station adjacent to the HDB Block 803 is as shown in Figure 8.

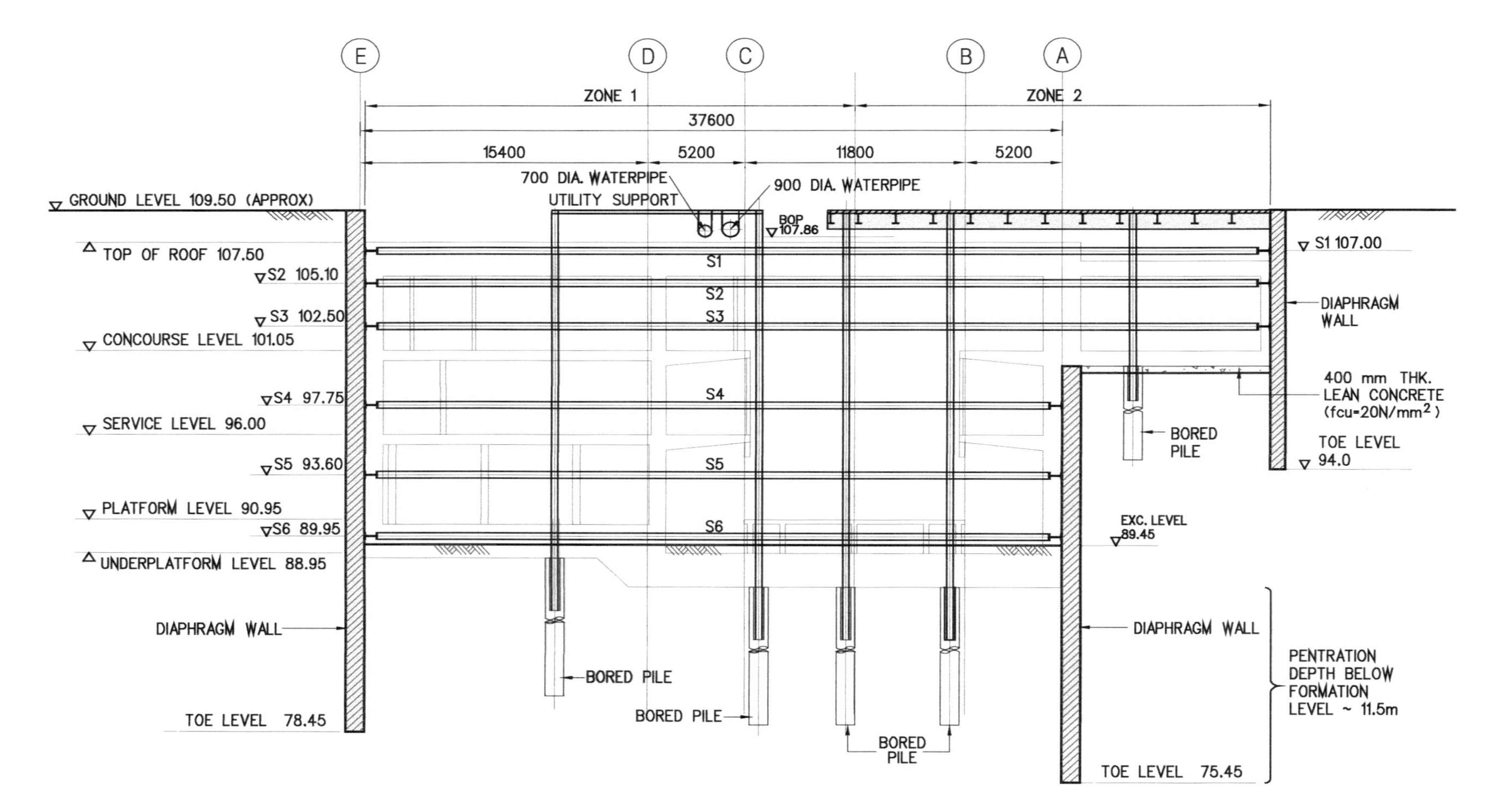

Figure 8. Typical sections of ERSS for TPW station adjacent to the HDB Block 803.

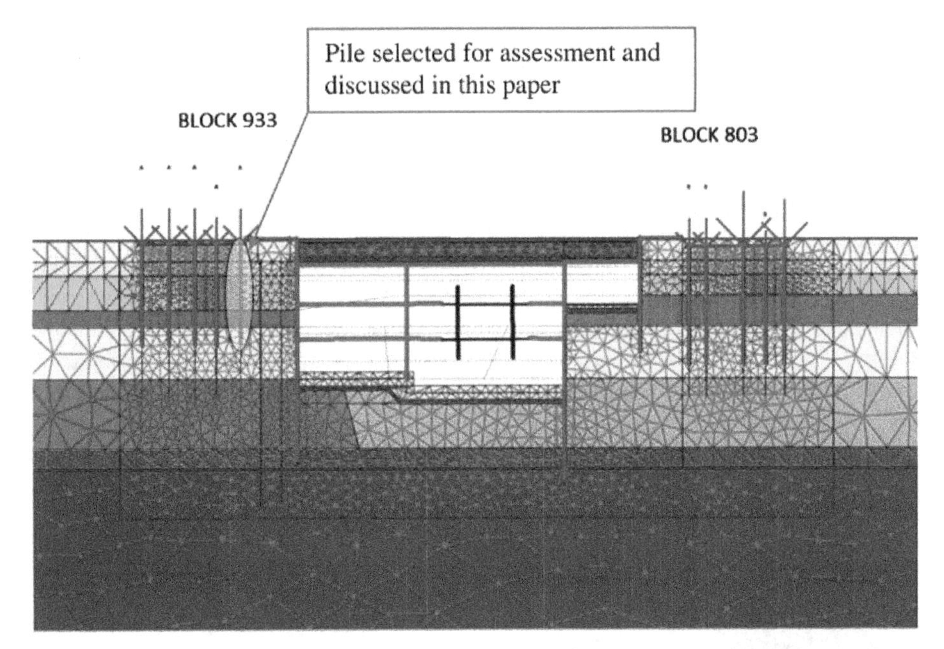

Figure 9. Finite element meshes showing the pile foundation arrangement adjacent to a deep excavation (encircled closest pile considered in this study).

For a typical case of piles supporting 9-storey HDB Blocks No. 803 and 933 adjacent to deep excavation for TPW Station, the finite element model is shown in Figure 9. This model is a 2D plane strain model with 15 noded isoparametric triangular elements for soil layers and 5 noded plate elements for all structural elements except for struts for which node-to-node anchor elements/plane truss element were adopted. The piles were also modelled as plate elements with their stiffnesses smeared for the average pile spacing and the soil–pile interaction was modelled by using the interface elements around the pile elements which were assigned with reduced soil strength properties (decreased by a strength reduction factor, R_f). In the stress analysis, both x and y displacements were set to zero at the bottom boundary, whereas at the truncated sides, only nodal displacements in the x-direction were set to zero.

The soil constitutive model used in the analysis was the Mohr–Coulomb model, an elastic perfectly plastic bilinear stress strain model. Undrained behaviour was set for all clay layers by choosing undrained material type, undrained elastic modulus, E_u, and undrained strength parameters of S_u and $\Phi_u = 0$ and drained behaviour was set with drained material type, effective elastic modulus, E′ and effective stress parameters viz, c' and φ' for all sandy soils. However, since OA soil behaviour is in the transition between drained and undrained behaviour, both these cases were considered thus leading to two sets of numerical analyses covering drained and undrained cases. In all cases of numerical analyses, pore water pressure of soil elements was determined by a steady-state 2D seepage analysis which was carried out prior to elasto-plastic stress analysis for each excavation/construction stage. Calculated pore water pressures in this way were used for the effective stress calculations and then used in the elasto-plastic analysis.

The pile foundation for the HDB Block No. 803 and 933 consists of 700 mm diameter bored piles at approximately 3.5 m centre-to-centre spacing. The lengths of the piles were not available in the as-built drawings retrieved from BCA. Initial estimate of the piles was based on the estimated working load from the columns. Later, geophysical survey was carried out to verify the estimated pile length. Finally, pile length of 18 m was adopted in the analyses and study. For ERSS design, Sandi, Shen, Leung, Liew, and Kho (2007) compared 2D and 3D FE analyses with field measurements and a range of smearing factors to be adopted in 2D FE analysis and concluded that using smearing factor of $3d_{pile}$ gives similar predictions using 3D numerical analysis. In 2D analyses, we adopt a smearing effect for the discrete structural elements like piles to simulate the discrete element using 2D plane strain analysis.

4. Results of numerical analyses

Figures 10 and 11 show the vertical and horizontal soil displacements, respectively, for the complete cycle of excavation and backfill. As shown in Figure 10, the downward displacement on both sides of the plate element, which represents the pile, is unsymmetrical. The vertical downward displacement on the side nearer to the excavation is larger than the vertical downward displacement on the other side further away from the excavation. This has resulted in the unsymmetrical changes in the shear stress on both sides of the plate element. This will be discussed in more detail in Section 5 of this paper. On the other hand, as shown in Figure 11, the horizontal displacement of the soil on the side of the plate element is almost similar. This is due to the reason that the pile element has been modelled as a thin plate element and thus has not caused any significant change in the stress field and vertical movement of the soil on both sides of the plate element. This has resulted in symmetrical decrease in the normal stress on the plate element.

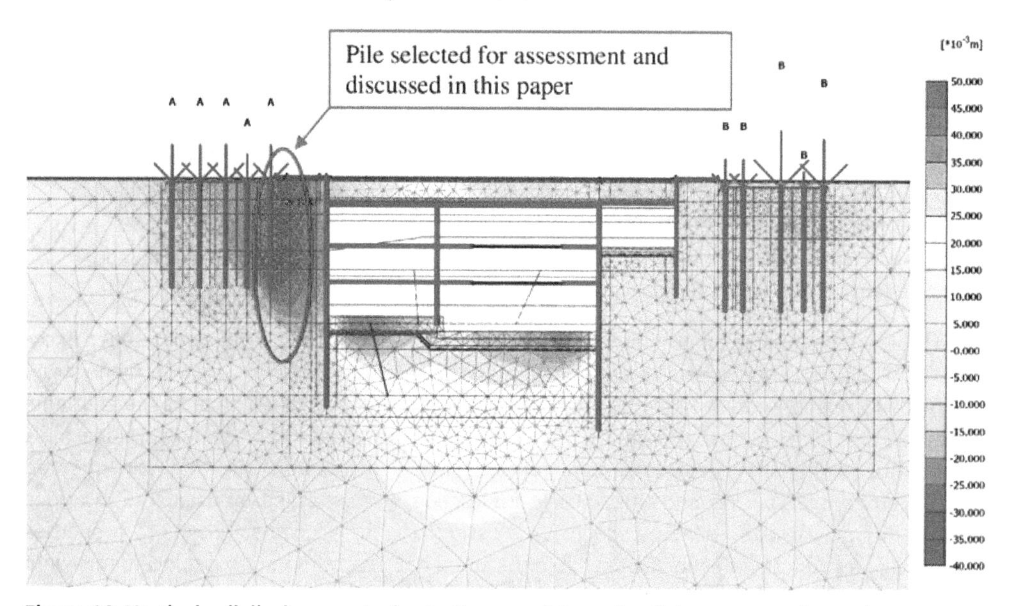

Figure 10. Vertical soil displacements due to the complete cycle of stage excavation and construction of the station structure.

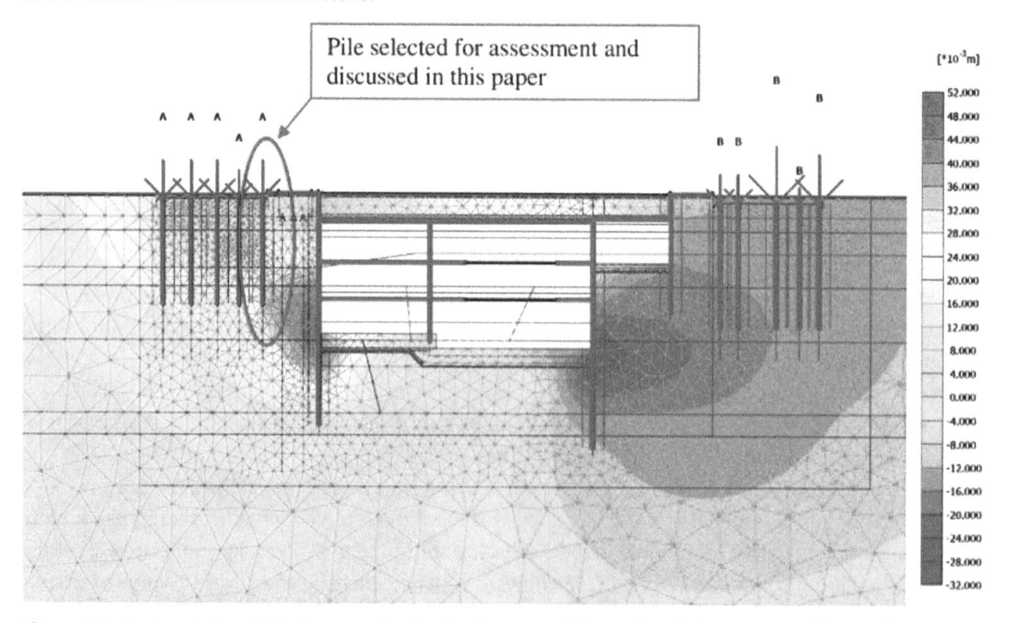

Figure 11. Horizontal soil displacements due to the complete cycle of stage excavation and construction of the station structure.

5. Reduction in skin friction resistance of the pile

For the pile close to the deep excavation, comparison of effective normal stress distribution on the pile shaft before and after excavation is shown in Figure 12. The kinks especially at elevations closer to the ground levels seen at the plot of effective normal stress after excavation are due to the pre-loading effect. The resultant of the normal stress on pile interfaces both on soil side and excavation side is same. However, the resultant of the normal stress before the excavation is 1,155 kN/m, whereas after excavation, it reduces to 878 kN/m, which is about 24% reduction from the value before excavation. The response of pile in terms of the normal stress acting on the pile is almost symmetrical due to the reason that the pile has been modelled as a plate element with smeared properties of the pile. This thin plate element has caused insignificant change in the stress field on both sides of the plate due to the low stiffness. In addition, the results also indicate that there is no build up of active and passive soil pressure on the two sides of plate element when soil movement occurred. This is likely due to the low stiffness of the plate element.

In the conventional method of checking the geotechnical capacity of bored pile, empirical relationship, $fs = Ks \times N$ is commonly used to determine the ultimate skin friction, where $N = SPT$ values and the coefficient, Ks, varies from 1.5 to 2.5 for stiff-to-hard cohesive soils, including Bukit Timah Granite and Jurong formation soils and 2–3 for dense and hard, cemented OA soils. For both the cases, the limiting values of fs are specified CP04:200 as 150 and 300 kPa respectively. Similar way of estimating the ultimate end bearing from the *SPT-N* values is also described in CP 04:2003 (2003). While adopting this empirical method for estimating the pile's ultimate geotechnical capacities, the same reduction factor for the skin friction due to adjacent deep excavation can be applied in order to include the effect of adjacent deep excavation on ultimate skin friction of the piles.

Comparison of developed skin friction on the both sides of the plate elements which simulate the pile shaft before and after excavation is shown in Figure 13. The sign convention for the representation of the graphs is positive for upward direction shear stress and negative for downward direction shear stress for the face near to excavation and is negative for upward direction shear stress and positive for downward direction shear stress for the face away from excavation. The shear stress

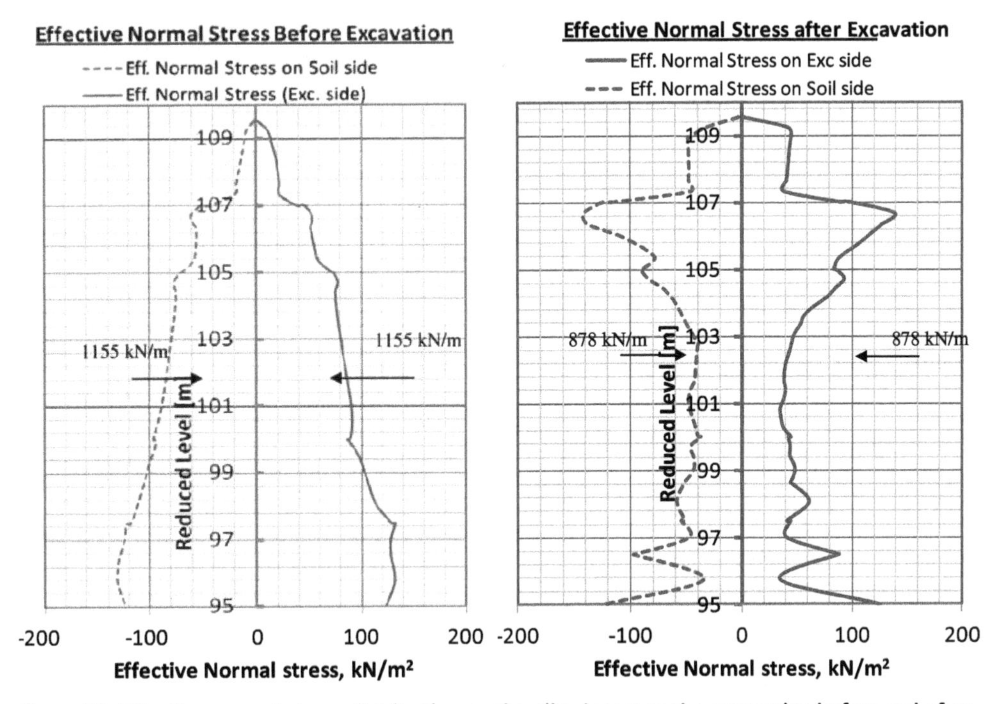

Figure 12. Effective normal stress distribution on the pile closest to the excavation before and after excavation.

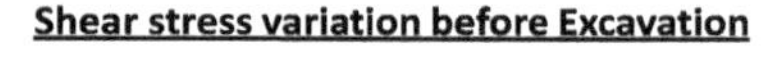

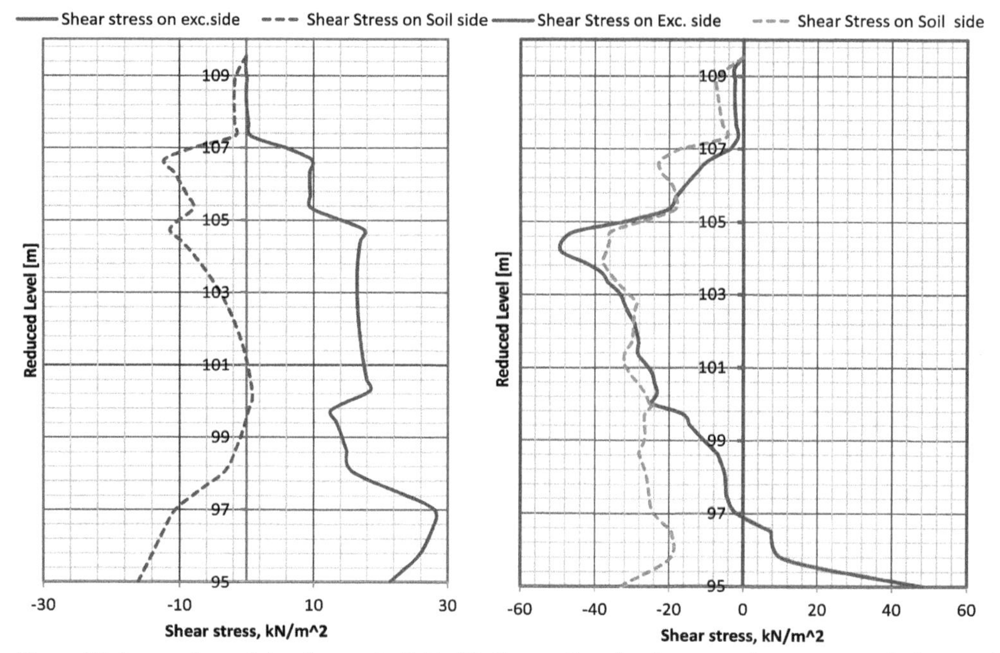

Figure 13. Comparison of development of skin friction on the pile closest to the excavation before and after excavation.

distribution on the two sides of the plate element of the pile before excavation is both showing upward direction but is unsymmetrical. The unsymmetrical distribution is due to the reason that the pile is on the edge of the series of piles for the building.

After excavation, there is soil movement towards the excavation and also subsurface soil settlement. The resulting soil movement due to the excavation is such that there is also similar change in the horizontal displacement on both sides of the plate element and there is more downward displacement of the soil on the side of the plate element closer to excavation than the other face away from excavation. As a result of this soil movement, there is an increase in the upward shear stress on the face of the plate element away from the excavation, while there is a decrease in the upward shear stress and increase in downward shear stress on the face of the plate element near to the excavation. This is likely to be caused by the settlement of the subsoil layers and induces negative skin friction on the pile. The effect of the change in shear stress acting on the pile on its end bearing pressure and overall geotechnical capacity is discussed in more detail in Sections 6 and 7.

6. Increase in end bearing pressure of the pile

In order to maintain the force equilibrium condition, the reduction in skin friction will obviously lead to increase in pile axial force and thus the pile end bearing pressure. Figure 14 shows the pile axial force variation with depth. Before excavation, the axial force is 112 kN/m at the toe of the plate element. After excavation, the axial force at the toe of the plate element increases to 275 kN/m due to the increase in down drag of the soil as a result of settlement. Assuming that the displacements and stress field changes in the soil due to deep excavation would not change the bearing capacity of the soil, it is necessary to check whether this greater end bearing pressure is within the allowable end bearing capacity of the soil layer at the pile toe level. The assessment of pile geotechnical capacity will be described in Section 7.

7. Assessment of pile structural and geotechnical capacity

For the pile structural capacity, it has been checked using the bending moment and shear force diagram obtained from the analyses and multiplied with the smearing factor and found to be acceptable. The details for the structural capacity check is not included in this paper as it is a straightforward

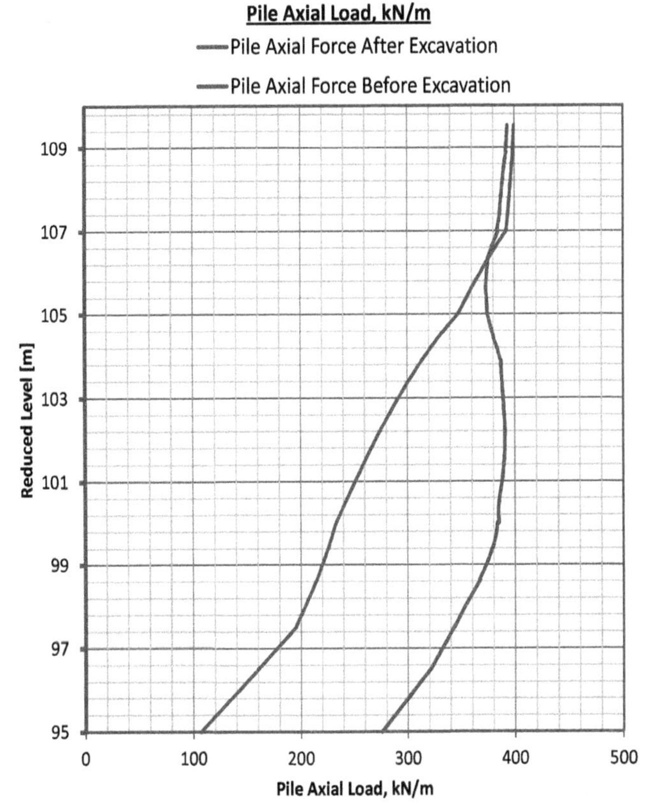

Figure 14. Variation of pile axial force with depth at the pile closest to the retaining wall before and after excavation.

procedure and commonly used by engineers. This section will focus on the assessment of geotechnical capacity of the pile as a result of the ground movement caused by deep excavation of TPW Station, which is important to ensure the safety of the building but not commonly checked by engineers probably due to the difficulties in interpreting the results from the analyses and understanding the response of the pile before and after excavation.

Table 3 shows the comparison of the pile shaft friction and factor of safety for the cases before and after excavation. The total mobilized skin friction per pile is calculated by multiplying the total shear stresses along the plate element on both faces by the smearing factor. The total mobilized skin friction was reduced from 1,004.5 to 490 kN. The average skin friction is obtained by dividing the total skin friction by the shaft area. The average skin friction is reduced from 30.6 to 15 kPa. After the excavation, the soil has been disturbed along the pile. The load transfer from the pile to soil will

Table 3. Summary table of comparison of pile shaft friction and factor of safety for the cases before and after excavation

Developed skin friction	Before excavation	After excavation
Skin friction mobilized	287 kN/m	140 kN/m
Skin friction per pile	287 × 3.5 kN	140 × 3.5 kN
	=1,004.5 kN	=490 kN
Average skin friction pressure	1,004.5/shaft area	490/shaft area
	=1,004.5/32.8	=490/32.8
	=30.6 kPa	=15 kPa
	Ult skin friction (2.5 N) = 125–250 kPa (Average *SPT* values are in the range of 50–100)	Shaft friction will be disturbed after excavation

be redistributed. The reduction in the shaft resistance will be compensated by the increase in the end bearing resistance. Hence, it is not necessary to evaluate the factor of safety of the pile in terms of shaft friction.

Table 4 shows the comparison of the pile end bearing and factor of safety for the cases before and after excavation. The end bearing force was obtained from the axial force at the toe of the plate element multiplied with the smearing factor. The end bearing force increases from 420 kN per pile before excavation to 963 kN per pile. This is corresponding to the increase of end bearing pressure on the soil from 1,010 to 2,502 kPa. The end bearing pressure is obtained by dividing the end bearing force per pile by the pile base area. The *SPT-N* value for the soil at the toe level of the pile is approximately 80. Based on the *SPT-N* value of 80 and limiting end bearing pressure, *Fb* = 60 N, the limiting end bearing pressure is 4,800 kPa. Thus, the factor of safety of the pile in terms of end bearing is adequate.

The above assessment for the geotechnical capacity of the pile has considered the increase in end bearing pressure of the piles predicted as by the analyses, and has calculated the factor of safety in terms of end bearing for the piles as a result of the station excavation. These show that the factor of safety is still sufficient to meet the minimum requirement.

Before the station is constructed, the piles will have a working load W applied at the pile head. This will be almost entirely resisted in skin friction as the OA strata are highly competent and the piles would predominantly be friction piles. The end bearing pressure at the base of the piles will be very small. During station excavation, there will be some relatively small deformations of the ground outside the diaphragm walls. As a result, the piles are likely to settle by a small amount and this will cause some redistribution of the friction and end bearing components of the pile load, with some increase in end bearing and a corresponding reduction in skin friction. However, the ultimate pile capacity has not been reduced. At the end of the station construction, the same working load W is applied to the pile (assuming no redistribution of load between adjacent piles); this is still resisted mainly by skin friction and by an increased amount of end bearing. The pile will have the same ultimate capacity. The only change is that the pile will simply have settled by a small amount, and there will have been a redistribution of the friction and end bearing components resisting the original pile working load.

Table 5 shows the summary table of the assessment of the pile. The pile lengths of 18 m have been considered for Blocks 803 and 933. In view of the uncertainty due to the unavailability of as-built information for the pile foundation, geophysical testing was essential to verify the assumed pile length. As the ground conditions are principally OA, the piles are acting as predominantly friction piles, with little or no load acting on the pile base. This means that the pile was mainly resisted by friction and has little resistance from end bearing in the original state. After excavation, the pile load transfer to soil merely change its path from via shaft friction to via end bearing at toe.

Table 4. Summary table of comparison of pile shaft friction and factor of safety for the cases before and after excavation

Developed end bearing pressure	Before excavation	After excavation
End bearing force/m	112 kN/m	275 kN/m
End bearing forces/pile	112 × 3.5 = 420 kN	275 × 3.5 = 963 kN
Developed end bearing pressure	392/pile area = 1,018 kPa	963/pile area = 2,502 kPa
Ultimate end bearing	Based on *N* = 80 and *Fb* = 60 N	Based on *N* = 80 and *Fb* = 60 N
	Fb = 4,800 kPa	*Fb* = 4,800 ka
FOS (end bearing)	4.7 (minimum and based on *N* = 80)	1.9 (minimum and based on *N* = 80)

Table 5. Summary table of impact on the pile foundation on HDB Block 803 and 933							
Building	**Distance from retaining wall (m)**	**Foundation type**	**Pile length from ground level (m)**	**Maximum pile/footing absolute movements (mm)**	**Maximum horizontal pile relative movement (mm)**	**Maximum bending moment due to excavation (kN/m)**	**Differential settlement angular distortion**
Block 933 (9 Storey)	7	Bored pile foundation	18 (GPR)	25(Hor)	15	24	1:2100
			15 (Est)	18.6 (Vert)			
Block 803 (9 Storey)	6.5	Bored pile foundation	18 (GPR)	18 (Hor)	5	10.6	<1:10000
			15 (Est)	8 (Vert)			

As shown in Table 5, Blocks 803 and 933 are predicted to settle by 8 and 19 mm, respectively, assuming the stiffness for the OA to correspond to $E_u/N = 3$. These settlement predictions are likely to be conservative because the stiffness of the OA is generally higher than given by $E_u/N = 3$. The inherent stiffness of the buildings will also lead to load redistribution between the piles and lead to smaller settlement than prediction.

8. Conclusions and recommendations

In conclusion, the damage assessment approach and procedure for buildings on pile foundation due to adjacent deep excavation or tunnelling as discussed in this paper is reasonable. The variations of the effective normal stress on the pile before and after the deep excavations are also examined and then an approach for assessing the effect on the ultimate skin friction and end bearing of the pile is presented. The redistribution of load transfer from the pile to the soil has also been highlighted. The analysis results show the phenomena of load transfer in pile from shaft friction to end bearing during excavation for the proposed MRT Station. The transfer of load from shaft friction to end bearing is associated with a small amount of settlement as reflected in the analysis.

The decrease in shaft friction and the increase in end bearing of pile do not compromise the overall capacity of the pile. The only change is the pile will settle by a small amount which has insignificant impact to the existing building. The predicted pile/footing settlement will likely be small, in the range of 5–25 mm in competent OA, which has insignificant impact to the buildings. Adopting $E_u = 3$ N for prediction of building settlement is reasonable. The piles have adequate FOS despite the transfer of resistance from shaft friction to end bearing.

Based on the assessment of pile response, predicted movement, structural and geotechnical capacities of the pile, it was found to be within the acceptable limit and the pile foundation has adequate factor of safety with the deep excavation in close proximity.

Using the proposed method, the complete damage assessment of buildings supported on pile foundations can be carried out. It is also shown that by using smearing factor of $3d_{pile}$, the 2D FEM analyses results are appropriate.

Acknowledgement
The authors would like to thank Mr Song Siak Keong of Land Transport Authority (LTA) for allowing the information for the project to be published in this paper.

Funding
The authors received no direct funding for this research.

Author details
C.G. Chinnaswamy[1]
E-mail: cgc@meinhardt-infra.com.sg
David N.G. Chew Chiat[1]
E-mail: ncc@meinhardt-infra.com.sg; davidng456@yahoo.com
[1] Meinhardt Infrastructure Pte Ltd, Singapore, Singapore.

References

Boscardin, M. D., & Cording, E. G. (1989). Building response to excavation-induced settlement. *Journal of Geotechnical Engineering, 115*, 1–21. http://dx.doi.org/10.1061/(ASCE)0733-9410(1989)115:1(1)

Burland, J.. (2008, 16 de Diciembre de). *The assessment of risk of damage to buildings due to tunneling and excavations.* Jornada Tecnica de Movimientos de Edificios Inducidos por Excavaciones, Barcelona.

Burland, J. B. (1997). Assessment of risk of damage to buildings due to tunneling and excavation. In Ishihara (Ed.), *Earthquake geotechnical engineering* (pp. 1189–1201). Rotterdam: Balkema.

Burland, J. B., & Wroth, C. P. (1974). Settlement of buildings and associated damage, SOA review. In *Conference on settlement of structures* (pp. 611–654). Cambridge: Pentech Press.

Chiam, S. L., Wong, K. S., Tan, T. S., Ni, Q., Khoo, K. S., & Chu, J. (2003). The old alluvium. In *Proceedings underground Singapore 2003* (pp. 409–440). Singapore: Nanyang Technological Singapore.

Chu, J., Goh, P. P., Pek, S. C., & Wong, I. H. (2003). Engineering properties of the old alluvium soil. In: *Proceedings underground Singapore 2003* (pp. 285–315). Singapore: Nanyang Technological Singapore.

Civil design criteria—Revision A7 for road & rail transit systems. (2008). Land Transport Authority (PED/DD/K9/106/A6).

CIRIA PR 30. (1996). *Prediction and effects of ground movements caused by tunneling in soft ground beneath urban areas* (Project Report 30). London: Construction Industry Research and Information Association.

CP 04:2003 (Singapore Standard). (2003). *Code of practice for foundations.*

Loganathan, N. (2011). *An innovative method for assessing tunneling induced risks to adjacent structures.* PB2009 William Barclay Parsons Fellowship Monograph 25. New York, NY: Parsons Brinckerhoff.

Potts, D. M., & Addenbrooke, T. I. (1996). The influence of an existing surface structure on the ground movements due to tunneling. In *International Symposium on Geotechnical Aspects of Underground Construction in Soft Ground* (pp. 573–578). Rotterdam: A A BALKEMA.

PWD. (1976). *Geology of the Republic of Singapore.* Singapore: Author.

Sandi, M. S., Shen, R. F., Leung, C. F., Liew, Y. K., & Kho, C. M. (2007). *Comparison of 2D and 3D FEA with measurements of pile response adjacent to deep excavation.* Underground Singapore 2007. Singapore.

Wong, K. S., Li, W., Shirlaw, J. N., Ong, J. C. W., Wen, D., & Hsu, J. C. W. (2001). Old alluvium: Engineering properties and braced excavation performance. In *Proceedings underground Singapore* (pp. 210–218). Singapore: Nanyang Technological Singapore.

12

Use of triangular membership function for prediction of compressive strength of concrete containing nanosilica

Sakshi Gupta[1*]
*Corresponding author: Sakshi Gupta, Department of Civil Engineering, ASET, Amity University, Gurgaon, Haryana, India
E-mail: ersakshigupta18@gmail.com

Reviewing editor: Raja Rizwan Hussain, King Saud University, Saudi Arabia

Abstract: In this paper, application of fuzzy logic technique using triangular membership function for developing models for predicting compressive strength of concrete with partial replacement of cement with nanosilica has been carried out. For this, the data have been taken from various literatures and help in optimizing the constituents available and reducing cost and efforts in studying design to develop mixes by predefining suitable range for experimenting. The use of nanostructured materials in concrete can add many benefits that are directly related to the durability of various cementitious materials, besides the fact that it is possible to reduce the quantities of cement in the composite. Successful prediction by the model indicates that fuzzy logic could be a useful modelling tool for engineers and research scientists in the area of cement and concrete. Compressive strength values of concrete can be predicted in fuzzy logic models without attempting any experiments in a quite short period of time with tiny error rates.

Subjects: Composites; Civil, Environmental and Geotechnical Engineering; Concrete & Cement; Structural Engineering; Mathematics & Statistics for Engineers; Nanoscience & Nanotechnology

Keywords: Fuzzy Logic; Nanosilica; Concrete; Compressive strength; prediction; Triangular membership function; Modelling

ABOUT THE AUTHOR

Sakshi Gupta is working as an assistant professor, Civil Engineering Department, ASET, Amity University, Manesar, Gurgaon, Haryana, India. Her areas of interest include the use of nanosilica in paste, mortar and concrete, its study of mechanical and durability properties, use of data mining techniques such as ANN and Fuzzy logic to correlate different properties of concrete and high-strength concrete, their performance and durability aspects, incorporating waste materials. In order to predict the effects of nanosilica on compressive strength values of concrete, models were carried out in fuzzy logic system. Successful prediction by the model indicates fuzzy logic could be a useful modelling tool for engineers and research scientists in the area of cement and concrete. Compressive strength values of concrete can be predicted without attempting any experiments in quite short period of time with tiny error rates. The present research work is one such effort towards attaining above goal.

PUBLIC INTEREST STATEMENT

This work deals with the optimization of the constituents available and reducing cost and efforts in studying design to develop mixes by predefining suitable range for experimenting. This has been done using the Fuzzy Logic tool in MATLAB where the parameters have been chosen from different literatures which are used in the prediction of the compressive strength of concrete containing nanosilica. This will help to know the best possible use and the amount of replacement of nanosilica with cement in concrete and knowing the optimized percentage replacement for a high compressive strength of concrete. This will in turn reduce the cost of the structure when the concrete containing nanosilica is employed in construction work.

1. Introduction

Concrete being one of the oldest materials in the construction industry is a mixture of paste and aggregates. It is the most widely used construction material due to its flowability in most complicated forms and its strength development characteristics when it hardens.

The necessity of the cementitious materials in the construction industry is nowadays beyond any doubt; however, their variety of applications must not hide their complexity. They are indeed composite materials with truly multi-scale internal structures that keep evolving over centuries. With the onset of nanotechnology, it allows engineers and architects to use various materials in structural applications that were once impossible. Nanotechnology creates new prospects to improve the material properties for civil construction. Attracting civil engineers to adopt nanotechnology could enable them to provide pioneering elucidation to the complicated problems of construction today. It is well known that materials such as concrete, being the core elements of construction industry, could be developed using nanotechnology. Nanosilica is typically a highly effective pozzolanic material consisting of very fine vitreous particles approximately 1,000 times smaller than the average cement particles. nS has proven to be an excellent admixture for cement to improve strength and durability and decrease permeability (Aitcin, Hershey, & Pinsonneault, 1981; ARI News, 2007).

The present study was envisaged to develop a relationship between various input parameters and an output parameter, i.e. 28-day compressive strength, using triangular membership function in fuzzy logic technique. The objective was to use the triangular membership function for prediction of compressive strength of concrete containing nanosilica, with data obtained from literature (Gupta, 2014).

Over the last two decades, different data mining methods, such as the fuzzy logic and artificial neural network, have become popular and have been used by many researchers for a variety of engineering applications. In daily life, information obtained is used to understand the surroundings to imbibe new things and to make plans for the future. Over the years, the ability to reason has been developed on the basis of evidence available to achieve the required goals. To deal with the problem of uncertainty, the theory of probability had been established and successfully applied to many areas of engineering and technology. The principal catalyst for introducing fuzzy theory is to represent the uncertain concepts. It does not need to handle the laborious mathematical models but only need to set a simple controlling method based on the engineering experiences. It is convenient and easy to use fuzzy logic models for numerical experiments to review the effects of each variable on the mix proportions (Akkurt, Tayfur, & Can, 2004; Demir, 2005; Topcu & Sarıdemir, 2008; Ünal, Demir, & Uygunoğlu, 2007).

2. Fuzzy logic

The concept of "fuzzy set" was preliminarily introduced by Zadeh (1965), who pioneered the development of fuzzy logic (FL) replacing Aristotelian logic which has two possibilities only. FL concept provides a natural and atypical way of dealing with the problems in which the origin of imprecision/unreliability is the absence of sharply defined criteria rather than the presence of random variables (Demir, 2005; Sen, 1998). Herein, the uncertainties do not mean random, probabilistic and stochastic variations, all of which are based on the numerical data. Fuzzy set theory provides a methodical calculus to deal with such information linguistically. Fuzzy approach performs numerical computation using linguistic labels stimulated by membership functions. Therefore, Zadeh introduced linguistic variables as variables whose values are sentences in a natural or artificial language. Although FL was brought forward by Zadeh (1965), the fuzzy concepts and systems attracted attention after a real control application conducted by Mamdani and Assilian in the year 1975 (Demir, 2005).

A general fuzzy inference system (FIS) has basically four components:

(1) Fuzzification
(2) Fuzzy rule base
(3) Fuzzy output engine
(4) Defuzzification

In Fuzzification, each piece of input data is converted to degrees of membership by a lookup in one or more several membership functions. Fuzzy rule base includes rules that have all possible fuzzy relation between inputs and outputs. These rules are expressed in the IF–THEN format. There are primarily two type of rule base: (1) Sugeno type and (2) Mamdani type. Fuzzy inference engine takes into consideration all the fuzzy rules in the fuzzy rule base and learns how to transform a set of inputs to the corresponding outputs. There are essentially two kinds of inference operators: minimization (min) and product (prod). Defuzzification converts the resulting fuzzy outputs from the fuzzy inference engine to a number. There are many defuzzification methods such as weighted average (wtaver) or weighted sum (wtsum). In the present study, the fuzzy model used is of Mamdani fuzzy rule type and the prod method was employed because of its more precise result methodology. For the defuzzification in the fuzzy model, weighted average method has been applied (Akbulut, Hasiloglu, & Pamukcu, 2004; Bouzoubaa & Lachemi, 2001; Ho & Zhang, 2001; Jang & Sun, 1995; Passino, 1998; Takagi & Sugeno, 1985).

The key idea in FL is the allowance of partial belongings of any object to different subsets of the universal set instead of belonging to a single set totally. Partial belonging to set can be described numerically by a membership function which assumes values between 0 and 1 contain. All the available inputs to a parameter are used to form their respective membership functions assigning a membership value of 1 to the crisp input in the database. The base of all the membership functions has been chosen to be the entire range of values to ensure that no region is left out of the functions. For convenience, functions have been assigned names on the basis of values of nodal points for the functions.

The accuracy of the predictions of a network was quantified by the root mean squared error difference (RMSE), between the measured and the predicted values and mean absolute error (MAE).

$$\mathrm{RMSE} = \sqrt{\frac{1}{N}\sum_{n=1}^{N}(actual - predicted)^2}$$

$$\mathrm{MAE} = \frac{1}{N}\sum_{n=1}^{N}(actual - predicted)$$

3. Database

The database for the FIS models was collected from available literature on concrete containing nanosilica, as summarized in Table 1.

Thus, large varieties of data were collected and in total 32 data-sets have been used with the following input and output variables (Beigi, Javad, Lotfi, Sadeghi, & Iman, 2013; Givi, Rashid, Aziz, & Salleh, 2010; Heidari & Tavakoli, 2013; Li, 2004; Li, Xiao, Yuan, Ou, & Ou, 2004; Ji, 2005; Jo, Kim, Tae, & Park, 2007; Nili, Ehsani, & Shabani, 2010; Said, Zeidan, Bassuoni, & Tian, 2012; Zhang & Li, 2010). The basic parameters considered in this study were cement content, fine aggregate content, coarse aggregate content, nanosilica content, diameter of nanosilica, water-to-binder ratio and superplasticizer dosage. The exclusion of one or more of concrete properties in some studies and the ambiguity of mixtures proportions and testing methods in others was responsible for setting the criteria for identification of data. The successful model to predict the 28-day compressive strength depends upon the magnitude of the training data using Triangular and Gaussian membership functions. The predicted results were compared with the values obtained experimentally.

Table 1. Details of data used in modelling

S. No.	Cement (kg/m³)	FA (kg/m³)	CA (kg/m³)	W/b ratio	SP (kg/m³)	nS (kg/m³)	D (nm)	28-d CS (MPa)	Researcher (Year)
1	396.6	826	722	0.37	7	16.5	15	75.2	Beigi et al. (2013)
2	380	826	722	0.35	7	33	15	86.1	
3	363.5	826	722	0.33	7	49.6	15	85.4	
4	318.4	840	1040	0.5	2.71	1.6	15	36.8	Heidari and Tavakoli (2013)
5	316.8	840	1040	0.5	4.75	3.2	15	40.2	
6	390	783	1175	0.4	1.78	23.4	35	70	Said et al. (2012)
7	390	774	1162	0.4	3.56	46.8	35	76	
8	390	769	1154	0.4	1.27	23.4	35	60	
9	390	762	1143	0.4	2.54	46.8	35	66	
10	356.4	650	1260	0.42	5.4	3.6	10	66.36	Zhang and Li (2010)
11	349.2	650	1260	0.42	7.2	10.8	10	61.16	
12	447.75	492	1148	0.4	0	2.25	80	39.2	Givi et al. (2010)
13	445.5	492	1148	0.4	0	4.5	80	40.3	
14	443.25	492	1148	0.4	0	6.75	80	41.2	
15	441	492	1148	0.4	0	9	80	38.1	
16	447.75	492	1148	0.4	0	2.25	15	42.7	
17	445.5	492	1148	0.4	0	4.5	15	43.6	
18	443.25	492	1148	0.4	0	6.75	15	42.9	
19	441	492	1148	0.4	0	9	15	39.7	
20	394	811	915	0.45	1.68	12	15	53.8	
21	388	811	915	0.45	2.32	24	15	56.5	
22	382	811	915	0.45	3	36	15	60	Nili et al. (2010)
23	247.5	625	0	0.5	4.5	7.5	40	54.3	Jo et al. (2007)
24	240.6	626	0	0.5	5.8	14.4	40	61.9	
25	241.8	627	0	0.5	7	23.2	40	68.2	
26	227.7	628	0	0.5	7.5	27.3	40	68.8	
27	370	647	1088	0.49	13.5	13.9	15	44	Ji (2005)
28	568.36	1757.8	0	0.5	8.85	17.5	15	32.9	Li et al. (2004)
29	556.64	1757.8	0	0.5	14.58	29.3	15	33.8	
30	527.34	1757.8	0	0.5	29.3	58.59	15	36.4	
31	556.64	1757.8	0	0.5	10.28	11.71	15	35.4	
32	480	647	1140	0.28	10	20	10	75.8	Li (2004)

Note: All types of SP have been considered to be same.

The variables used are as follows:

- Cement
- FA (fine aggregates)
- CA (coarse aggregates)
- W/b ratio
- SP (superplasticizer)
- nS (nanosilica)
- Diameter of nanosilica

The ranges of various input and output parameters used in FL technique are given in Table 2.

Table 2. Input and output variables

Variables	Parameter	Abbreviation	Database range	
			Minimum	Maximum
Input	Cement (kg/m3)	Cement	227.70	568.36
	Fine aggregate (kg/m3)	FA	492	1757.80
	Coarse aggregate (kg/m3)	CA	0	1260
	Water to binder ratio	W/b ratio	0.28	0.50
	Superplasticizer (kg/m3)	SP	0	29.30
	Nanosilica (kg/m3)	nS	1.60	58.59
	Diameter of nanosilica (nm)	D	10	80
Output	28-day compressive strength (MPa)	28-d CS	32.90	86.10

4. Application of FL technique

Fuzzy modelling is a system identification task involving two phases: structure identification and parameter prediction. Structure identification consists of the issues such as selecting the relevant input variables, choosing a specific type of FIS, determining the number of fuzzy rules and their antecedents and consequents, and determining the type and number of membership functions. Parameter prediction is the determination of aimed values response to evident input values of embodied model. For this aim, in the study, 32 data results were used in the processes of Mamdani-type Fuzzy interference model in the FL system. Training meant to present the network with the experimental data and have it learn, or modify its weights, such that it correctly reproduces the strength behaviour of mix. However, training the network successfully required many choices and training experiences. It was observed that an individual's membership in a fuzzy set admit some uncertainty and hence, it is said that its membership is a matter of degree of association.

5. Results and discussions

The various membership functions for different parameters are presented in Figure 1 (a–d) and the ruler view is shown in Figure 2. The if-then rule base is applied in this case. Membership functions are the building blocks of the fuzzy set theory. Membership functions were first prepared for each input and output data depending upon their ranges and variability. According to the importance, the shape of the membership function here was decided to be triangular. Figure 3 gives the surface diagrams between various parameters. The Rule Viewer is used as a diagnostic; it can show, for example, which rules are active, or how individual membership function shapes are influencing the results to predict the model. The Surface Viewer in the form of surface diagrams is used to display the dependency of one of the outputs on any one or two of the inputs—that is, it generates and plots an output surface map for the system. The MATLAB FL toolbox is used in the modelling where it generates a plot of the output surface of a given FIS using the first two inputs and the output. To compare the performance of models, graph between actual and predicted values is plotted and is represented by Figure 4. Table 3 gives the actual and the predicted values of the 28-day compressive strength of concrete containing different percentage replacement of cement with nanosilica. Table 4 gives the statistical parameters for the FL model, i.e. the correlation coefficient (CC), the root mean squared error (RMSE) and the mean absolute error (MAE). The best measure of model fit depends on the researcher's objectives, and more than one are often useful. The RMSE will always be larger or equal to MAE; the greater the difference between them, the greater the variance in the individual errors in the data-set. Since the correlation coefficient of the given set of data is 0.968581, i.e. it is near to 1, it indicates that the variables actual and predicted 28-day compressive strength have a positive correlation. This means that if one variable moves a given amount, the second moves proportionally in the same direction, with the strength of the correlation growing as the number approaches one. Minimizing the MAE is the key criterion of the development of the FL model and getting it close to zero which is an ideal condition, practically never possible. The MAE is representing the difference between the actual and the predicted observations of the model and is used to determine the extent to which the model fits the data and whether the removal or some

Table 3. Actual and predicted values of 28-day compressive strength

S. No.	Actual 28-d CS (MPa)	Predicted 28-d CS (MPa)
1	75.2	72.5
2	86.1	77.6
3	85.4	76.6
4	36.8	42.8
5	40.2	52.9
6	70	65.4
7	76	63.2
8	60	58.7
9	66	64.6
10	66.36	63
11	61.16	61.3
12	39.2	43.1
13	40.3	46.2
14	41.2	50.1
15	38.1	44.9
16	42.7	46.3
17	43.6	44.1
18	42.9	53.9
19	39.7	43.6
20	53.8	56.9
21	56.5	58
22	60	59.4
23	54.3	58.9
24	61.9	60.3
25	68.2	63.9
26	68.8	65
27	44	44.6
28	32.9	38
29	33.8	39.3
30	36.4	40.1
31	35.4	39.5
32	75.8	70.3

Table 4. Statistical parameters for the fuzzy logic model

S. No.	Statistical parameter	Value
1	Correlation coefficient (CC)	0.968581
2	Mean absolute error (MAE)	1.00875
3	Root mean square error (RMSE)	5.769963

explanatory variables, simplifying the model, is possible without significantly harming the model's predictive ability. Compared to the similar MAE, RMSE amplifies and severely punishes large errors and can be used to distinguish model performance. The values of MAE and RMSE were found out to be 1.008750 and 5.769963, respectively. The two biggest advantages of MAE or RMSE are that they provide a quadratic loss function and that they are also measures of the uncertainty in forecasting. Lower values of RMSE indicate better fit. RMSE is a good measure of how accurately the model predicts the response.

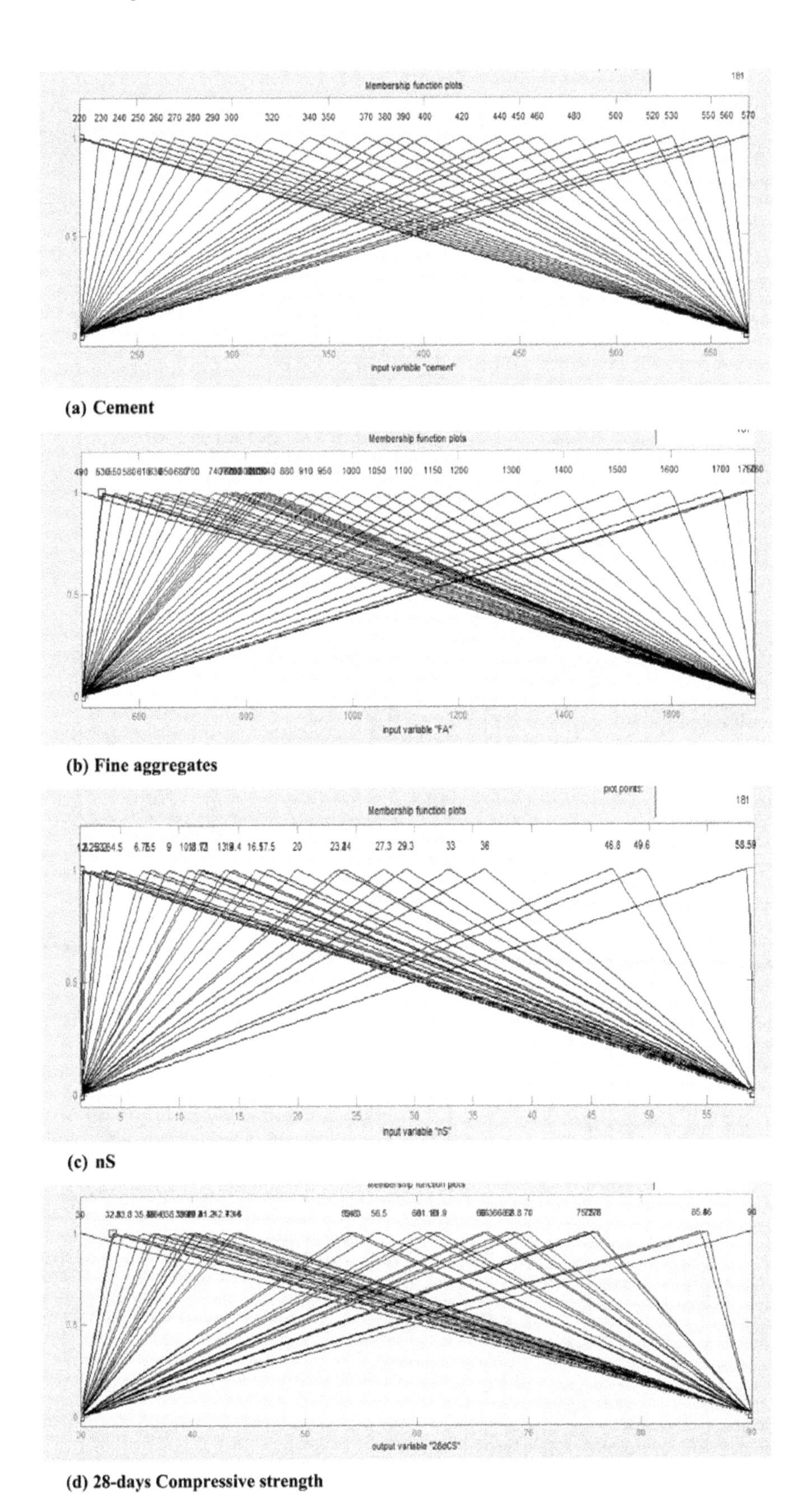

(a) Cement

(b) Fine aggregates

(c) nS

(d) 28-days Compressive strength

Figure 1. Membership functions for inputs and output (a) cement (b) fine aggregates (c) nanosilica (nS) (d) 28-day compressive strength.

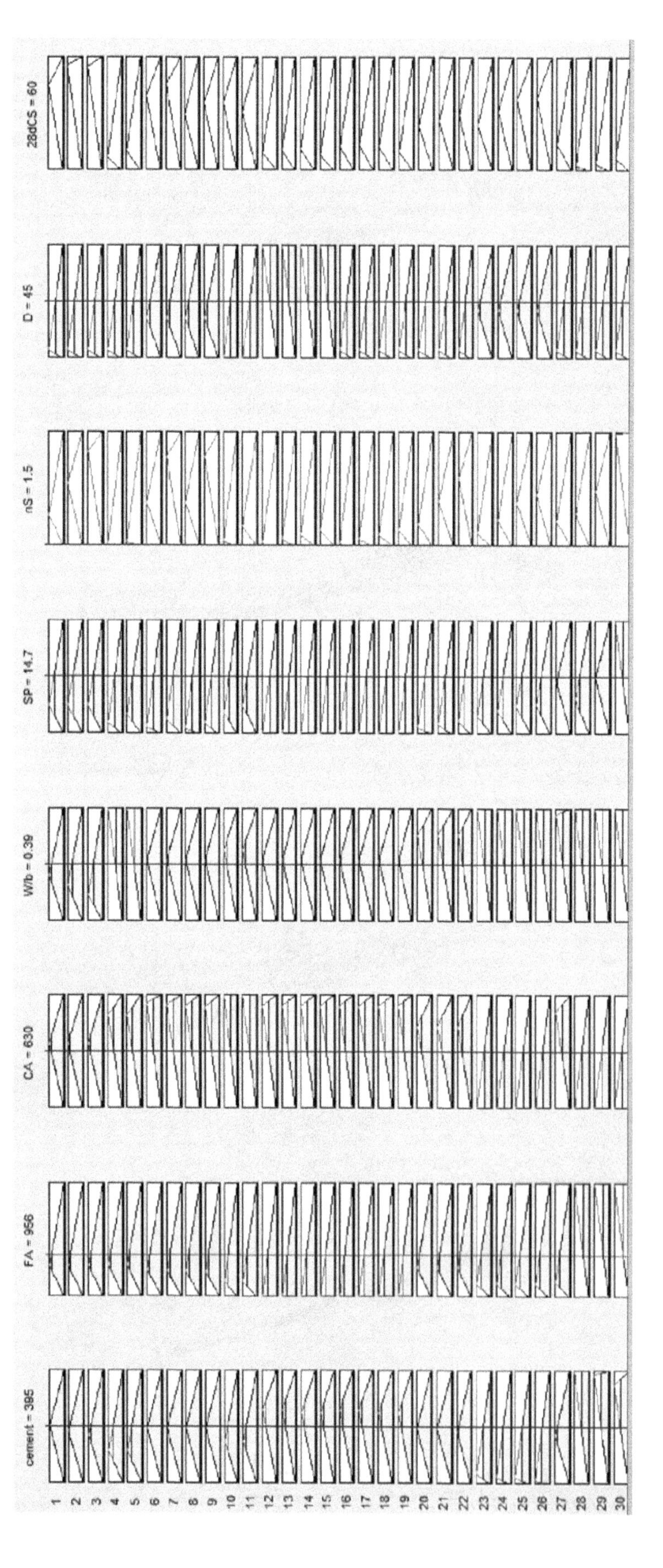

Figure 2. Rule viewer.

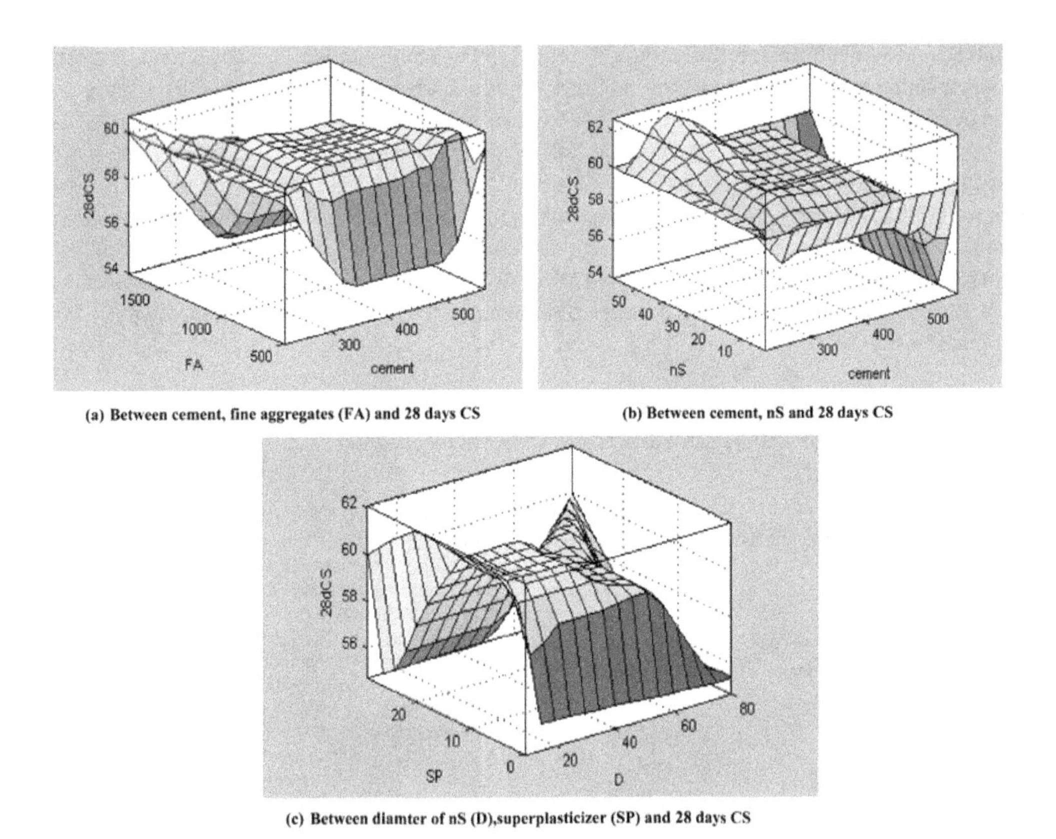

Figure 3. Surface diagrams (a) between cement, fine aggregates (FA) and 28-day CS (b) between cement, nS and 28-day CS (c) between diameter of nS (D) superplasticizer (SP) and 28-day CS.

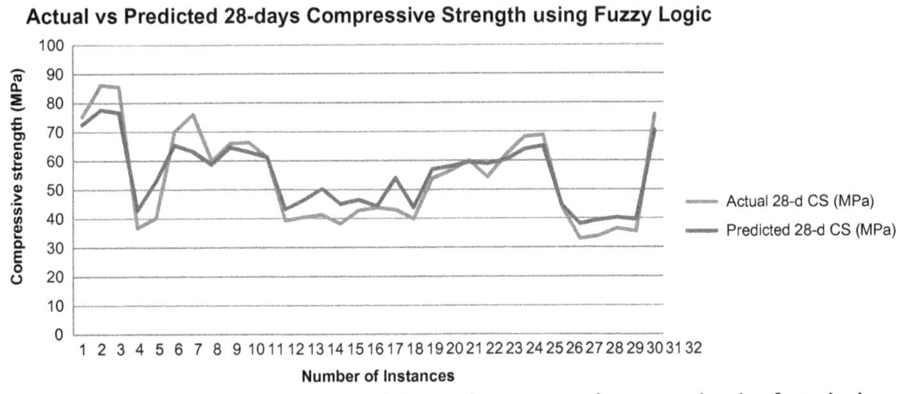

Figure 4. Actual vs. predicted results of the 28-day compressive strength using fuzzy logic approach.

6. Conclusions

In the present study, an FL prediction model for 28-day compressive strength has been developed. In order to predict the effects of nanosilica on compressive strength values of concrete without attempting any experiments, the models were carried out in FL system. A successfully trained model is characterized by its ability to predict strength values for the data it was trained on. The models were trained with input and output experimental data. The values are very closer to the experimental data obtained from FL models. CC, RMSE and MAE are statistical values that are calculated for comparing experimental data with FL model. The correlation coefficient is 0.968581 which is approaching 1 indicates that the correlation between actual and predicted values is strong. As a result, compressive strength values of concrete can be predicted in FL models without attempting any experiments in a quite short period of time with tiny error rates. The values of MAE and RMSE were found out to be

1.008750 and 5.769963, respectively. The difference between MAE and RMSE values is small indicating lesser variance in the individual errors in the data-set chosen. The FL technique could be improved upon by combining it with another method, i.e. artificial neural network and genetic algorithm for its optimization purpose. Also, the FIS could be further improved on using larger data-sets and more parameters. Also, in this work, the triangular membership function is used in the prediction of compressive strength of concrete containing nanosilica. This can be further done by considering other membership functions such as Gaussian, two-sided Gaussian, π-shaped, S-shaped, trapezoidal, sigmoid and bell shaped and also different rule base. As the number of data increases, there will be more accurate prediction and have a better correlation of the actual and predicted values of the compressive strength.

Thus, Successful prediction by the model indicates that FL could be a useful modelling tool for engineers and research scientists in the area of cement and concrete.

Funding
The authors received no direct funding for this research.

Author details
Sakshi Gupta[1]
E-mail: ersakshigupta18@gmail.com
[1] Department of Civil Engineering, ASET, Amity University, Gurgaon, Haryana, India.

References

Aitcin, P. C., Hershey, P. A., & Pinsonneault. (1981). Effect of the addition of condensed silica fume on the compressive strength of mortars and concrete. *American Ceramic Society, 22*, 286–290.

Akbulut, S., Hasiloglu, A. S., & Pamukcu, S. (2004). Data generation for shear modulus and damping ratio in reinforced sands using adaptive neuro-fuzzy inference system. *Soil Dynamics and Earthquake Engineering, 24*, 805–814. http://dx.doi.org/10.1016/j.soildyn.2004.04.006

Akkurt, S., Tayfur, G., & Can, S. (2004). Fuzzy logic model for the prediction of cement compressive strength. *Cement and Concrete Research, 34*, 1429–1433. http://dx.doi.org/10.1016/j.cemconres.2004.01.020

ARI News. (2007, June). *Nanotechnology in construction—One of the top ten answers to world's biggest problems.* Retrieved from http://www.aggregateresearch.com/articles/6279/Nanotechnology-in-Construction---one-of-the-top-ten-answers-to-worlds-biggest-problems.aspx

Beigi, M. H., Javad, B., Lotfi, O. O., Sadeghi, N. A., & Iman, M. N. (2013). An experimental survey on combined effects of fibers and nanosilica on the mechanical, rheological, and durability properties of self-compacting concrete. *Materials & Design, 50*, 1019–1029.

Bouzoubaa, N., & Lachemi, M. (2001). Self-Compacting concrete incorporating high volumes of class F fly ash Preliminary results. *Cement and Concrete Research, 31*, 413–420.

Demir, F. (2005). A new way of prediction elastic modulus of normal and high strength concrete-fuzzy logic. *Cement and Concrete Research, 35*, 1531–1538. http://dx.doi.org/10.1016/j.cemconres.2005.01.001

Givi, A. N., Rashid, S. A., Aziz, F. N. A., & Salleh, M. A. M. (2010). Experimental investigation of the size effects of SiO_2 nano-particles on the mechanical properties of binary blended concrete. *Composites Part B: Engineering, 41*, 673–677. http://dx.doi.org/10.1016/j.compositesb.2010.08.003

Gupta, S. (2014). A review on the use of nano-silica in cementitious compositions. *International Journal of Concrete Technology, 1*(1), 1–15.

Heidari, A., & Tavakoli, D. (2013). A study of the mechanical properties of ground ceramic powder concrete incorporating nano-SiO_2 particles. *Construction and Building Materials, 38*, 255–264. http://dx.doi.org/10.1016/j.conbuildmat.2012.07.110

Ho, D. W. C., & Zhang, P. A. (2001). Fuzzy wavelet networks for function learning. *IEEE Transactions on Fuzzy Systems, 9*, 200–211. http://dx.doi.org/10.1109/91.917126

Jang, J. S. R., & Sun, C. T. (1995). Neuro-fuzzy modeling and control. *Proceedings of the IEEE, 83*, 378–406. http://dx.doi.org/10.1109/5.364486

Ji, T. (2005). Preliminary study on the water permeability and microstructure of concrete incorporating nano-SiO_2. *Cement and Concrete Research, 35*, 943–947.

Jo, B. W., Kim, C. H., Tae, G., & Park, J. B. (2007). Characteristics of cement mortar with nano-SiO_2 particles. *Construction and Building Materials, 21*, 1351–1355. http://dx.doi.org/10.1016/j.conbuildmat.2005.12.020

Li, G. (2004). Properties of high-volume fly ash concrete incorporating nano-SiO_2. *Cement and Concrete Research, 34*, 1043–1049. http://dx.doi.org/10.1016/j.cemconres.2003.11.013

Li, H., Xiao, H., Yuan, J., Ou, J., & Ou, J. (2004). Microstructure of cement mortar with nano-particles. *Composites Part B: Engineering, 35*, 185–189. http://dx.doi.org/10.1016/S1359-8368(03)00052-0

Nili, M., Ehsani, A., & Shabani, K. (2010, June 28–30). Influence of nano-SiO_2 and micro-silica on concrete performance. In *Second International Conference on Sustainable Construction Materials and Technologies*. Università Politecnica delle Marche, Ancona, Italy.

Passino, M. (1998). Stable fuzzy logic design of point to point control for mechanical systems. In S. Yurkovich (Ed.), *Fuzzy control*. Menlo Park, CA: Addison-Wesley Longman.

Said, A. M., Zeidan, M. S., Bassuoni, M. T., & Tian, Y. (2012). Properties of concrete incorporating nano-silica. *Construction and Building Materials, 36*, 834–844.

Şen, Z. (1998). Fuzzy algorithm for estimation of solar irradiation from sunshine duration. *Solar Energy, 63*, 39–49. http://dx.doi.org/10.1016/S0038-092X(98)00043-7

Takagi, T., & Sugeno, M. (1985). Fuzzy identification of systems and its applications to modeling and control. *IEEE Transactions on Systems Man and Cybernetics, 15*, 116–132. http://dx.doi.org/10.1109/TSMC.1985.6313399

Topçu, İlker Bekir, & Sarıdemir, M. (2008). Prediction of mechanical properties of recycled aggregate concretes containing silica fume using artificial neural networks and fuzzy logic. *Computational Materials Science, 42*, 74–82. http://dx.doi.org/10.1016/j.commatsci.2007.06.011

Ünal, O., Demir, F., & Uygunoğlu, T. (2007). Fuzzy logic approach to predict stress–strain curves of steel fiber-reinforced concretes in compression. *Building and Environment*, 42, 3589–3595. http://dx.doi.org/10.1016/j.buildenv.2006.10.023

Zadeh, L. A. (1965). Fuzzy sets. *Information and Control, 8*, 338–353. http://dx.doi.org/10.1016/S0019-9958(65)90241-X

Zhang, M. H., & Li, H. (2010). Pore structure and chloride permeability of concrete containing nano-particles for pavement. *Construction and Building Materials, 25*, 608–616.

Permissions

All chapters in this book were first published in CE, by Cogent OA; hereby published with permission under the Creative Commons Attribution License or equivalent. Every chapter published in this book has been scrutinized by our experts. Their significance has been extensively debated. The topics covered herein carry significant findings which will fuel the growth of the discipline. They may even be implemented as practical applications or may be referred to as a beginning point for another development.

The contributors of this book come from diverse backgrounds, making this book a truly international effort. This book will bring forth new frontiers with its revolutionizing research information and detailed analysis of the nascent developments around the world.

We would like to thank all the contributing authors for lending their expertise to make the book truly unique. They have played a crucial role in the development of this book. Without their invaluable contributions this book wouldn't have been possible. They have made vital efforts to compile up to date information on the varied aspects of this subject to make this book a valuable addition to the collection of many professionals and students.

This book was conceptualized with the vision of imparting up-to-date information and advanced data in this field. To ensure the same, a matchless editorial board was set up. Every individual on the board went through rigorous rounds of assessment to prove their worth. After which they invested a large part of their time researching and compiling the most relevant data for our readers.

The editorial board has been involved in producing this book since its inception. They have spent rigorous hours researching and exploring the diverse topics which have resulted in the successful publishing of this book. They have passed on their knowledge of decades through this book. To expedite this challenging task, the publisher supported the team at every step. A small team of assistant editors was also appointed to further simplify the editing procedure and attain best results for the readers.

Apart from the editorial board, the designing team has also invested a significant amount of their time in understanding the subject and creating the most relevant covers. They scrutinized every image to scout for the most suitable representation of the subject and create an appropriate cover for the book.

The publishing team has been an ardent support to the editorial, designing and production team. Their endless efforts to recruit the best for this project, has resulted in the accomplishment of this book. They are a veteran in the field of academics and their pool of knowledge is as vast as their experience in printing. Their expertise and guidance has proved useful at every step. Their uncompromising quality standards have made this book an exceptional effort. Their encouragement from time to time has been an inspiration for everyone.

The publisher and the editorial board hope that this book will prove to be a valuable piece of knowledge for researchers, students, practitioners and scholars across the globe.

List of Contributors

Abdolhossein Baghlani, Mohsen Sattari and Mohammad Hadi Makiabadi
Faculty of Civil and Environmental Engineering, Shiraz University of Technology, Shiraz, Iran

Lina I. Shbeeb
Faculty of Engineering, Ahliyya Amman University, Amman, Jordan

Wáel H. Awad
Faculty of Engineering Technology, Al Balqá Applied University, Amman, Jordan

Hao Lei and Jeffrey Taylor
Department of Civil and Environmental Engineering, University of Utah, Salt Lake City, UT 84112, USA

Xuesong Zhou
School of Sustainable Engineering and the Built Environment, Arizona State University, Tempe, AZ 85287, USA

George F. List
Department of Civil, Construction, and Environmental Engineering, North Carolina State University, Raleigh, NC 27695-7908, USA

Anagi M. Balachandra
Metna Co., 1926 Turner St., Lansing, MI 48906, USA

Libya Ahmed Sbia and Parviz Soroushian
Department of Civil and Environmental Engineering, Michigan State University, 3546 Engineering Building, E. Lansing, MI 48824-1226, USA

Amirpasha Peyvandi
Bridge Department, HNTB Corporation, 10000 Perkins Rowe, Suite No. 640, Baton Rouge, LA, 70810, USA

Mohammad Ebrahim Banihabib
Department of Irrigation and Drainage Engineering, University College of Aburaihan, University of Tehran, 20th km Imam Reza Road, P.O. Box 11365/4117, Pakdasht, Tehran, Iran

Nimish Dharmadhikari
Indian Nations Council of Governments, Tulsa, OK 74103 USA

EunSu Lee
Upper Great Plains Transportation Institute, North Dakota State University, Fargo, ND 58102, USA

Youssef I. Hafez
Royal Commission Yanbu Colleges and Institutes, Yanbu University college, Yanbu, Saudi Arabia

Abdullah Fettahoglu
Department of Civil Engineering, Bursa Orhangazi University, Yildirim 16310, Turkey

Monika Suchowska-Kisielewicz, Andrzej Jedrczak and Zofia Sadecka
Institute of Environmental Engineering, University of Zielona Gora, Licealna 9, 65-417 Zielona Gora, Poland

Xuesong Zhou
School of Sustainable Engineering and the Built Environment, Arizona State University, Tempe, AZ 85287, USA

Jeffrey Taylor
Department of Civil and Environmental Engineering, University of Utah, Salt Lake City, UT 84112-0561, USA

C.G. Chinnaswamy and David N.G. Chew Chiat
Meinhardt Infrastructure Pte Ltd, Singapore, Singapore

Sakshi Gupta
Department of Civil Engineering, ASET, Amity University, Gurgaon, Haryana, India

Abdolhossein Baghlani, Mohsen Sattari and Mohammad Hadi Makiabadi
Faculty of Civil and Environmental Engineering, Shiraz University of Technology, Shiraz, Iran

Lina I. Shbeeb
Faculty of Engineering, Ahliyya Amman University, Amman, Jordan

Wáel H. Awad
Faculty of Engineering Technology, Al Balqa'Applied University, Amman, Jordan

Hao Lei and Jeffrey Taylor
Department of Civil and Environmental Engineering, University of Utah, Salt Lake City, UT 84112, USA

Xuesong Zhou
School of Sustainable Engineering and the Built Environment, Arizona State University, Tempe, AZ 85287, USA

George F. List
Department of Civil, Construction, and Environmental Engineering, North Carolina State University, Raleigh, NC 27695-7908, USA

Anagi M. Balachandra
Metna Co., 1926 Turner St., Lansing, MI 48906, USA

Libya Ahmed Sbia and Parviz Soroushian
Department of Civil and Environmental Engineering, Michigan State University, 3546 Engineering Building, E. Lansing, MI 48824-1226, USA

Amirpasha Peyvandi
Bridge Department, HNTB Corporation, 10000 Perkins Rowe, Suite No. 640, Baton Rouge, LA, 70810, USA

Mohammad Ebrahim Banihabib
Department of Irrigation and Drainage Engineering, University College of Aburaihan, University of Tehran, 20th km Imam Reza Road, P.O. Box 11365/4117, Pakdasht, Tehran, Iran

Nimish Dharmadhikari
Indian Nations Council of Governments, Tulsa, OK 74103 USA

EunSu Lee
Upper Great Plains Transportation Institute, North Dakota State University, Fargo, ND 58102, USA

Youssef I. Hafez
Royal Commission Yanbu Colleges and Institutes, Yanbu University college, Yanbu, Saudi Arabia

Abdullah Fettahoglu
Department of Civil Engineering, Bursa Orhangazi University, Yildirim 16310, Turkey

Monika Suchowska-Kisielewicz, Andrzej Jedrczak and Zofia Sadecka
Institute of Environmental Engineering, University of Zielona Gora, Licealna 9, 65-417 Zielona Gora, Poland

Xuesong Zhou
School of Sustainable Engineering and the Built Environment, Arizona State University, Tempe, AZ 85287, USA

Jeffrey Taylor
Department of Civil and Environmental Engineering, University of Utah, Salt Lake City, UT 84112-0561, USA

C.G. Chinnaswamy and David N.G. Chew Chiat
Meinhardt Infrastructure Pte Ltd, Singapore, Singapore

Sakshi Gupta
Department of Civil Engineering, ASET, Amity University, Gurgaon, Haryana, India

Index